Classics in Cartography

Classics in Cartography

Reflections on Influential Articles from *Cartographica*

Edited by

Martin Dodge
Department of Geography,
School of Environment and Development,
University of Manchester, UK

With a foreword by Jeremy W. Crampton,
editor of the journal *Cartographica*

WILEY-BLACKWELL

A John Wiley & Sons, Ltd., Publication

This edition first published 2011, © 2011 by John Wiley & Sons Ltd

Wiley-Blackwell is an imprint of John Wiley & Sons, formed by the merger of Wiley's global Scientific, Technical and Medical business with Blackwell Publishing.

Registered office:
John Wiley & Sons Ltd, The Atrium, Southern Gate, Chichester, West Sussex, PO19 8SQ, UK

Editorial Offices:
9600 Garsington Road, Oxford, OX4 2DQ, UK
The Atrium, Southern Gate, Chichester, West Sussex, PO19 8SQ, UK
111 River Street, Hoboken, NJ 07030-5774, USA

For details of our global editorial offices, for customer services and for information about how to apply for permission to reuse the copyright material in this book please see our website at www.wiley.com/wiley-blackwell

The right of the author to be identified as the author of this work has been asserted in accordance with the UK Copyright, Designs and Patents Act 1988.

Library of Congress Cataloging-in-Publication Data

Classics in cartography : reflections on influential articles from cartographica / edited by Martin Dodge.
    p. cm.
 Includes index.
 ISBN 978-0-470-68174-9 (cloth)
 1. Cartography. I. Dodge, Martin, 1971-
 GA101.5.C53 2010
 526–dc22

                                                   2010027896

A catalogue record for this book is available from the British Library.

This book is published in the following electronic formats: ePDF: 978-0-470-66947-1;
Wiley Online Library: 978-0-470-66948-8

Set in 10.5/12.5 pt Minion by Thomson Digital, Noida, India
Printed and bound in Singapore by Markono Print Media Pte Ltd

First Impression   2011

Copyright Notice

Chapters 2, 4, 6, 8, 10, 12, 14, 16, 18 and 20 are reprinted by permission of the University of Toronto Press, <www.utpress.com>.

Contents

Contributors' Biographies

Jeremy W. Crampton
Associate Professor at Georgia State University in Atlanta, USA

Jeremy's research interests lie in the cartographic politics of identity, critical approaches to cartography, the biopolitics of race and the work of Michel Foucault. His latest book is *Mapping* (Wiley-Blackwell, 2010). He has served as editor of the journal *Cartographica* since 2008.

Ferenc Csillag
Deceased, June 2005

Martin Dodge
Lecturer in Human Geography, School of Environment and Development, University of Manchester, UK

Martin's research focuses on conceptualizing the socio-spatial power of digital technologies and urban infrastructures, virtual geographies, and the theorization of visual representations, cartographic knowledges and novel methods of geographic visualization. He was curator of the well known Web-based *Atlas of Cyberspaces* and has co-authored three books covering aspects of the spatiality of computer technology: *Mapping Cyberspace* (Routledge, 2000), *Atlas of Cyberspace* (Addison-Wesley, 2001) and *Code/Space* (MIT Press, 2010). He has also co-edited two books, *Geographic Visualisation* (Wiley, 2008) and *Rethinking Maps* (Routledge, 2009), focused on the social and cultural meanings of new kinds of mapping practice. He is a member of the editorial board of the journal *Cartographica*.

David H. Douglas
Independent Scholar

David obtained his undergraduate degree at the Royal Military College of Canada in 1963. He served as an officer and pilot in the Royal Canadian Air Force until 1966, after which he obtained a Master's degree in Geography at Carleton University in Ottawa. On completing his degree he was offered a post in the Department of Geography at the University of Ottawa, which he held from 1970 until the end of 1999. After that he held a three year position in the Department of Surveying at the University of Gävle in Sweden.

His research has involved the compression of cartographic data in lines and surfaces, along with work on simple projections for quantitative mapping, polygon topology for dasymetric and choropleth maps, and shortest path algorithms.

Matthew H. Edney
Osher Professor in the History of Cartography at the University of Southern Maine, USA

Matthew directs the History of Cartography Project, University of Wisconsin-Madison. He has also taught at SUNY-Binghamton and the University of Michigan, Ann Arbor. Broadly interested in the nature and history of cartography, his research currently focuses on eighteenth century mapping, especially of British North America.

Sarah Elwood
Associate Professor of Geography at the University of Washington, USA

Sarah's work intersects critical Geographic Information Systems (GIS), and urban and political geography. She studies the social and political impacts of spatial technologies such as GIS, and the changing practices and politics of local activism, community organizing and other modes of civic engagement. Her current research focuses on the ever-expanding range of interactive Web-based technologies that are enabling collection, compilation, mapping and dissemination of geographic information by vast numbers of people.

John Fels
Adjunct Associate Professor at North Carolina State University, USA

John has worked as a professional cartographer with the Ontario Ministry of Natural Resources in Canada and as a freelance cartographic designer and consultant. He developed and taught the core Design curriculum in the Cartography Program at Sir Sandford Fleming College, Ontario, and is currently Adjunct Associate Professor in the Graduate GIS Faculty at North Carolina State University. He is the author of the *North Carolina Watersheds* map and co-author of *The Power of Maps* (Guilford Press, 1992), *The Natures of Maps* (University of Chicago Press, 2009) and *Rethinking the Power of Maps* (Guilford Press, 2010).

Rina Ghose
Associate Professor of Geography at the University of Wisconsin-Milwaukee, USA

Rina's research involves critical GIS, political economy and urban geography. She examines the implementation and use of spatial technologies and their socio-political impacts. She has conducted longitudinal research on how spatial knowledge is used by grassroots community organizations and activists as well as powerful actors and networks in shaping the inner-city planning process. Currently she is examining the impact of neoliberal policies upon public participation GIS.

Leonard Guelke
Retired Scholar

Len graduated with a BSc in Geography from the University of Cape Town, South Africa, in 1961. The curriculum of the time had a strong emphasis on cartography, something the leading thinkers on the nature of geography all agreed was a central component of geographic inquiry. This preparation helped him land his first job as a cartographic compiler and editor with Thomas Nelson & Sons of Edinburgh, which was followed by a two year assignment (1965–1967) as coordinator of the Atlas of Alberta. He obtained a PhD in Historical Geography from the University of Toronto in 1974. Although cartography was not specifically the focus of his graduate studies, on the basis of his earlier education and work experience he was deemed sufficiently well qualified to be appointed a faculty member at the University of Waterloo responsible for teaching cartography. This position stimulated a period, from 1975–1990, of cartographic research and the active participation in the Canadian Cartographic Association. Len retired in 2005.

Mordechai (Muki) Haklay
Senior Lecturer in GIS at University College London (UCL), UK

Muki's research focuses on usability engineering aspects of geospatial technologies, public access to environmental information and participatory GIS. He holds a BSc in Computer Science and Geography and an MA in Geography from the Hebrew University in Jerusalem, together with a PhD in Geography from UCL. He is a member of the editorial board of the journal *Cartographica*.

J. Brian Harley
Deceased, December 1991

Catherine Emma (Kate) Jones
Lecturer in Human Geography at the University of Portsmouth, UK

Kate completed her PhD at University College London in 2008 in the area of Health Geography. She is a specialist in GIS for collaborative research in social and urban geography.

David M. Mark
Professor of Geography at the University of Buffalo, the State University of New York and Director of the Buffalo site of the National Centre for Geographic Information and Analysis (NCGIA), USA

David completed his PhD in Geography at Simon Fraser University (Burnaby, Canada) in 1977, and joined the University at Buffalo in 1981. He has written or co-authored more than 220 publications, including 80 refereed articles and four edited books. His research interests include ontology of the geospatial domain, geographic cognition, cultural differences in geographic concepts, geographic information systems, human-computer interaction and digital elevation models.

Jeremy Mennis
Associate Professor of Geography and Urban Studies at Temple University, Philadelphia, USA

Jeremy received his PhD in Geography from Pennsylvania State University in 2001. His research has focused on spatio-temporal data models and semantic GIS representation. Current research is investigating how social and geographic environments influence crime and health behaviours.

Mark Monmonier
Distinguished Professor of Geography in the Maxwell School of Citizenship and Public Affairs at Syracuse University, USA

Mark is author of sixteen books, including *How to Lie with Maps* (University of Chicago Press, 1991, 1996), *Coast Lines: How Mapmakers Frame the World and Chart Environmental Change* (University of Chicago Press, 2008), and *No Dig, No Fly, No Go: How Maps Restrict and Control* (University of Chicago Press, 2010). His awards include a Guggenheim Fellowship (1984), the American Geographical Society's O.M. Miller Cartographic Medal (2001) and the German Cartographic Society's Mercator Medal (2009).

Donna J. Peuquet
Professor of Geography at Pennsylvania State University, USA

Donna's research interests are in the areas of geographic knowledge representation, knowledge discovery, spatio-temporal data models, spatial cognition, AI approaches to knowledge representation, geocomputation and GIS design. Since the early 1990s, her work has centred on the representation of time and temporal dynamics, including database, visual and cognitive representation and how these interrelate.

Thomas K. Poiker
Retired Scholar

Thomas Poiker, formerly Peucker, grew up in Austria, studied in Germany (Geography) and taught (Economic Geography, Geographic Information Systems (GIS), Cartography and – in the Computing Science Program – Computer Graphics) at Simon Fraser University in Burnaby, British Columbia, Canada. His research area was GIS, especially digital terrain models (DTMs) where he developed the triangulated irregular network structure. He is known for the first text in GIS and some articles in GIS, especially DTMs. He developed an online program (UniGIS, two years, 12 courses) which he directed from 1998 to his retirement in 2007.

Matthew Sparke
Professor of Geography and International Studies at the University of Washington, USA

Funded by a National Science Foundation CAREER grant, Matthew's recent research and teaching have been about globalisation, neoliberal governance, and the impact of transnational market ties on the geography of politics. He is the author of *In the Space of*

Theory: Postfoundational Geographies of the Nation-State (University of Minnesota Press, 2005), and *Introduction to Globalisation* (Blackwell, forthcoming), as well as over 50 articles, book chapters and reviews. In 2007 he received the University of Washington's Distinguished Teacher Award. His work on the politics of cartography is now leading into new research on the geography of global health, including attention to both neogeography Web 2.0 technologies used for risk management in rich countries, and the collective remapping of the globe itself – sometimes also enabled by volunteers – as a space of shared vulnerability and health citizenship.

Denis Wood
Independent Scholar

Denis holds a PhD in Geography from Clark University where he studied map-making under George McCleary. He was curator for the award-winning Power of Maps exhibition for the Smithsonian, and writes widely about maps. His most recent book is *Rethinking the Power of Maps* (Guilford Press, 2010). A former Professor of Design at North Carolina State University, Denis is currently an independent scholar living in Raleigh, North Carolina.

Foreword

Cartographica: The International Journal for Geographic Information and Geovisualization (to give its full title) is one of the longest standing peer review journals for the publication of cartography and mapping and geographic information systems (GIS).

The journal was founded by Bernard Gutsell (1914–2010) and his wife Barbara. As a young man in his native England, Gutsell was a squadron leader in the Royal Air Force during World War II. It might be that he found his love of geography and maps at that time. After the war, Gutsell moved to Canada, where he got a job with the government in the Geographical Bureau, later joining York University in Toronto. He retired from York in 1989. To some degree his career parallels that of another cartographer, the American Arthur Robinson. Like Gutsell, Robinson was involved in World War II, although his job was more directly cartographic, since he was in charge of the Map division of the Office of Strategic Services (OSS), which became the CIA after the war. It would be interesting to compare their careers further, but they do highlight the deep interconnections between cartography and government.

In the early 1960s the Gutsells asked for funds to start a journal from Canada's National Research Council (NRC) and the Canada Council. The sponsors asked him to produce Volume 1 before they were prepared to issue any funds. Since Gutsell had no money to do this, he actually started the journal with Volume 2! This was May 1965. It was only a few years later that Volume 1 appeared (labelled '1964'), in response to many enquiries from people asking where they could get a copy.

At this time the journal was simply called *The Cartographer*, and although it has gone through several name changes, one constant of its first thirty years was Gutsell's editorship. His involvement is generation-spanning. When he started the journal I was a baby; when I published my first articles in the journal in the early 1990s he was still editor. He stepped down in 1994. Due to his long tenure I'm still only the fifth editor of the journal after 45 years (and the first from outside Canada, although like Gutsell I am originally British). Following Gutsell, the journal was edited by Michael Coulson (1994–1999), Brian Klinkenberg (1999–2004) and Peter Keller (2004–2007). Other significant figures include Roger Wheate and Cliff Wood, who served as co-editors from 2004–2010. In addition, Ed Dahl must be mentioned in a number of roles, such as Associate Editor (1981–1994) and Board member (1994–2007). It was Ed who arranged the responses to the Harley article I discuss later in this book.

The very idea of classics in cartography might seem anachronistic in an age when cartography has become GIS, and GIS itself is either going to have to revolutionise or be subsumed by so-called Volunteered Geographic Information (VGI) or the geospatial Web and their corporate ilk such as Google Earth. It raises the question not only of what constitutes a classic (Something cited a lot? Something 'old'? Something cited a lot at first but not much now?) but of what cartography actually is (and whether the answer to that is historically variable or constant). Perhaps, in fact, a classic is something which *changes the definition of cartography.*

Indeed, Denis Wood's manifesto cry that 'Cartography is Dead – Thank God!' might at first glance appear to be the *sine quo non* judgement upon cartography. But this idea bears further examination. Wood does not celebrate the end of maps and mapping, but rather of a certain species of cartographic enquiry (academic, dry, irrelevant to real-world map use) that he sees as all too prevalent – and after all he is on the editorial board of *Cartographica*. Wood's point is subtle, it is not maps that have betrayed us, rather we have betrayed maps. We have flogged them to death and analysed them as if they were disturbed mental patients on the psychiatric couch. We have prosecuted them for war crimes, for supporting militaristic conquests and colonial exploitation, for propping up ministers and monarchs. And perhaps most indefensibly we have forgotten the beauty and wonder of maps, not to mention their sheer power amongst the general public.

No doubt Wood has a point, and his own work on map art has done much to correct these imbalances. But cartography, as a study of maps and mapping, is a product of modernity, and like most disciplines has undergone shifts in emphasis. Cartography embraced the scientific reason of the European Enlightenment as a practice of mapping and surveying in the sixteenth and seventeenth centuries, and as an academic discipline in the late nineteenth and early twentieth centuries. As this book neatly demonstrates, 'classics' come in a number of forms, and the classical scientific side of the discipline is well represented – notably the very influential work of Douglas and Peucker (now Poiker), which has been cited well over 1000 times according to Google Scholar (unfortunately academic citation databases have not indexed that year of *The Canadian Cartographer*, as the journal was then known). We also see from the selection how the study of mapping has evolved. Harley's article, which has been cited about half as much, represents a very different tradition, that of map critique. Newer concerns such as participatory GIS (PGIS) and experiential mapping are also included (see the Introduction for a fuller discussion of the choices included).

So classics can be thought of as articles that attract attention, whether formally through citations or more informally by word of mouth, that serve to shift the discipline and cause us to rethink maps.

Classics in cartography also raises the question of the relation and importance of the field in the larger sense. To Wood, it's dead, but I think that a little ungenerous to those of us still interested in thinking about mapping (and by 'us' I don't just mean academic cartographers, but map artists, geographers, philosophers, historians, political activists and the like). My old professor at Penn State, Peter Gould, used to say that geography was a great place to begin, but a bad place to remain. The same is probably true of cartography. Cartography is strongest (and I think to me this is what 'classics in cartography' ultimately means) when it reaches out and joins with these other forms of questioning. Cartography for cartography's sake is probably not going to light up the world. But cartography for art's sake, for philosophy's sake or for politics' sake, now that's something.

Jeremy W. Crampton
Editor, *Cartographica* 2008–2010

Acknowledgements

I would like to thank Rachael Ballard, Fiona Woods, Izzy Canning and colleagues at Wiley-Blackwell for their sterling work in supporting the development and production of the book.

For useful suggestions and ideas at points along the way, thanks to Jeremy Crampton, Muki Haklay, David O'Sullivan, Sara Fabrikant and Chris Perkins. I am very grateful to Graham Bowden, in the Cartographic Unit at the University of Manchester, for all his hard work on the illustrations. Morag Robson and Donna Pope were also helpful for gaining access to the original bound volumes of *Cartographica* held in the John Rylands University Library.

I also acknowledge Pete Fisher and his earlier book on GIS 'classics' for sparking the initial idea for this volume.

Martin Dodge

1

What are the 'Classic' Articles in Cartography?

Martin Dodge

School of Environment and Development, University of Manchester, UK

1.1 Outline of the Book

The intention of *Classics in Cartography* is to provide an intellectually-driven reinterpretation of a selection of some of the most influential articles from the last thirty years of academic cartography research. The ten chosen 'classic' articles were written by a range of the leading academic cartographers, geographers and allied scholars. They were all published in the international peer-reviewed journal *Cartographica*.

While the ten 'classic' articles are diverse in their agendas and approaches, they are all thought provoking texts that demonstrate how different aspects of mapping work as a mode spatial representation; they also shed light how different cartographic practices have been conceptualised by academic researchers. They are reprinted in full in this volume and, importantly, they are accompanied by newly commissioned reflective essays by the original authors (or other eminent researchers) to provide fresh interpretation on the meaning of the ideas presented and their wider, lasting impact on cartographic scholarship. Moreover, these essays give insights into how academic ideas emerge and some present a personal perspective on the nature of scholarly research. As such it is hoped that they will furnish current and future researchers with insights into how influential academic ideas come about and circulate as catalysts that can codify and instigate important areas of research within cartography and generate novel theoretical perspectives on mapping. While the focus on past 'classics' is perhaps rather backward looking in an era of such rapid social and technical change in cartography, it can be counter-argued that today there is real intellectual value in historical reflection because of the ways it helps us to understand better the present context for cartographic studies and to better inform future strategies for more innovative, creative mapping research (Dodge, Perkins and Kitchin, 2009; Kitchin and Dodge, 2007).

The book's intellectual focus on reflecting on 'classic' work in cartographic research, as opposed to GIScience or geovisualization is a conscious decision (see Dodge, McDerby and Turner, 2008; Fisher, 2006 for coverage of these allied fields). There is a strong case that cartography, broadly conceived, has become a newly reinvigorated topic in recent years, and that mapping has growing relevance to many scholars and students across the social sciences and humanities disciplines (Dodge and Perkins, 2008). The turn towards the 'visual' and 'spatial' in many large social science disciplines (such as anthropology, literary studies, sociology, history and communications) means there is extensive interest in spatial representations and mapping practice in its many forms (Warf and Arias, 2008). Meanwhile, mapping approaches are also proving instrumentally powerful in the information sciences, bio-informatics and human-computer studies as the basis for novel knowledge discovery strategies (Börner *et al.*, 2009). There is also much more lively engagement with cartography beyond academia, with growing artistic interest, numerous exciting participatory mapping projects and, of course, mass consumer enrolment of interactive spatial media on the Web, on mobile phones and in-car satellite navigation systems to solve myriad daily tasks (Crampton, 2009; Elwood, 2010).

So, looking beyond the core readership in cartography and GIScience, it is hoped that *Classics in Cartography* will have utility more widely across the sciences, social sciences and humanities, meeting the needs of a range researchers and postgraduate students interested in maps. It provides a new route into the wealth of significant cartographic literature, a unified and coherent way to bring a range of important mapping theories to the attention of a wide range of people looking to intellectually inform their mapping practice. The combination of 'classic' articles with new interpretation, which includes the significant work of many of the most well known cartographic scholars, makes this a uniquely useful book.

1.2 Delimiting the Cartographic 'Classics'

At the heart of the academic discipline of cartography are a set of theoretical frameworks and empirical findings that provide the intellectual basis for understanding the nature of maps and the work they do in the world. While such theories and findings are often the incremental product of the collective thought of many scholars, there are also signature pieces of writing that become recognized as 'classics' because of the way in which they push forward understanding or praxis by a significant degree. Such books and articles, through dint of their novel insights, analytic rigour or breadth of scholarship, gain recognition as foundational touchstones for students and academic researchers in cartography.

However, the task of drawing up a *short* and *definitive* list of such 'classic' work for any academic discipline that would achieve widespread agreement is an almost impossible one. The idiosyncratic interests, personal biases, partial knowledge and political agendas of the list maker will always mean the selection is less than perfect. To begin there are multiple dimensions upon which 'classic' status can be defined and the judgements made are almost always subjective. Perhaps most obviously a 'classic' might be delineated in terms of the degree of novelty and originality in the material: being first to publish can often be crucial in claiming rights to found a field of research. Additionally, 'classic' status might be judged by the impact the paper or book has in terms of setting on-going research agendas and acting as the initiator of something bigger – it is a 'classic' not so much for what it is but because of what it caused. Along a different track, it could be argued that some writing is rightly regarded as 'classic' because it is an archetypal model or stylish synthesis of a large and important body of knowledge, it elegantly encapsulates an argument better than rest, and the quality of expression and depth of scholarly interpretation means it becomes widely referenced as the definitive source. Such articles and books can also be powerful in pedagogic terms – giving students and the next generation of academics their 'route maps' into ideas and interpretation of the literature. So 'classics' are classic because teachers and textbooks cite them as such. The longevity of the work can also award 'classic' stature as ageless pieces that every serious student and new scholar must read (although many do not!). A piece of work can also be elevated to the prominence of a 'classic' because it provides a convenient shorthand signifier for a much large body of scholarship by one academic or research group; it becomes the totemic masterwork of a lifetime of research. This is particularly the case where scholarly reputations inflate and evolve after the death of the person concerned. One could argue, for example, that J.B. Harley's 'classic' article *Deconstructing the Map* (Chapter 16), which was published shortly before his sudden death in 1991, has subsequently been cited oftentimes as a summary of his larger body of work on the politics of maps.

'Classics' can also emerge because what they say becomes the centre of controversy, either by accident or deliberate design by the author. Such pieces can spark a flurry of responses and commentaries in journals – and now online discussions and blog posts – and also generate an inflated citation score. While sometimes pieces can become a 'classics' because they got things wrong and are seen as prime exemplars of how misguided scholars were in the past. Others become elevated as talismans of failed paradigms or as placeholders for politically unacceptable viewpoints of previous generations (e.g. in political geography dealing with the overt colonial ideologies of past in Halford Mackinder's writing with its infamous 'Heartlands' mapping, Blouet, 2005; or the racist agenda underlying the cartographic analysis of W.Z. Ripley, Winlow, 2006).

This kind of revisionism also begs the questions, is 'classic' a permanent state – once its achieved, does it remain forever more? Perhaps it is less so now given the extent to which theories seem to change with fashion and the rapidity of cycling through research agendas in contemporary social science scholarship. Consequently, 'classic' status must be regarded as provisional: a touchstone piece for the in vogue paradigm can become moribund as the core research agenda shifts and it is superseded by other, better – or perhaps just different – work. And, one of the interesting academic games is to try to find such 'lost classics' and resurrect them to bolster a newly emerging perspective.

Beyond these intellectual issues, subjective judgements and temporal fluctuations, there is a panoply of projects that seek to 'scientifically' assess the most significant scholarly work using citations counts, impact factors, h-scores and an assortment of other quantitatively derived metrics. Such calculative 'classics' seem to offer objectivity, but this is very much a veneer that masks a whole host of messy realities, fallacies and

contingencies with quantitative approaches, particularly relating to relative compara-
bility through time and across subject areas. As anyone who has used citations knows,
the major databases recording them are also incomplete, with varying coverage over
time, by language, publishing formats and academic disciplines. The partiality of the
data sources is easily highlighted in their inconsistencies when comparing citation
scores for the same article across the three main databases (e.g. citations to my 2007
paper *Rethinking Maps*: ThompsonISI's Web of Science: 13; Google Scholar: 32;
Elsevier's Scopus: 17). Moreover, the practices and intellectual significance of citations
varies across scholarly domains, which means measuring 'classics' absolutely, in
quantitative terms, across subjects areas is unworkable.

  Yet these acknowledged flaws in citations do not stop a significant degree of
fascination with such metrics by academics (particularly, perhaps, by those who seem
to have high scores or want higher ones!), by promotion committees, grant giving
bodies and government funding agencies. Increasingly over last decade, quantitative
assessment of the significance of published work has figured in efforts to systematically
profile academics, allocate funding amongst departments and rank institutions in the
name of improving quality, rewarding so-called research excellence and achieving
greater value for money. It is interesting to ponder how cartographic research – with a
relatively small core body of active scholars and particular publishing practices and
arguably peripheral outlets – fares in these kinds of citational games.

  High citation scores are no guarantee of scholarly quality or intellectual significance;
they are at best a popularity indicator. Controversial articles get cited just for being
controversial, not necessarily because they are good. Poor quality work can pile up
citations simply because its visible or easily accessible. Other articles can accrue citations
because of assiduous self-promotion by authors and through lazy citation practices by
fellow academics (too many of us cite articles or book as 'tokens' of credibility without
ever having read them).

  An interesting alternative to the admittedly subjective individual list-maker and
attempts at objective metrics would be to try a robust qualitative approach using a
sample survey method. Here a kind of opinion poll would be taken of a wide sample of
scholars in a subject area who would be asked to select their 'classics'. Combining
multiple selections could generate a consensual view of 'classics', assuming a broadly
representative sample of people willing to participate in the poll could be achieved.
While this was not done for this book – which is primarily the result of subjective list-
making – it would be a potentially worthwhile exercise to try to draw a more definitive
list of cartography's 'classics' by such a polling approach.

1.3 Why Re-publish 'Classics'?

Notwithstanding the problematic task of identifying 'classics', there is certainly an
intellectual tradition of evaluating developments in scholarship in terms of such
touchstone articles and books, through their re-reading and re-interpretation. Such
reflection and appraisal is conducted in a range of forms of academic publishing in many
disciplines. For example, leading disciplinary journals, such as *Progress in Human*

Geography, regularly consider citation 'classics' with the goal of 'reflecting on books and other works that have more than stood the test of space and time in shaping the discipline and practice of human geography'. Over the years a number of books reprising 'classics' from a specific sources – journals, institutions and research groups – are put together: Peter Fisher, for example, edited a large collection of *Classics from IJGIS* (2006), which reprinted nineteen important articles, accompanied by reflective essays, from the *International Journal of Geographical Information Science*. There is also a practice of publishing festschrifts of the 'classics' work of eminent scholars as a reflection and celebration of their contribution to the intellectual progress of disciplines. Examples include *The New Nature of Maps*, a posthumously edited collection of Brian Harley's later theoretical papers on cartography put together by Paul Laxton (2001), and *Land and Life*, edited by John Leighly (1976), presented significant work of Carl Sauer.

In pedagogic terms a range of different kinds of collections of 'classics' are a staple of academic publishers. Produced chiefly as primers for undergraduate students, examples include: *Human Geography: An Essential Anthology* (1996) edited by John Agnew and colleagues. A large number of reader style volumes that try to provide synoptic coverage of a body of literature by excerpting from 'classic' work often with editorial interpretation have been produced over the last decade as well (e.g. Moseley *et al.*'s *Introductory Reader in Human Geography*, 2007). Other books focus solely on the interpretation of 'classics', without actually reprinting or excerpting the originals: Hubbard, Kitchin and Valentine (2008), for example, edited *Key Texts in Human Geography*, a successful collection of essays appraising twenty-six of the most significant books for the discipline. More broadly there are now host of disciplinary dictionaries, introductory handbooks and larger encyclopaedia projects that seek to codify an academic field, in part, by identifying and evaluating 'classic' material.

Such anthologies, readers and key texts type books, while often aimed at students, are also important because the selection decisions of the academic editors ineluctably also provide signifiers in terms of 'what matters' intellectually. They influence directly what student's see, read and come to regard as the canons of the discipline. More broadly, it might be argued that there is scholarly value in this raft of reflective publishing, in that it can help cut through 'information overload' generated by bibliographic databases and ready online access to e-journals and digital books. A coherently edited collection can save much time and effort for students and academics in tracking down a set of the landmark articles that are otherwise scattered across many decades worth of journal issues (much of which can now be quickly downloaded but without means or indicators to determine what is worth reading). In summary, the wide range of competing and complimentary reflective efforts does, in some senses, establish the merit of considering 'classics' in cartography.

1.4 Choosing the 'Classics' in Cartography from *Cartographica*

The task of drawing up top ten lists of academic articles is at one level a rather worthless exercise, yet it can focus minds and provide a starting point for considering what are the most significant works and why. Also, given peoples liking for such 'list-o-mania', it can also offers up some prurient entertainment, talking points at seminars and something to argue over at conference lunch lines!

In the initial planning of this book it was decided to use the 'top ten' approach[1] as an organizing principle and also to focus on a single core disciplinary journal as the source.[2] In drawing up a shortlist of 'classics' from the wide array of high-quality material published in the four decades worth of *Cartographica*, an effort was made to select articles that have provided significant contributions to advance cartographic thought and praxis. This was guided, in part, by looking at citations counts of articles but this was not in itself seen as a sufficient or always reliable indicator of significance. Therefore, a sizeable degree of subjective editorial judgement was also involved in the final selection of 'classics'.

The decision to focus solely on the journal *Cartographica* as the source rather than to look more widely across other peer review outlets was a conscious choice and offers several advantages. From the mid-1960s *Cartographica* has grown into a key academic guiding force in cartography, reflected in both the breath and quality of the research articles it has published over the years. It now enjoys a well established international reputation for publishing innovative peer-reviewed work, with material submitted from a broad range of scholars in cartography, GIScience, human geography and allied disciplines. It has been the outlet of choice for many landmark pieces of research along with some larger monographs, particularly during the 1980s. It has published several hundred peer-reviewed research articles covering a huge breadth of ideas and approaches to mapping. It now has extensive historical archives, having been produced continuously since 1964, and is currently published on a quarterly basis by University of Toronto Press. (Further discussion of the historical context of *Cartographica* is provided in Jeremy Crampton's foreword.)

Using peer-reviewed articles only from *Cartographica* was felt to be a strength in the book's intellectual design, not a weakness. It has positive virtues of concentrating the choice on a cogent and high-quality range of material. It would be fair to say none of the other cartography journals has the same *international* reputation for publishing innovative research and for intellectual leadership. While the use of *Cartographica* alone is an arbitrary decision, widening sourcing to look across multiple cartography journals would have increased the diversity of selected material but would not have necessarily identified a higher quality set of 'classics' or improved the insight offered by this volume.

Looking for 'classics' across more journals would have made selection decisions that much harder and also raised thorny boundary issues – what is cartography and what is not. (In some senses this was side-stepped here – if it was published in *Cartographica*, it

---

[1] Of course ten is an arbitrary but universal feature of the mania for lists. Psychologically, ten seems like a good number – large enough to be useful but not too long to be unwieldy and impossible to grasp. It does not mean there are only ten 'classic' articles in *Cartographica* – it could easily have been fifteen or twenty. However, ten also fits pragmatic demands of publishers for a scale of book that is commercially viable.

[2] I am very grateful to the assistance of Jeremy Crampton in formulating the book and in the process of selecting 'classic' material for inclusion. Note, Jeremy currently serves as editor of *Cartographica*. I serve on the editorial board of the journal.

counts as cartography!). Defining the scope of any academic field is a problematic task because where to draw the intellectual boundaries cannot be objectively decided. What lies inside the disciplines borders is subjective, related to individual's interests and experience. (Indeed, is everything published in *Cartographica* actually about cartography? This depends on your viewpoint.). How do you police the scholarly margins and handle overlaps with other subjects, particularly given the broad nature of mapping. Cartographic research has many shared interests with a raft of quantitative methods, regional geography, surveying, along with graphic design, aesthetics, statistical analysis, photogrammetry, to name but a few fields. The rise of research focused on GIS techniques and concepts since the late 1980s, and latterly GIScience has come to eclipse cartography and absorb much of mapping research. This changing domain focus is reflected in the changed names of cartography journals to include GIS in their titles and to encompass the topic as a key remit for relevance (for example, *The American Cartographer* changed to *CaGIS*, Cartography and Geographic Information Systems, in 1990).

In the initial intellectual design of the book it was not apparent that there were any attempts to produce a list of the most highly cited work across cartographic literature in general. However, under the editorship of Jeremy Crampton, a useful 'top ten' list of the articles published in *Cartographica* based on a citation metric was drawn up not long ago (Table 1.1). This was done, in large part, as a promotional device for the journal, with the ten articles being 'given away' as free downloadable pdfs on the University of Toronto Press website.[3] The 'accuracy' of the citation data is open to question (particularly given the partial coverage of the source) but the selection does identify an interesting range of work, including pieces by scholars that would be widely acknowledged in cartography. The interpretative text accompanying this list notes how a couple of pieces 'focus on the representation of elevation data, presciently foreshadowing today's interest in virtual globes and terrain representation' and, more generally, the selection highlights the 'deep ties between cartography and GIS' (www. utpjournals.com/carto/CARTO_post.pdf).

While this list, derived solely from citation counts, was a helpful initial point of reference for identifying the 'classics' for this book, it did not by itself seem sufficiently robust to use in totality. While there is a 40% overlap between this list and the ten 'classics' eventually chosen for this collection, it was felt that 'raw' citation counts resulted in a rather unbalanced selection, that in particular tended to favour technical and algorithmic pieces above more philosophical, political and design-orientated work. Some papers seemed to speak much more to a GIS audience rather than scholarship around cartographic representations and mapping practice. The goal for this book was different from the needs of commercial promotion for the journal, seeking a broader view of cartographic research and a greater diversity of authorship to capture wider range of approaches and intellectual perspectives. For example, the 2007 citation-based 'top ten' list featured two papers by both J.B. Harley and David Mark – this kind of

---

[3] Online at http://utpjournals.metapress.com/content/120327/. It would be interesting to see if this promotional exercise increased the reading and citing of these selected pieces and, as such, re-enforced their 'classic' status.

Table 1.1 A top ten listing of the most highly cited articles published in the journal *Cartographica*. It was drawn by Jeremy Crampton based on Elsevier's Scopus citation database in December 2007. Articles ranked according to number of citations

	Author(s)	Date	Title	Volume
1	Harley, J.B.	1989	Deconstructing the Map	26(2)
2	Mark, D.M.	1984	Automated Detection of Drainage Networks From Digital Elevation Models	21(2/3)
3	Langran, G. and Chrisman, N.R.	1988	A Framework for Temporal Geographic Information	25(3)
4	Peuquet, D.J.	1984	A Conceptual Framework and Comparison of Spatial Data Models	21(4)
5	Mark, D.M. and Csillag, F.	1989	The Nature of Boundaries on 'Area Class' Maps	26(1)
6	Monmonier, M.	1990	Strategies for the Visualisation of Geographic Time-Series Data	27(1)
7	Blakemore, M.	1984	Generalisation and Error in Spatial Data Bases	21(2/3)
8	Harley, J.B.	1990	Cartography, Ethics and Social Theory	27(2)
9	Carter, J.R.	1992	The Effect of Spatial Precision on the Calculation of Slope and Aspect Using Gridded DEMs	29(1)
10	Kumler, M.P.	1994	An Intensive Comparison of Triangulated Irregular Networks and Digital Elevation Models	31(2)

duplication was avoided. The citation-based list also seemed to be missing a number of noteworthy scholars who have written interesting and insightful pieces for *Cartographica*, such as Denis Wood and Matthew Edney.

Furthermore, the time span of the citation-based 'top ten' list was overly narrow, with all the papers identified having been published in a ten year window from 1984 to 1994 (with three published in a single year, 1984). Given that *Cartographica* started publishing in 1965 (known then as *The Cartographer*) and continues to the present day, it was felt a more even distribution of 'classics' through time would be appropriate and hopefully capture more of the changing nature of cartographic scholarship during the past forty plus years. In particular, the fact that the 'top ten' citational list included no 'classic' from the last fifteen years seemed odd (one might argue this is statistically correct but not intellectually justifiable). The underlying citational statistics themselves (generated with Elsevier's Scopus database) also have a question mark against them, as they apparently missed out Douglas and Peucker's 1973 article, which has highest citation count by significant margin, approximately double J.B. Harley's, according to Google Scholar (Table 1.2).

The rationale then for the 'classic' articles actually selected for inclusion *Classics in Cartography* was somewhat of a subjective fudge! (Table 1.2) It was felt necessary to look beyond citations to choose other less visible articles that are intellectually significant – overlooked 'classics' perhaps. The book needed wider coverage through time and with a broader range of recognizable authors. The selection tried to incorporate articles which

Table 1.2 The ten 'classic' peer-reviewed articles from the journal *Cartographica* selected for this volume. Articles ranked according to their date of publication

	Author(s)	Date	Title	Volume	Citations[a]
1	Douglas, D.H. and Peucker, T.K.	1973	Algorithms for the Reduction of the Number of Points Required to Represent a Digitised Line or its Caricature	10(2)	1088
2	Guelke, L.	1976	Cartographic Communication and Geographic Understanding	13(2)	18
3	Peuquet, D.J.	1984	A Conceptual Framework and Comparison of Spatial Data Models	21(4)	256
4	Wood, D. and Fels, J.	1986	Designs on Signs: Myth and Meaning in Maps	23(3)	73
5	Mark, D.M. and Csillag, F.	1989	The Nature of Boundaries on 'Area Class' Maps	26(1)	106
6	Harley, J.B.	1989	Deconstructing the Map	26(2)	555
7	Monmonier, M.	1990	Strategies for the Visualisation of Geographic Time-Series Data	27(1)	91
8	Edney, M.H.	1993	Cartography Without 'Progress'	30(2/3)	43
9	Sparke, M.	1995	Between Demythologising and Deconstructing the Map	32(1)	28
10	Elwood, S. and Ghose, R.	2001	PPGIS in Community Development Planning	38(3/4)	26

[a]Citations according to Google Scholar, March 2010.

changed the purview and remit of sub-disciplinary fields of cartography. The result is a selection of articles that represents a wide range of cartographic interests, and includes work by many of the most influential scholars of the last few decades. It is undoubtedly an idiosyncratic list, and is perhaps too eclectic, but hopefully it is also a plausible list of the 'classics' in scholarly cartographic research that have appeared in the journal *Cartographica*.

The selected 'classics' have a widely variable range of citation counts, from over one thousand to a couple of articles with under fifty. For example, there is the foundational work of Douglas and Peucker on cartographic generalization that has been cited 1088 times according to Google Scholar (March 2010) and is referenced in myriad of places across the Web. The second most-cited article in this collection is quite different, but no less influential, being a very well known piece by Brian Harley, often credited with changing the way we think about maps in social terms. Harley's article from 1989 has over five hundred citations and remains influential across the social sciences.

While they cover a fairly wide time span – nearly thirty years – there are unfortunately none identified and selected from the very earliest volumes of *Cartographica*. Equally, the latest issues are not as well represented as needed, with the most recent article selected as a 'classic' being Elwood and Ghose's 2001 research on PPGIS. Looking at the dates of the pieces listed in Table 1.2, it is still somewhat evident of a bulge of material selected as 'classics' from the later half of the 1980s. Perhaps this does represent a purple patch for cartographic scholarship that became eclipsed by the growth in GIS research in subsequent years.

In terms of structuring the selected 'classics' in this volume, we did not want to present them in chronological order (or the reverse, from new to old) as this can imply simple linear progress and a naive narrative of improvement over time. The reality is research from different time periods is different and cannot be easily evaluated as better or worse based on when it was published. Also, we did not want to present the ten 'classics' ranked by their citation count. Again this implies a spurious hierarchy in scholarly worth and reifies the citation, which, as discussed above, is a problematic measure of quality. Instead, it was decided that *Classics in Cartography* would be organized thematically, allowing readers to see links between key ideas and debates in academic cartography. The ten articles were grouped into three broad sections: epistemological practice; ontological understanding; and politics and society.

In terms of the editorial process all ten 'classic' articles from *Cartographica* are reprinted in full here. They have all been reformatted for consistency and to remove variability of layout and referencing style evident in the original versions published in the journal. The degree of standardization, particularly the switch from footnote citations to Harvard style referencing in a couple of articles, has necessitated some very minor changes to the texts themselves. Nearly all the original illustrations have been faithfully redrawn for this book by Graham Bowden (Cartographic Unit, University of Manchester) to ensure higher quality reproduction than the pdf scans available from the University of Toronto Press.

1.5 The Significance of Scholarly Reflection

What is the value in reflecting on 'classic' articles? It is perhaps one of the distinctive practices of scholarship that much effort is expended in understanding how we work, where ideas come from and what influences the development of a discipline. This can be regarded as mere academic navel gazing or dismissed as vainglorious posturing. (And, of course, there are elements of self-indulgence, vanity and ego involved.) However, I would strongly argue that the reflective approach taken by *Classics in Cartography* has significance because it exposes how knowledge construction proceeds. It 'lifts the lid' on the messy practices of research and the uncertainties in the writing process. It highlights how provisional ideas and theories often are. This back story is typically left out of the finished product – the edited, proofed and professionally formatted article in the journal appears concrete and complete. The published article is like the ship in the bottle, upright and ready; what a reflective approach offers is the chance to narrate the unwritten story of how the ship actually got into the bottle. In some regards this is often as interesting (and perhaps sometimes more interesting) as the finished product. Such reflection can provide new insights when read against the original articles, especially for younger scholars or for outsiders who are coming to the material for first time.

Now is also a good time to contemplate and reflect upon the nature of cartographic scholarship. It seems contemporary academia is trapped by ever increasing pressures to publish, the speeded-up cycles of submission and review, the 'salami slicing' of research to generate multiple 'new' papers that are thin on original material, the growth in the size of journals and a proliferation of outlets, the pressure to measure the impact of articles, the need to be setting the next agenda and to demonstrate an international profile. It might be argued that these processes, working in concert, are diminishing the intellectual value of peer-reviewed articles today. The rapidity of circulation of research findings and the relentless attempts to promote new approaches mean it is increasingly hard to keep up and identify what is important. (Often it seems we run and run in our academic 'hamster wheels' without moving forward). The rate and volume of published research needed to be a 'successful' scholar today does not necessary promote quality or allow the time for genuine innovation. As such, it could be argued, that the time is now ripe for more measured reflection and consideration of the past ideas in cartography, rather than running blindly into the future.

For this volume, thoughtful reflection was sought on how scholarly ideas in cartography originate, circulate and come to be influential. Authors of the new essays were requested to be personal, anecdotal and opinionated if they wanted to. And in many cases they have grasped this opportunity and are genuinely introspective about the uncertain processes of authorship. They were also tasked to make their reflection essays accessible to readers without deep theoretical background in the particular field. The added value of the reflective essays are also significant in how they highlight the position and biography of the authors, which is useful for seeing how now-established scholars were working earlier in their careers. Many of the essays discuss the wider significance of the original *Cartographica* articles in relationship to cartography debates and issues current at the time of writing, and consider how the value of the article has changed – and continues to influence – the way scholars understand the map. Some also consider whether the original article is right to be regarded as a 'classic'. Box 1.1 is a listing of the types of themes that the authors were asked to consider in their reflective essays. This was not a rigid template and the style and substance of essays is quite variable, which is inevitable and also refreshingly realistic.

In terms of the preparation of the new reflection essays, six out ten were written by the original authors (or joint authors) of the *Cartographica* articles. In the case of David Mark and Ferenc Csillag's 1989 'classic' article, *The Nature of Boundaries on 'Area Class' Maps*, the reflection essay was of necessity put together solely by David Mark, as Ferenc Csillag passed away in 2005. The remaining three reflection essays were not written by the original authors, but by other eminently qualified scholars; they obviously have a different tone and perspective. J.B. Harley died in 1991 and the task of re-interpreting his 'classic' article, *Deconstructing the Map*, was taken up enthusiastically by Jeremy Crampton. Donna Peuquet was asked to reflect upon her 1994 article, *A Conceptual Framework and Comparison of Spatial Data Model*, but was unfortunately unable to participate due to pressure of other writing commitments; Jeremy Mennis gamely stepped up to the task of writing the essay. Lastly, the reflection essay for Len Guelke's 'overlooked classic' from 1976, *Cartographic Communication and Geographic Understanding*, was undertaken insightfully by Muki Haklay and Kate Jones; Len, himself, did not feel up to the task as he is happily ensconced in other more interesting retirement projects.

Box 1.1 Themes for Author Consideration

- Genesis of the theoretical ideas in the paper.

- Context in which the paper was written, reviewed and published. Discuss the wider significance of the paper in relationship to cartography debates and issues current at the time of writing.

- Consider the initial impact and reception of the paper at the time of publication.

- How has the paper influenced ideas over time? Evaluate how it has subsequently been cited, critiqued and incorporated into academic cartography discourse.

- Re-interpret your paper with the power of hindsight and reflect on the lasting validity of the paper's main arguments, methods and sources of evidence. Do you stand by the arguments and ideas presented in the paper? What did the paper do well and what were its weaker points? Does it contain mistakes and things you would now change? How have your ideas changed subsequent to publication?

- Do you agree that it is a 'classic?' Why do you think it's proved to be influential?

- Discuss how the broader field of cartography that your paper relates to has subsequently developed. Has it grown? Or shrunk in importance? Has it taken unexpected directions? Is it now overlooked perhaps?

1.6 Epilogue

Over the past couple decades there has been a sustained scholarly engagement in thinking about the ontological basis of cartographic representation and exploring new epistemologies of mapping. Moreover, there is a burgeoning interest from many scientists, social scientists and humanities scholars in theorizing the nature of cartography and productively applying mapping and geographic visualization to solve research problems they face. This coupled with tremendous socio-technical developments in the production of cartographic representations has led to a widening and more vibrant array of different kinds of mapping employed by scholars. The goal in editing *Classics in Cartography* is to further advance theoretical understanding of cartography in terms of social science scholarship by reflecting on some of the significant ways maps have been researched. Hopefully, the reflection essays, in combination with the original *Cartographica* articles, work as a useful set of intellectual signposts, particularly for postgraduates and new researchers, in understanding the evolution of 'classic' cartographic theories and for developing new mapping ideas.

References

Agnew, J., Livingstone, D.N. and Roger, A. (1996) *Human Geography: An Essential Anthology*, Blackwell, Oxford.

Blouet, B.W. (2005) *Global Geostrategy: Mackinder and the Defence of the West*, Routledge, London.

Börner, K., Bettencourt, L.M.A., Gerstein, M. and Uzzo, S.M. (2009) *Knowledge Management and Visualization Tools in Support of Discovery*. NSF Workshop Report (http://vw.slis.indiana.edu/cdi2008/NSF-Report.pdf).

Crampton, J.W. (2009) Maps 2.0. *Progress in Human Geography*, **33**(1), 91–100.

Dodge, M. and Perkins, C. (2008) Reclaiming the map: British Geography and ambivalent cartographic practice. *Environment and Planning A*, **40**(6), 1271–1276.

Dodge, M., McDerby, M. and Turner, M. (2008) *Geographic Visualization: Concepts, Tools and Applications*, John Wiley & Sons Ltd, Chichester.

Dodge, M., Perkins, C. and Kitchin, R. (2009) Mapping modes, methods and moments: a manifesto for map studies, in *Rethinking Maps: New Frontiers in Cartographic Theory* (eds. M. Dodge, R. Kitchin and C. Perkins), Routledge, London, pp. 220–243.

Elwood, S. (2010) Geographic information science: emerging research on the societal implications of the geospatial web. *Progress in Human Geography*, **34**(3), 349–357.

Fisher, P. (2006) *Classics from IJGIS: Twenty Years of the International Journal of Geographical Information Science*, CRC Press, Boca Raton, FL.

Hubbard, P., Kitchin, R. and Valentine, G. (2008) *Key Texts in Human Geography*, Sage, London.

Kitchin, R. and Dodge, M. (2007) Rethinking maps. *Progress in Human Geography*, **31**(3), 331–344.

Laxton, P. (2001) *The New Nature of Maps: Essays in the History of Cartography*, The Johns Hopkins University Press, Baltimore, MD.

Leighly, J. (1976) *Land and Life: A Selection of the Influential Writings of Carl Ortwin Sauer*, University of California Press, Berkeley, CA.

Moseley, W., Lanegran, D.A. and Pandit, K. (2007) *The Introductory Reader in Human Geography: Contemporary Debates and Classic Writings*, Blackwell, Oxford.

Warf, B. and Arias, S. (2008) *The Spatial Turn*, Routledge, New York.

Winlow, H. (2006) Mapping moral geographies: W.Z. Ripley's races of Europe and the U.S. *Annals of the Association of American Geographers*, **96**(1), 119–141.

SECTION ONE
EPISTEMOLOGICAL PRACTICE

2

Algorithms for the Reduction of the Number of Points Required to Represent a Digitized Line or its Caricature

David H. Douglas and Thomas K. Peucker[1]

Abstract

All digitizing methods, as a general rule, record lines with far more data than is necessary for accurate graphic reproduction or for computer analysis. Two algorithms to reduce the number of points required to represent the line and, if desired, produce caricatures, are presented and compared with the most promising methods so far suggested. Line reduction will form a major part of automated generalization.

Lines from maps and photographs are recorded numerically for cartographic manipulation to facilitate their reproduction at different scales and projections, and to allow map compilation with other geographic databases. Usually lines are approximated by straight line segments, the end points of which are recorded by a pair of coordinates in either polar or orthogonal measure. The other more important methods by which lines are recorded are chain encoding and skeleton encoding. Chains approximate lines by a sequence of end-to-end vectors, where the length and direction of the vectors are selected from a fixed, usually four or eight, number of possibilities (Freeman, 1961). Skeleton encoding is directed more at recording closed areas or polygons by filling the area with circles or rhombi of different sizes. The lines forming the boundaries are recorded by implication (Pfaltz and Rosenfeld, 1967). The conversion of graphic data to

[1] Originally published: 1973, *Cartographica (The Canadian Cartographer)*, **10**(2), 112–122.

At the time of publication: Douglas was a Lecturer in Geography, University of Ottawa and on study leave at Simon Fraser University, British Columbia; Peucker was Associate Professor at Simon Fraser University, British Columbia.

Classics in Cartography: Reflections on Influential Articles from Cartographica   Edited by Martin Dodge
© 2011 John Wiley & Sons, Ltd

computer readable numerical forms is effected with a coordinate digitizer, a bit plane scanner or an automatic line follower. A coordinate digitizer converts a pointer's location on a table to x–y values which can be written on punched cards or magnetic devices. Polar coordinate digitizers, which consist of a slide in a rotating anchor head, record a radius and an angle from a base vector. Another digitizing device consists of a pointer suspended from a pair of retracting wires which activate potentiometers. Conversion of values in one recording coordinate system to another can be performed easily with small computer programs.

Drum scanners superimpose a vast and very fine grid over the document to be digitized recording a 'yes–no' or 'on–off' value for each cell location, depending on whether that cell covers a line or not. A trade-off is introduced between the fineness of the mesh, implying more computer processing time to reduce the data to forms which are easily handled, and coarseness of the image recorded. On the other hand, the mesh density, being dependent on hardware, is fixed at the time of manufacture and is usually set to be somewhat smaller than the minimum line width. In all cases, the reduction of a bit plane scan, in which lines are represented by clouds of cells containing numerous discontinuities, to chain or vector encoded lines, is a complex process requiring processing time and resources which could only be described as being quite substantial.

With a coordinate digitizer lines may be recorded in point mode, time or increment automatic modes. Lines recorded in point mode are effectively generalized by the operator who subjectively selects points which best approximate the line to the degree he desires. This presumes, amongst other things, that he is his own customer. Point digitizing is extremely tedious, however, and is unsuitable for anything but the simplest data sets, such as the generalized outlines of counties or census tracts. Most coordinate x–y digitizers on the market possess, as options, time or increment automatic recording modes. Points are recorded automatically in a given time interval, or after the cursor has moved a preset distance along the x and/or y axis. The prime limiting factor on the speed of recording is the speed of the output device. Magnetic tape transports which record up to 300 characters per second are commonly available, allowing up to 20 or 30 points to be recorded each second. To record coastlines, contour lines or other lines of high frequency oscillation it is evident that the minimum speed required, given the speed at which an operator can follow a line, is in the order of 5 to 10 points per second, which effectively eliminates paper tape and punched cards as output media. Digitizing onto magnetic tape has more than its share of problems, primarily because there are no foolproof means to ensure the data are correctly recorded at the time of digitizing, and because of the inordinately frequent occurrence of non-confirmable digitizing errors, such as line ends which should, but do not meet, . . . lines recorded twice and so forth. The editing procedures necessary are time consuming and clumsy. These problems have been met by elaborate online procedures where a mini-computer interfaced to the digitizing table oversees the whole operation, checks and double checks the data recorded, closes loops and signals when it senses a great many errors, such as cursor movement too fast to be accurate (Bolye, 1970).

All digitizing methods, except perhaps for the possible exclusion of point digitizing on a coordinate digitizer, record, as a general rule, far more points than necessary to reproduce the line on most graphic devices, even at the scale and resolution of the original line. The elimination of data representing unnecessary points, such as duplicates, and points along a straight line can be of significance, simply because of the diminished storage requirements. As well, the operating speed of many spatial analysis programs and the plotting speed of many graphic devices are related inversely to the number of points to be processed or plotted. Reduction of a line by elimination of unnecessary points representing it assumes a more positive advantage if the line is to be abstracted or caricatured purposely, if the scale of reproduction is to be smaller, or if the output device, such as some Cathode Ray Tube plotters, has a cruder resolution than represented by the original digitized line. Lines which have a higher frequency of oscillation than can be represented within the resolution capability of the graphic device become fuzzy and weak (and are similar in effect to the data clouds recorded by a bit plane scanner). Figure 2.1 illustrates line data at the resolution of recording, its reproduction on the computer printer, and the reproduction of a greatly generalized version of the line. Given the crudeness of the printer as a graphic device it is evident that the simplified version of the line is preferable to the unsimplified one, mainly because of the elimination of most of the double lines and data clouds. Since this line was better represented by 25 points than it was by the original 140, obviously some computer pre-processing was justified.

There have been a great many approaches suggested and algorithms programmed to reduce the number of points required to represent numerically recorded lines. Some of these are in regular use within planning agencies and cartographic units. Not all of the methods have been exhaustively tested to measure or judge their cartographic useful-ness and there have been few, if any, studies to compare the methods with each other. The methods can be classed broadly into the categories of: elimination of points along the line by one or more of a multitude of criteria; approximation of the line with a mathematical function; and deletion of specific cartographic features represented by the

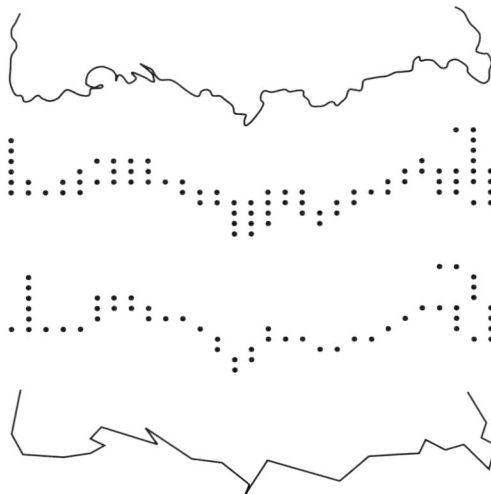

Figure 2.1 Line represented by 140 points on the plotter and the printer, and the same line represented by 25 points.

line. Of these categories, it would seem that the last one would come closest to duplicating the task as performed by an experienced cartographer as he generalizes.

The cartographer attempts to maintain the character and overall impression of an empirically defined, or hand drawn line by selective deletion of some of the details. A fjorded coast is represented by only a few of the actual number of fjords, a delta by only a few of the actual number of channels and so forth. The automation of this approach would rely, therefore, on the ability to programme the computer to recognize specific cartographic features. One attempt is based on an interactive computer programme which has the ability to 'learn' from the actions of an operator (Clement, 1973). The operator generalizes a line plotted on a cathode ray screen by signalling the deletion or maintenance of points. As the computer 'learns' from what the operator selects it attempts to recognize similar features on its own. This system at its present level of development concentrates on the angular and length relationships of a very small number of segments, but the number of possible ways to represent a single simple class of feature, such as a peninsula, is simply staggering. This interactive system, therefore, represents but a small step towards the solution of a fantastically complex problem.

The second group seeks to approximate the points along a line with mathematical functions. This can be done for the whole line at once or it can be done in some piece-wise order, taking a small number of connected points at a time. There are several different methods fitting into the latter category. One developed by A.R. Boyle for the Hydrographic Survey of Canada (1972) computes a first order least squares line through a fixed number of points and then steps forward in that direction by a predetermined distance. Two other approaches begin by defining the ends of segments as averages of a fixed number of points along the line. Koeman and Vander Weiden (1970) suggest taking the mean while Jancaitis and Junkins (1973) take the distance-weighted centroid. When these central points are joined the results simulate a piece-wise approximation with functions of the first order. It must be mentioned, however, that the stated purpose of Jancaitis and Junkins (1973) was to smooth and not necessarily to reduce the line.

The resulting data sets of extracted functions are economical in terms of storage space required, but are relatively time consuming in the processing stage. The greater the number of points, the more costly and complex the operation. These functions reproduce lines which are typically much smoother than the lines they represent. In the main they are probably much better suited for smoothing than reduction and have to be considered of limited value for generalizing. Functions extracted in a piece-wise fashion tend to under-represent erratic curves and over-represent smoother curves. Methods which look for central tendencies are inclined to depress the effect of extreme points. Unfortunately, these are often the very points which give character to the line.

Of the group of methods which eliminate points, some concentrate on the points which are to be deleted while others are directed towards selecting those points which are to be maintained. The algorithms directed at deleting points are usually the simplest. In the case of data recorded by time-automatic digitizing, a simple test to drop those closer than one resolution unit can eliminate a large percentage of the points recorded. This method can be extended by purposely decreasing the numerical resolution or by establishing a threshold distance. Points closer than this distance to neighbours are dropped (Tobler 1966). For chain encoding, a simple compression on the basis of consecutively equal vectors can also result in significant savings. This can be extended as well for other types of encoding by dropping points whenever the direction of the line is not changed through a threshold angle by the segments subtended on it. The underlying purpose of these methods is to eliminate wasted data space but, since the line plotted after this kind of processing would look very much the same as it would before, it cannot represent a significant step towards automated generalization.

The simplest and most often used method of line reduction is to delete all but every nth point along the line where n is a fixed integer based upon the desired degree of reduction (Experimental Cartographic Unit, 1971). The method does not require much in the way of computing resources and it furnishes acceptable results if the digitizing was extremely dense. The primary disadvantage is the frequent elimination or misrepresentation of important features along the line, such as promontories, indentations, sharp angles and so forth. A secondary limitation is that straight lines are still over-represented. These shortcomings are made obvious in Figure 2.2.

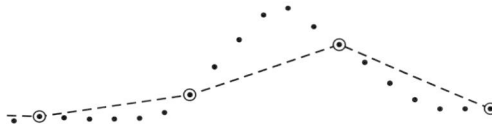

Figure 2.2 Line reduction by the selection of every sixth point.

The alternative to deleting points is to select them. In the special case of monotonically increasing lines (for instance, just one value of 'y' for every 'x'), crests and troughs may be selected. The obvious disadvantage here is the omission of points where there is a change of direction but which nonetheless are not crests or troughs. For irregular planar curves, the problem is more difficult. Jarvis converts the Cartesian to polar coordinates and then looks for crests and troughs (Jarvis 1971). This is useful for curves which can be made monotonic by this conversion but, as for Cartesian measure, the solution cannot be considered general.

One alternative to line generalization which seemed to hold conceptual promise was that method provided by the German firm A.E.G., which supplied the Experimental Cartographic Unit with its GEAGRAPH 4000 plotter, and was described by Lang (1969). This method was reported as producing acceptable results but was eventually rejected as a general purpose technique by the Experimental Cartographic Unit on the grounds that it required far too much computer time for the online processing system being operated at the time. The objective of the procedure was to delete points if they were found to lie within a tolerance distance of a straight line segment being tested to represent a portion of the line. From one representative point it constructs straight lines to subsequent points until one point between the representative point and the subpoint is further away from the line linking the two than a pre-set tolerance value. As soon as this condition is satisfied, the point before the subpoint becomes a new representative point and the procedure is repeated. The method gives acceptable results in the case of smooth curves but it does not detect the best representative points on sharp curves and the results are particularly unsatisfying where sharp angles are numerous.

The methods proposed in this paper are based on a concept somewhat similar to the pre-set tolerance ideas described by Lang but concentrate rather on the selection of points rather than on their deletion. Approaches to a computerized solution to many problems begin with an examination of the way one would solve them subjectively. Consider the line represented by points illustrated in Figure 2.3(a). One might choose the encircled points as those which represent the original line to our own requirements of accuracy. Perhaps the reason we would select these points and not others might be illuminated by examining the simpler situations in Figure 2.3(b) and (c). Starting with the obligation to begin with the end points, the question might be: 'Why would there be a compulsion to insert a point C in (b), where no such compulsion would exist in (c)?' The perpendicular distance of C from the segment A–B may provide a clue. This suggests that an arbitrary maximum distance could be established. If no point along the line is further than this distance from the straight line segment connecting its end points, then the straight line segment will suffice to represent the original line. If this condition is not satisfied, then another point along the curved line must be selected and the same test would be carried out with the new segments. The next question is: 'What point along the curved line should he selected to become the end point of the two new straight segments created?' The obvious answer is the furthest point from the straight segment. Although it is possible that this point may be embedded in a long smooth curve, it is more likely that it is the apex of a relatively sharp angle. As well, this point has already been identified as a result of the distance search; therefore, the benefits associated with its selection far outweigh the possible attraction of selecting some other representative point. In the case of closed loops, where the first and the last point do not define a line, then the maximum perpendicular distance from the segment is replaced with the maximum distance from the point. The same process would be repeated with the new segments created until the maximum distance requirement is satisfied for all straight segments.

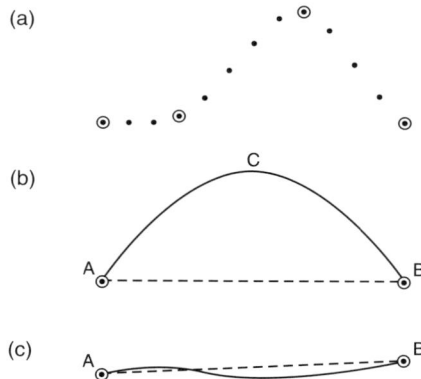

Figure 2.3   Subjective selection of representative points.

Two different procedures embodying these principles have been encoded in FORTRAN IV and tested. In addition, Method 2 has been encoded as a recursive function in ALGOL W (Clement, 1973).

Method one begins by defining the first point on the line as an anchor and the last as a floating point. These two points define a straight segment. The intervening points along the curved line are examined to find the one with the greatest perpendicular distance between it and the straight line defined by the anchor and the floater. If this distance is less than the maximum tolerance distance, then the straight segment is deemed suitable to represent the whole line. In the case where the condition is not met, the point lying furthest away becomes the new floating point. As the cycle is repeated the floating point advances toward the anchor. When the maximum distance requirement *is* met the anchor is moved to the floater and the last point on the line is reassigned as the new floating point. The repeat of this latter operation comprises the outer cycle of the process. The points which had been assigned as anchor points comprise the generalized line.

Method two is exactly the same as method one except that note is taken of all points which have been assigned as floaters on previous inner cycles. These are stacked in a vector. After the anchor point is moved to the floating point, the new floating point is selected from the top of this stack, thereby avoiding the necessity of re-examining all the points between the floater and the end of the line. This procedure usually results in the selection of a slightly greater number of points than Method 1, but takes approximately 5% of the computing time and is thought to produce better caricatures. This method can also be thought of as taking a logically hierarchical approach to line reduction. On one cycle extreme points are selected and these tested to see if they suffice. If they do not, intermediate points are taken and the same question asked about each of the two new segments produced, and then each of the four new segments are examined, . . . and so on as if in a branching tree. Each branch is terminated when the offset tolerance criterion is satisfied.

To enable valid comparisons four separate subroutines were written on the basis of the procedure described by Lang. One was an exact duplication of that procedure while the other three were combinations of two incorporated modifications.

The programme Lang describes starts by assigning the first point as the anchor and the third as a floater. The second is tested to see if it lies within tolerance distance of the segment defined by the anchor and the floater. If it does, the fourth is assigned as the floater and the second and third arc examined and so on. The first floating point defining a segment which does not allow all intervening points to satisfy the tolerance criteria causes the anchor to move to the point before the floating point. Since selection of the point immediately before the floating point has no cartographic justification, the first modification of the procedure has the anchor point move to the point furthest from the segment. The reasoning behind selecting the furthest point is that it is the one most likely to subtend a sharp angle and would, therefore, have the best chance of properly representing the line. The second modification attempts to cut computing time by avoiding unnecessary repeated calculations of distance. From Figure 2.4, it is clear that in most cases, the sum of the distances $a + b + c$ is greater than the greatest distance that P_1, P_2, or P_3 lies from the segment P_0P_4. In other words, if $a + b + c$ is less than the tolerance distance then d also would be less than the tolerance. Only one distance, rather

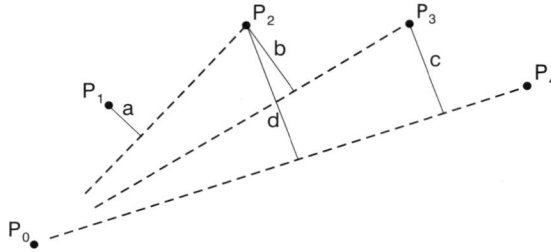

Figure 2.4   Running registers of accumulated offset distances.

than all of the intervening ones, has to be calculated on each cycle. The inner cycle, intended to find the point lying furthest from the segment, is invoked only in the cases where the accumulated total is greater than the tolerance. Positive and negative accumulations are kept in separate registers to avoid subtractions from their absolute magnitudes in the case of double curves. The maintenance of these running registers is particularly useful when series of points lie along straight lines.

The first modification, which attempts to select a point which is more rationally defined than simple convenience, has the expected result of approximately doubling the number of points selected and the processing time required to isolate them. The second modification definitely reduces the time required to process a given line, especially if a great many points are deleted because they lie along relatively straight segments.

All procedures were tested and compared, both for their ability to remove unnecessary points, that is with the offset tolerance set to be less than the resolution of the plotting device (Figure 2.5), and their ability to produce caricatured representations (Figure 2.6). All were judged to produce satisfactory results for simple line reduction; however; the versions of the A.E.G. procedure without the modification to pick the furthest point from the tested segment did not produce satisfactory caricatures because of the tendency to omit and cut corners. The methods presented in this paper were tested with substantial data sets and found to be operationally suitable both for simple reduction and in the production of satisfactory abstractions (Figures 2.5 and 2.6).

Detailed comparisons in computing time required for each subroutine were made on the basis of a three inch square and a three inch diameter circle, each made up of 4000 points evenly distributed along its periphery. It was felt that the square would give ample opportunity to demonstrate the power of each routine in the case where many points along a line are to be deleted, whereas the circle would be more representative in sinuosity of drawn or empirically recorded lines as far as this timing test was concerned. Table 2.1 presents results by the established offset tolerance in number of points selected and the central processing unit time in seconds required for the reduction procedure (IBM 370/155 under O/S MVT).

Figure 2.7 illustrates plotted results for a 1/10 inch offset tolerance with the points selected by each routine marked with a heavy dot. The fact that the routines selected different points and different numbers of points is not unusual or unexpected and similar differences occur in the case of sinuous lines. The shortcoming of the unmodified A.E.G. procedure is evident in the case of points selected to represent the

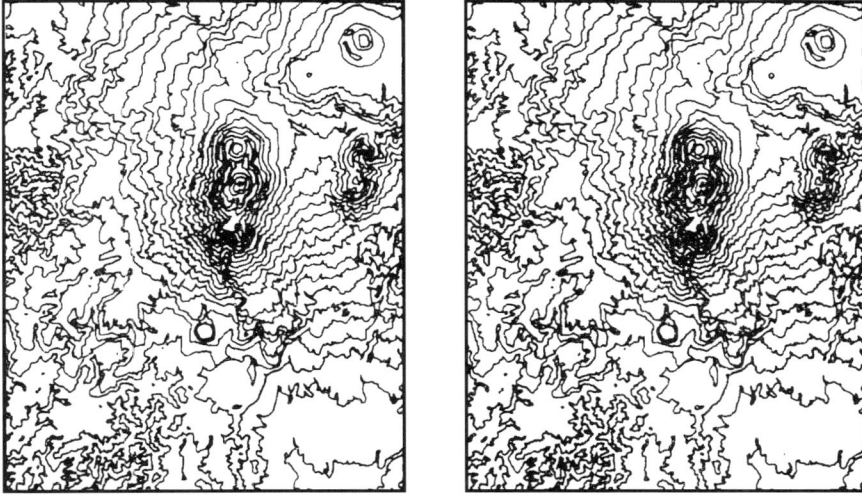

Figure 2.5 Contours plotted from the original digitized, unduplicated at 0.001 resolution, 41 311 points (left), and from 7782 points (right) reduced by Method 2 with a tolerance set to half the resolution of the plotter. The reduction procedure added 16.5 seconds to the 64 seconds required to read and write the data to plot the map on the left. The images may be compared with a simple stereoscope.

square, . . . which were just less than one tolerance unit from the corners for all but the first and last point.

Each routine selected five points to represent the square and each took approximately the same time regardless of the tolerance, except with the second modification. In this case the first step off the straight line caused the inner cycle to be invoked, which found the new anchor point on the first iteration. More iterations were required for the other tolerance limits.

In the case of the circle, an increase in tolerance limit caused a decrease in the number of points found to represent it for all methods. Those methods, which push the

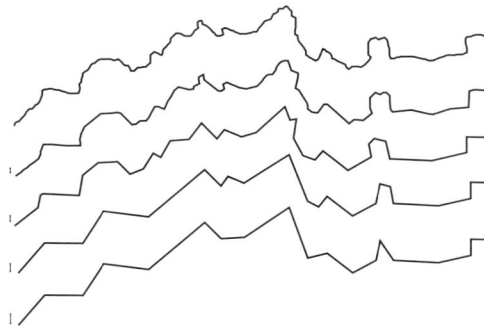

Figure 2.6 Line reduced and caricatured by Method 2. The tolerance value employed is shown to scale at the left of each caricature, which was reduced from the original data represented by the top line.

Table 2.1 Processing time (in seconds) required to reduce a three inch circle and a three inch square, made up of 4000 points each evenly spaced along the perimeter, to the number of points indicated with the given offset tolerance.

| | Offset Tolerance (inches) | | | | | | | | | | | |
| | 0.001 | | 0.005 | | 0.01 | | 0.05 | | 0.1 | | 0.5 | |
SQUARE 4000 points	Points	Time	Points	Time	Points	Time	Points	Time	Points	Time	Points	Time
A.E.G. procedure	5	88.4	5	88.6	5	87.3	5	88.3	5	86.4	5	86.9
A.E.G. plus Mod. 1	5	87.8	5	88.9	5	88.6	5	88.3	5	89.8	5	113.9
A.E.G. plus Mod. 2	5	22.6	5	44.5	5	46.0	5	45.8	5	44.5	5	44.7
A.E.G. plus Mods. 1 and 2	5	22.8	5	22.5	5	22.4	5	22.9	5	23.1	5	33.4
Method 1	5	0.7	5	0.7	5	0.8	5	0.8	5	0.8	5	0.7
Method 2	5	0.6	5	0.6	5	0.6	5	0.5	5	0.5	5	0.6
CIRCLE 4000 points												
A.E.G. procedure	88	5.5	40	11.1	29	14.9	14	32.6	10	42.2	5	97.1
A.E.G. plus Mod. 1	171	10.4	77	20.4	55	28.6	25	60.4	18	84.8	5	109.4
A.E.G. plus Mod. 2	88	5.4	40	10.8	29	15.4	14	32.7	10	41.6	5	92.2
A.E.G. plus Mods. 1 and 2	171	10.6	77	21.5	55	30.0	25	60.9	18	87.8	8	170.2
Method 1	127	25.1	56	10.4	39	7.5	18	3.7	13	2.7	6	1.0
Method 2	129	1.8	65	1.5	33	1.2	17	0.9	17	1.0	5	0.6

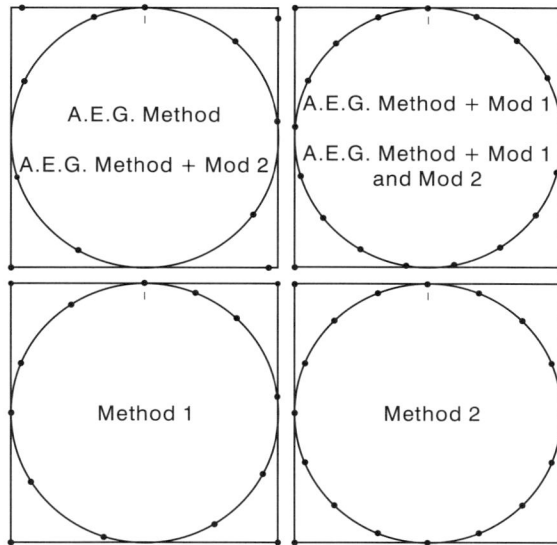

Figure 2.7 Plotted results for a circle and a square each made up of 4000 points around its perimeter. The dots on the boundaries indicate the selected points by the indicated procedures. The tolerance value of 0.1 inch is illustrated to scale at the top centre of each diagram.

examination segment ahead of the anchor points, that is the A.E.G. method with none, one or two modifications, take longer to perform as the offset tolerance is increased. This, therefore, comprises the main reason that they have to be considered unsuitable in an operational context. These procedures are fastest if they are unable to delete any points, because in such cases they would have to examine only one point to come to that decision. On the other hand, if a great number of points are found to be deletable, increasingly large inner cycles are invoked for each advance of the floating point. The two methods presented in this paper work in entirely the opposite way and are fastest in the case of lines which are found to be representable with a smaller number of points. Presumably this is the object of the effort. In all cases, Method 2 is seen to take as little as 1% of the time required by the others.

The prime purpose of the routines discussed here is to reduce the number of points required to represent a line and to produce abstractions, or caricatures, of the line in cases where these will suffice. In many cases these could be considered to be perfectly adequate generalization procedures. While the scope of generalization is no doubt much broader, line reduction by means such as those described here represents an important portion of that topic.

References

Boyle, A.R. (1970) *Computer Aided Compilation*. Hydrographic Conference, Ottawa, January.
Clement, A.H. (1973) *The Application of Interactive Graphics and Pattern Recognition to the Reduction of Map Outlines*. Unpublished Master's Thesis, University of British Columbia.

Experimental Cartographic Unit (1971) *Automatic Cartography and Planning*, Architectural Press, London.

Freeman, H. (1961) On the encoding of arbitrary geometric configurations. *Institute of Radio Engineers Transactions on Electronic Computers*, **EC-10**, 260–268.

Jancaitis, J.R. and Junkins, J.L. (1973) *Mathematical Techniques for Cartography*. Final Contract Report for US Army Engineers Topographic Laboratory, Fort Belvoir, Virginia, Contract No. DAAK02-72-C-0256, February, pp. 15–20.

Jarvis, C.L. (1971) A method for fitting polygons to figure boundary data. *The Australian Computer Journal*, **3**, 50–54.

Koeman, C. and Vander Weiden, F.L. (1970) The application of computation and automatic drawing instruments to structural generalization. *Cartographic Journal*, **7**(1), 47–49.

Lang, T. (1969) Rules for robot draughtsmen. *Geographical Magazine*, **XLII**(1), 50–51.

Pfaltz, J.R. and Rosenfeld, A. (1967) Computer representation of planar regions by their skeletons. *Communications of the ACM*, **10**(2), 122, 25.

Tobler, W.R. (1966) *Numerical map generalization, Michigan Inter-University Community of Mathematical Geographers, Discussion Paper No. 8*, Department of Geography, University of Michigan.

3

Reflection Essay: *Algorithms for the Reduction of the Number of Points Required to Represent a Digitized Line or its Caricature*

Tom Poiker and David H. Douglas[1]

In the early 1970s, there were few geographers who worked with computers. Tom had taken a two-week course in Fortran programming, two months before he moved to Simon Fraser University, and when he mentioned that fact to his colleagues in the Geography Department, he was immediately declared 'the other computer expert in the Arts Faculty'. And that stuck. He never was able to pursue the topic he wanted to follow up from his dissertation (location theory and the geography of education). In 1970, Tom organised a session for 'Computers in Geography' at the annual meeting of the Canadian Association of Geographers and David came to the session and we started talking.

This was also a period when the limiting factor in the development of a field that dealt with computers was not our imagination, not even the available hardware (we simply assumed that by the time the article was published, the hardware would be available) but the tremendous restriction in memory. In cartography, the data volume that developed very quickly – with major governmental projects – was way beyond any computer capacity that we dared to imagine. Besides, most automatic data collection systems were so wasteful in the number of units which they collected that a selection would also speed up any processing of the data sets afterwards.

A case in point: the Central Intelligence Agency (CIA) had developed the 'World Data Bank II', one of the first data sets for geographic coordinates of the world coast lines, main rivers and national boundaries (state boundaries for the USA). The data set was

[1] Simon Fraser University and University of Ottawa, both retired.

Classics in Cartography: Reflections on Influential Articles from Cartographica   Edited by Martin Dodge
© 2011 John Wiley & Sons, Ltd

about five million points and every point was stored as 20 bytes. The data set was sold for \$200 – if memory serves right. When Tom tried to use it, the Computing Centre at Simon Fraser University refused outright to deal with anything that had hundred million bytes. So Tom had an idea: we create a new data set that uses the start and end of every line with the original 20 bytes but every intermediate point with just two bytes by converting absolute coordinates to relative coordinates. Relative coordinates were given as deviation from the previous point, a process that needed less than ± 128 geographic seconds. For the very few cases where this was not the case, we interpolated. The Computing Centre was so enthusiastic about this conversion that they wrote a spooling routine for us to treat one line at a time.

Even though Tom never described this development in a publication because it didn't seem a large enough project, he talked about it at conferences to friends and colleagues who in turn asked for copies of the data sets. There was never the thought to charge for the data set but was the CIA content with the transfer? Eventually, Tom called the CIA data division and asked for clearance. The data guys passed the request on to management and management passed it on to their lawyers. So, one day, Tom got a call from a CIA lawyer. After getting an explanation of the situation, the lawyer first decided that this was a new data set and therefore belonged to Tom. Tom protested and argued that the CIA collected the data, Tom only changed the structure. So the lawyer decided that Tom could distribute the data, for free, if the term CIA was used with the data name. At the end of the conversation, the lawyer said – prophetically – 'I believe that this is not the last time that we will discuss the issue of data ownership'.

David had his own ideas for reducing the size of line data sets. Government agencies were collecting line data automatically at a density that was beyond the capacity of most potential users. There was very little technology available to reduce the number of points and what was available was not good, we decided. David described his approach to Tom and asked at the same time to be accepted into the PhD program in the Geography Department of Simon Fraser University.

Shortly after David moved to Simon Fraser University, we started to work on a publication of the procedure. Tom thought that the whole idea ought to be based on some more theory but was slow in developing conceptual ideas. So David suggested that we finish the paper without the theory or he would write the paper alone. So we finished the paper and submitted it to *Cartographica* (then known as *The Canadian Cartographer*) (Douglas and Peucker, 1973).

It has to be said that we were not the only ones who saw the need for data processing routines that save storage capacity in spatial data. Line generalization was approached with moving averages (Tobler, 1964) and ellipses (Perkal, 1966). As a matter of fact, two months before our article was released in print, the computer scientist Ramer (1972) published his article that described basically the same method. And we were told that the method was presented at a conference and described in a book at about the same time (Duda and Hart, 1973). It is quite frequent that the same discovery happens at different places more or less at the same time because the problem poses itself at that particular juncture. But that does not diminish the merit of any of the developers.

However, it is not usual to publish two articles that describe the same algorithm. When the first one appears, the second is usually withdrawn. So why did the Editor of *The Canadian Cartographer* publish our article nevertheless? When we discovered Ramer's article and informed our editor, he answered that both articles were very differently written and were for two very different audiences and he therefore felt there was no harm in publishing the idea twice.

3.1 Improving on Our Explanation

When reading our original article over again – after 36 years – we both noticed that we actually gave a description of the algorithm that was quite inferior to how we described it later, at conferences or to students and friends. So, let us give this 'improved' description first.

- Given a line by a large number of points (Figure 3.1),

    - connect the two endpoints by a straight line (Figure 3.2);
    - find the point furthest away from the connecting line (1 in Figure 3.3) and make this the first point to be retained – apart from the start and the end.
    - create two new lines that have the new point as the end for the first and the start for the second line and repeat the process for both sub-lines (creating 2 and 3 as new points, Figure 3.3).
    - stop the process when the distance to the furthest point is less than a pre-determined threshold (Figure 3.4).

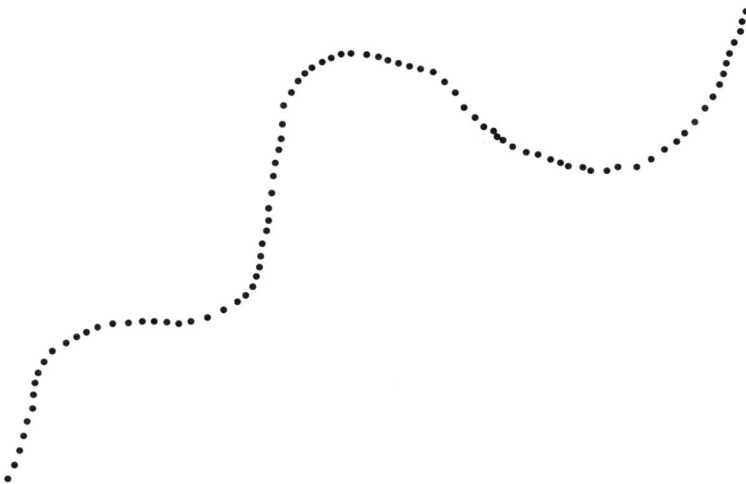

Figure 3.1

Reflection Essay

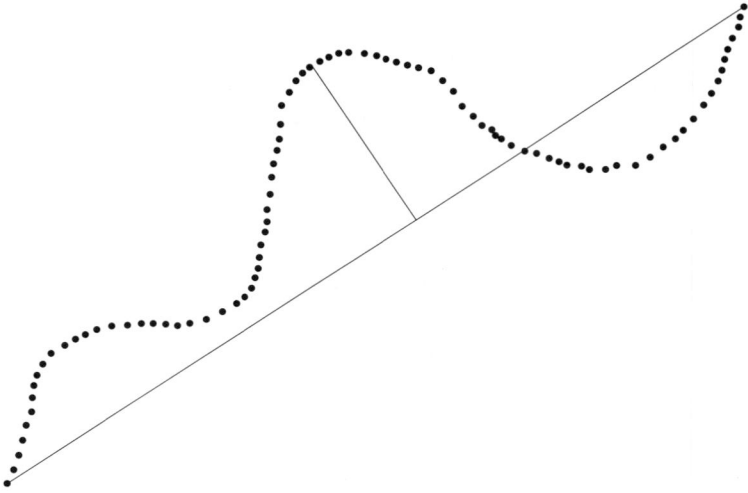

Figure 3.2

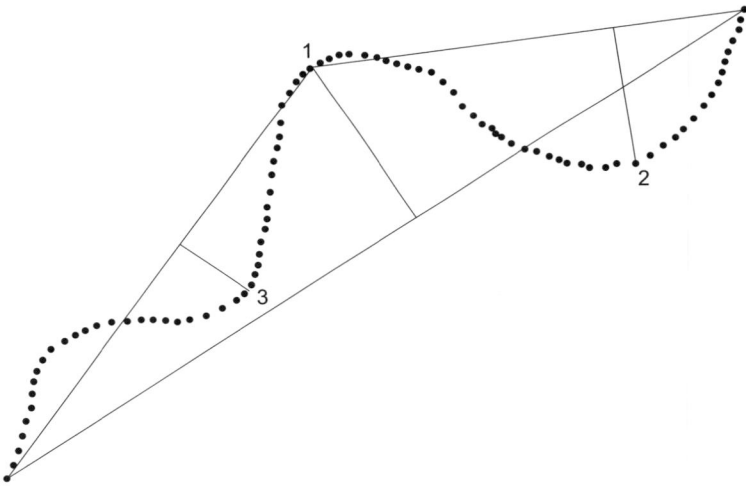

Figure 3.3

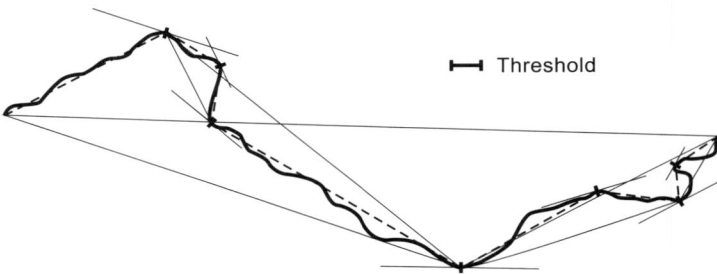

Figure 3.4

The routine is recursive, which means that whenever a maximum point is found, it becomes the endpoint and the starting point for two repetitions of the same routine. At the time that David programmed the algorithm, recursion was not very well known and not easily implemented with the available computer languages, especially Fortran, the most frequently used programming language of the time. But David found a very elegant solution by imitating a 'stack', another relatively new program component.

3.2 Intellectual Impact and Legacy

Both the technique and the article in *The Canadian Cartographer* have had considerable impacts, not all of which we are able to document authentically. The technique's impact, which we are happy to share with the other inventors, was accepted in – amongst others – cartographic software (later called Geographic Information Systems (GIS)), in graphics, for example as a basis of the .jpeg image compressor (source: verbal communication by George Jencks, University of Kansas), and it was also said that NASA was experimenting with it for the reduction of the data volume to send an audio signal, such as speech, by processing the stream (source: George Jencks and Waldo Tobler, then at the University of California, Santa Barbara). And there might be other applications that are kept confidential as commercial secrets.

These adoptions of the algorithm happened at different levels. Firstly, as straight implementations into cartographic – and GIS – software. There was always the question why we did not patent the algorithm. Well, firstly, academics don't regularly patent their ideas, especially when they are as young as we were at the time. But, more importantly, we knew very quickly that we were not the only ones with the idea. Besides, most companies that implemented the algorithm claimed that they had the idea themselves. Secondly, as adoptions into graphics environments. This is also the area where most of the 'expansions' and 'improvements' happened (see the work of John Hershberger and Snoeyink (1992) for a good example). We did not pay much attention to these 'improvements', since they suggested speed-ups of the algorithm that were way below the speed-up that computers were showing over the years and the expansions of data memory that were making these algorithms much less important than they were in the 1970s. Lastly, as expansions of the concepts into other fields: simplifications of surfaces (Fowler and Little, 1979), all the way to the simplification of sound surfaces (Jencks and Tobler, see above).

According to our last consultation of *Google Scholar*, the article has been cited over 1000 times, which makes it the most cited article by both authors but also one of the best cited articles – if not the best cited – in cartography. It is difficult to describe the individual publications that cited the article. For most it seemed that the article personified the digital cartographic world that it was mandatory to cite it, as soon as the article mentioned anything digital.

But there are other reactions to the article that are less than expected. For a long time, a geographer from the University of Wisconsin would start dancing when he saw one of us. When asked why he did that, he would answer 'this is the Douglas–Peucker routine'.

And when Tom got a visit by two graduates from the graduate programme in GIS at the University of Edinburgh, where the article was required reading, one of them asked: 'is Douglas Poiker your brother?'.

The fact that Tom changed his name from Peucker to Poiker in 1979 was less a confusion than the question of who invented the algorithm. In many cases, Tom was referred to as the inventor – being the senior academic. He always pointed at the sequence of authorship – David being the senior author – which referred to the first author as the one who had most of the ideas. It was somewhat depressing for Tom to so often be recognized for the algorithm – 'glad to meet you, Mr. Po..., oh Poiker as in Douglas–Peucker algorithm, right?' – and not for the Triangulated Irregular Network (TIN), which he considers his major achievement. After all, TIN is a whole data structure with a philosophy behind it, it was all his idea, and so on (Peucker *et al.*, 1978). In a non-published article he argues that an academic has a 10% chance of being remembered for what they would like to be remembered.

Eventually, Tom developed the conceptual framework for the routine (and the article) and published it as a separate paper (Peucker, 1976). The basic idea was that every line can be defined by a box that has a general direction, a length and a width (Figure 3.5). Lines can be given a hierarchical structure of boxes that allows for very fast procedures by finding the important points, without having to search through every single point along the line. That, of course, was also the idea that made the Douglas–Peucker algorithm so interesting. One of the more compelling applications of the concept is for the intersection of pairs of lines: intersects, not the lines but their boxes. The area that is common to both boxes represents the most likely area of intersection. Both lines are shortened to the area of intersection and two new boxes are created (Figure 3.6). This process is repeated until the remaining number of points is below a threshold and then the intersection can be performed iteratively. Tom never received any comments to this conceptual article and hardly ever saw it referenced.

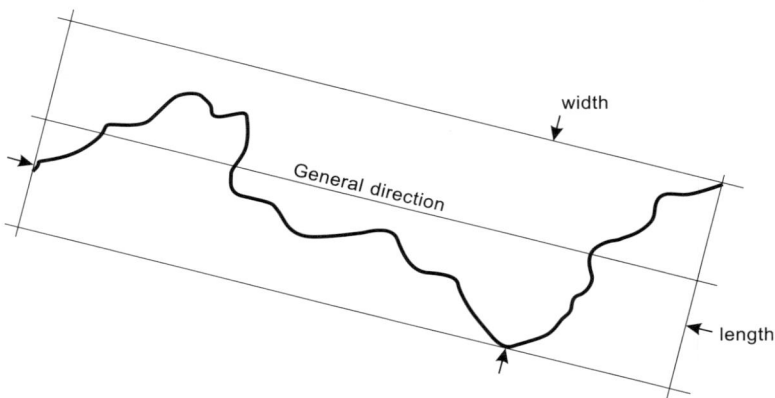

Figure 3.5   A line can be described by a box with a length, a width and a direction.

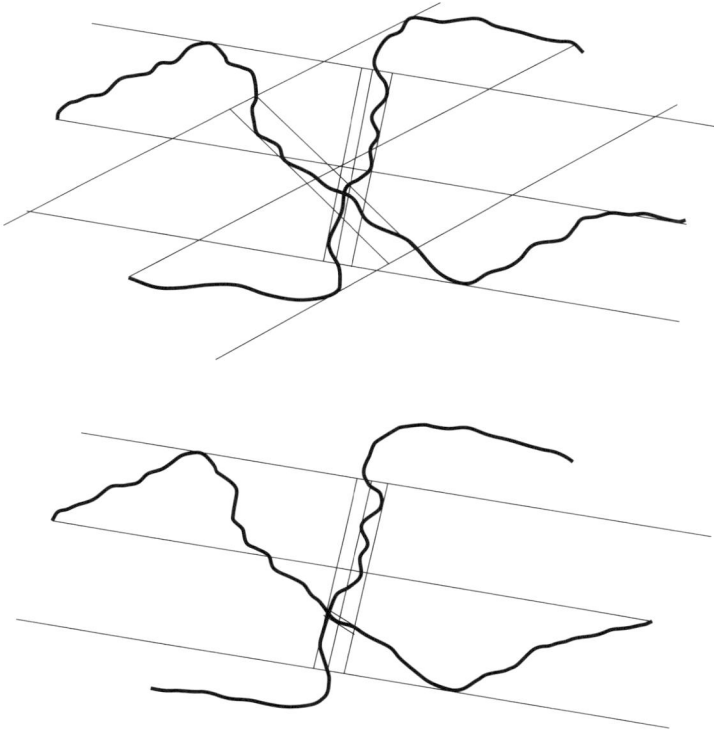

Figure 3.6 Intersection of two lines.

The 1970s was the decade in which academics and scientists developed their own GIS. Naturally, these 'systems' were very small and dedicated to a specific number of jobs. In the 1980s, the first integrated GIS came on the market and programming for geographic data became the domain of large companies that employed many computer scientists and few geographers. The GIS geographers became users with expertise in the application of these large systems. Some of them expanded the philosophical and conceptual side of the field (see several of the other *Cartographica* articles in this volume, including Chapters 8 and 12) but the development of 'algorithms' had basically passed, at least in academic geography.

However, the 'Douglas–Peucker algorithm' still seems to be alive. After 36 years, the interest has not abated and that is surprising. We still get papers to review, especially ones with long titles that have the term 'algorithm' in them.

Further Reading

The NCGIA Core Curriculum, 1990, is a very good and comprehensive overview of the whole field, (www.geog.ubc.ca/courses/klink/gis.notes/ncgia/).

References

Douglas, D.H. and Peucker, T.K. (1973) Algorithms for the reduction of the number of points required to represent a digitized line or its caricature. *The Canadian Cartographer*, **10**(2), 112–122.

Duda, R.O. and Hart, P.E. (1973) *Pattern Classification and Scene Analysis*, John Wiley & Sons, Inc., New York.

Fowler, R.J. and Little, J.J. (1979) Automatic extraction of irregular network digital terrain models. *Computer Graphics*, **13**, 199–207.

Hershberger, J. and Snoeyink, J. (1992) Speeding up the Douglas–Peucker line-simplification algorithm. *Proceeding, Fifth Symposium on Spatial Data Handling*, pp. 134–143. UBC Technical Report TR-92-07 (www.cs.ubc.ca/cgi-bin/tr/1992/TR-92-07).

Perkal, J. (1966) An attempt at objective generalization, Discussion Paper No. 10, Inter-University Community of Mathematical Geographers, Ann Arbor, Michigan.

Peucker, T.K. (1976) A theory of the cartographic line. *International Cartographic Yearbook*, **16**, 134–143.

Peucker, T.K., Fowler, R.J., Little, J.J. and Mark, D.M. (1978) Digital representation of three-dimensional surfaces by triangulated irregular networks (TIN). *Proceedings, Digital Terrain Model Symposium*, May.

Ramer, U. (1972) An iterative procedure for the polygonal approximation of plane curves. *Computer Graphics and Image Processing*, **1**, 244–256.

Tobler, W.R. (1964) An experiment in the computer generalization of maps. Technical Report No. 1, Office of Naval Research, Task No. 389-137

4

The Nature of Boundaries on 'Area-Class' Maps

David M. Mark and Ferenc Csillag[1]

Abstract

Appropriate generalization methods for geographic data must depend upon the kind of feature being generalized. Most research on cartographic line generalization has concentrated on linear features, such as coastlines, rivers and roads; however, methods for the generalization of such linear geographic features may not be appropriate for the generalization of other types of cartographic lines. In this paper, we present a model of another type of cartographic line, namely boundaries between categories or classes which occur over contiguous regions of geographic space. We focus our attention on 'natural' area-class data sets such as soil maps. In the model, such boundary lines are far more similar (mathematically) to elevation contours than they are to coastlines and rivers. Appropriate generalization methods may involve construction of surfaces representing probability of class membership, generalization or smoothing of such surfaces and 'contouring' the probabilities to find boundaries.

4.1 Introduction

Generalization is a major concern in cartography. Our thesis here is that appropriate generalization methods must depend upon the kind of line or feature being generalized. Most research on cartographic line generalization has ignored this point; in practice, most work has concentrated on the generalization of linear *features*, such as coastlines,

[1] Originally published: 1989, *Cartographica*, **26**(1), 65–78.

At the time of publication: Mark was a professor in the Department of Geography at the State University of New York at Buffalo; Csilliag was with the Research Institute for Soil Science and Agricultural Chemistry of the Hungarian Academy of Sciences, Budapest.

rivers and roads, which are observable as linear (or quasi-linear) objects on the earth's surface. Techniques for the generalization of linear geographic features may not be appropriate for the generalization of other types of cartographic lines. As a familiar example, it has long been recognized (for example, Pannekoek, 1962) that individual contour lines should not be generalized independently, since they represent a continuous surface – instead, the underlying topographic surface should be constructed and generalized, and this generalized surface should then be contoured (for recent discussions, see Weibel, 1987; Weibel *et al.*, 1987; Brassel and Weibel, 1988).

Geographic generalization must recognize the intimate association between non-spatial generalization in category space and cartographic generalization in two-dimensional space. This view of map generalization as applied geography is well established in traditional cartography (Pannekoek, 1962). Problems with the simulation of natural resources data (see discussion below) are in part due to the fact that traditional soil maps include the effects of geographical generalization, whereas the simulation models typically do not.

In this paper, we present a conceptual model of cartographic boundary lines of a particular kind, namely those which lie between categories or classes which occur over connected regions of geographic space. We then show how this model can be used as a basis for the generalization of such boundary lines. We focus our attention on the sorts of boundaries that occur between attribute classes on maps for natural resources or ecological phenomena.

Boundaries on a soils map seem to be a typical example of this kind of line, but boundaries of climatic regions are very similar in character. For reasons that will be discussed below, we propose that such boundary lines are far more similar (mathematically and geographically) to elevation contours than they are to linear features such as coastlines and rivers. Appropriate generalization methods may involve construction of surfaces representing probability of class membership, generalization of such surfaces, and either 'contouring' the probabilities to find boundaries or fitting boundaries to the probability surfaces in other ways. Thus, we present models for the generalization of the phenomena which underlie the maps, rather than focussing on the lines themselves.

4.2 Cartographic Background

In his classic paper, 'A Theory of the Cartographic Line' (Peucker, 1975), Thomas Poiker provided a conceptual model upon which to base algorithms for computerized line handling in cartography. Poiker noted that a drafted line, to be visible, must have a finite, non-zero width; he represented this aspect of cartographic lines through the use of recursive bands similar to Ballard's 'strip trees' (Ballard, 1981). Despite the importance of Poiker's paper in the development of analytical cartography, his choice of articles (definite–indefinite) in the title was unfortunate; whereas 'A' theory admits the possibility of future advances, 'the' cartographic line implies that cartographic lines represent a single phenomenon or feature-class, and that they can all be manipulated and generalized in the same way.

Several distinct types of geographic features are frequently represented by lines on maps (Table 4.1). Normally, lines of type 1 are not generalized at all, whereas lines of types 3 and 4b have been the examples used in most cartographic generalization research (Buttenfield, 1985; McMaster, 1986, 1987). Categorical boundaries (type 5) have received little attention in the cartographic literature, and will be focused upon here.

Table 4.1 A typology of geographic lines

Real-world feature is:	Example
1 Mathematical line	Latitude, longitude
2 Legislated line	Some political boundaries
3 Linear feature	
3a linear feature	Railway, road
3b line-like feature (variable width)	Stream, river
4 Zero-set of surface	
4a well defined surface	Contour
4b complex and/or dynamic	Shoreline
5 Area-class boundary	Climatic region boundary, vegetation boundary, soils boundary

An important kind of geographic information represents the distribution of categories: for every location (x, y) in some region of the plane, there is either a class to which an observation made at that point would belong, or a probability that the observation falls into each class within some set of classes. Following Bunge (1966: 14–23), we will call such data *area-class data*, and maps displaying such data *area-class maps*. Chrisman (1982: 16) has used the term 'categorical coverage' to refer to similar spatial data. However, we prefer Bunge's term 'area-class' data or map because the term is shorter, clearer and has priority in the literature.

The way in which geographic data are represented in many geographic information systems (GIS) may be the factor responsible for confusion about relations amongst choropleth maps, coverages and other terms. Fundamentally, GIS may be divided into two broad types, based on their basic data models: vector systems represent spatial information as points, lines and regions in what is essentially a simple object-orientated approach; raster-based systems, on the other hand, divide space into units (pixels, grid cells) which are independent of the distribution of any phenomena, and then list the objects or attributes found there. In a sense, each model misrepresents certain types of geographic data. In a vector system, all lines are represented in the same way (as vectors of coordinate pairs termed *chains* by NCDCDS, 1988), whether the lines represent linear features or class boundaries. Similarly, in a pure raster model, linear features are represented by (large) collections or sets of pixels, rather than as simple objects.

In light of this, it is not surprising that articles in the GIS literature have contradicted cartographic tradition by using the term 'choropleth map' to refer to any polygonal cellular data displays, including soils maps as well as shaded statistical units

(Burrough, 1986; Goodchild and Dubuc, 1987). The cartographic literature normally restricts the term 'choropleth' to situations in which the polygons are determined independently of the phenomenon, and in which the categories are based on sub-dividing the range of a single quantitative spatial variable. The difference between a soils map and a choropleth map (*sensu strictu*) lies mainly in the nature and origins of the lines which form their boundaries; this difference is central to the thesis of this paper.

Uncertainty in the locations of area-class boundaries, when it has been examined at all, usually has been addressed as an issue of accuracy. The US National Committee for Digital Cartographic Data Standards (NCDCDS, 1988) has provided the following definitions of accuracy and resolution:

- *Accuracy*: The closeness of results of observations, computations or estimates to the true values or the values accepted as being true (NCDCDS, 1988: 28);

- *Resolution*: The minimum difference between two independently measured or computed values which can be distinguished by measurement or analytical methods being considered or used (NCDCDS, 1988: 30).

Both accuracy and resolution can be expressed either in a geographic spatial domain (cartographic), or in the attribute or non-spatial domain. *Accuracy* is higher as 'error' is reduced, whereas *resolution* has to do with the sizes of the smallest features or objects that can be detected. In 'category space', the resolution of a classification scheme would tend to increase with the number of possible classes; in a remotely-sensed image, resolution is closely related to the size of the pixels. Spatial resolution for irregular (vector) observations is more difficult to measure, although Tobler (1984, 1988) has provided some suggestions.

Even if users recognize that the accuracy of cartographic products is not absolute, the users may still have an impression that 'spatial resolution' and 'attribute accuracy' are distinct (perhaps even independent) concepts: it is assumed that the resolution provided by the map-scale is adjusted (somehow) to the represented attribute. By taking some extreme examples, however, the link between spatial resolution and attribute accuracy becomes obvious. What is the accuracy of a forest-class at a resolution of 10 cm? What is the accuracy of a soil map at that resolution? What kind of vegetation classes (e.g. from weed species to global life-zones or biomes) can be represented at a resolution of 10 km with 80% accuracy? What resolution provides the best attribute accuracy? Soils data often are entered into geographic information systems as components in natural resource studies. A danger is that, in a GIS, any coverage can be overlain on any other, even in cases where differences in the accuracy or resolution of the component layers make the results meaningless.

4.3 Models of Space and Cartographic Representation

Many authors have treated the 'raster' and 'vector' models as simply alternative data structures for GIS, or even as convertible via reformatting techniques' (Peuquet, 1981a,

1981b). However, these two approaches are based on fundamentally distinct models of geographic space. The raster model is equivalent to the spatial occupancy model used in computational vision. Space is viewed as a 'container', possibly empty, and is divided up into resolution units (or *resels*, as Tobler, 1984, has called generalized resolution elements) without regard for any objects, attributes or properties of the space or things in it. The vector model is commonly used to represent geographic objects or features; in this case, geographic location itself becomes an attribute of the object. In a sense, each model is a dual of the other – they are alternative ways of looking at the same phenomena. Neither model is 'correct', but rather they are just different, although one may be substantially better than the other for certain phenomena or for particular types of modelling, just as in physics light may be treated as waves or particles (photons). Raster/vector differences obscure some of the crucial relationships between geometry, scale, resolution and cartographic generalization.

Traditionally, the basic properties of 'area-class' maps have been separated into attributes and geometry. Such a separation is consistent with an entity relationship model of phenomena, with geometry defining the objects, which then have attributes and relationships. Lakoff (1987: 158–160) recently criticised this E-R model, which is based on classical set theory and an objectivist paradigm, because it does not fit the way the human mind typically represents categories (see further discussion of category theory below). However, it does fit the models which scientists, engineers and planners use in their models of the world. Furthermore, the organization of academic disciplines promotes this separation, since, historically, some disciplines (such as pedology or forestry) have been concerned with the classification of phenomena, whereas others (such as cartography and surveying) have focused their interest on the data representation problem. For many types of spatial information, however, geometry and attribute are intimately intertwined, and any treatment of one in isolation from the other will have a high risk of misrepresenting the phenomenon.

There is a substantial difference between accuracy concepts in the vector and raster models: when the vector model considers uncertainty at all, it normally considers equal probability surfaces, that is, a probability density function can be assigned to the region near a line representing a boundary (Chrisman, 1982a, 1982b). In the raster model, one may instead represent uncertainty as a probability associated with a cell's attribute, or as a vector of probabilities across all possible attributes (Goodchild and Wang, 1988). Note that in the vector model, the uncertainty is commonly attached to the spatial dimension, whereas in the raster model it is attached to the attribute. The inclusion of spatial constraints in attribute classification schemes would be a major step in the development of geographic (cartographic) data models – in such an approach, the focus is on defining the boundaries themselves.

The differences outlined above are thought to be responsible for the lack of unified simulation studies applicable to both raster and vector situations. Researchers must be aware of the limitations of both data models: whereas the vector model is confined by constant neighbourhood functions over a pre-classified region, the raster model, even if class memberships are generated, operates on classes which often are non-spatial.

4.4 A Model of Area-Class Data

The following model of an area-class phenomenon is used in the discussions here. Firstly, we assume the existence of a classification system that, in common with many actual classification schemes used in the natural sciences, has the following characteristics:

1. The system defines a discrete number of mutually-exclusive and collectively-exhaustive classes. For example, given mean monthly values for temperature and precipitation at a site, Koeppen's climate classification scheme can assign the station to a class (see a physical geography text, such as Strahler and Strahler, 1984: 158–163). Or, given a quantitative description of a soil profile (vertical sequence of soil samples), including measures of chemistry, texture and colour, a set of rules would unambiguously assign that profile to a soil class. Thus the class that a location is assigned to is a deterministic function of the quantitative description of variables observed at that site. (Lakoff's 1987 thesis would argue that real-world phenomena do not have inherent classes, but that the human mind tends to divide such phenomena into classes in ways which are relatively consistent across individual minds. Here, we are modelling practice in the natural sciences, not general human cognition; future work to address more cognitively-based categories would be of interest.)

2. Since atmospheric variables can be observed by point or small-area instruments, and a soil profile can be taken for a vertical column at most a few centimetres in horizontal extent (sufficiently small to be considered a point in geographic space), class can be 'determined' at any point.

3. Determination of each variable composing the quantitative description of the class is subject to measurement and sampling errors; thus there is some level of uncertainty associated with the at-a-point determination of class.

Next, note that each of the variables which forms an input to the climatic or soil classification system exhibits spatial variability (Csillag, 1987); most (but not all) vary continuously, most are strongly and positively autocorrelated in space, and most variables are strongly correlated with each other. General 'theories' of soil development, such as the one introduced almost a century ago by the Russian soil scientist Dokuchaev, which underlie many classification schemes, assume that soil properties are some function of five independent soil-forming variables: parent material, climate, vegetation, topography and time. Each quantitative soil characteristic may be considered to be some function of these five factors. Even if the soil category can be determined with absolute certainty ($p = 1.0$) for every soil sample taken, spatial variability for independent variables over a grid cell (pixel) perhaps 30 metres on a side will nevertheless lead to non-zero probabilities for more than one class occurring within some cells.

   In the remainder of this paper, our examples will concentrate on soil. However, the principles clearly apply to climate data very well, and should also be relevant for finding boundaries amongst categories of vegetation, or for certain human activities or artefacts.

4.5 Digital Representations of 'Soil Type'

One objective of soils mapping could be to construct a raster soils image. The simple version of this is to determine the most probable soil type for each pixel or grid cell. A more ambitious objective is to determine for each grid cell a vector containing the probabilities that a soil profile from a randomly-selected point interior to that cell will belong to each member of the set of soil types and to then use those in subsequent analysis, display and modelling. In a vector representation, the objective would be to construct a vector soils map (Goodchild and Dubuc, 1987; Goodchild and Wang, 1988). This vector-based objective has two parts. The first part involves the construction of a set of space-filling (irregular) polygons, with a soil type assigned to each polygon, such that the risk that a 'soil classification error' (an event in which a soil sample from a point belongs to a different soil type than that of the enclosing polygon) is minimized. The second part of this problem is to attach to each boundary line or to each polygon some measure from which the probability of a soil classification error can be determined at any given (x, y).

If we want to use simulation in order to estimate the error associated with a soils map prepared by standard methods, either on the map or when the map has been digitized and stored in a vector-based GIS, then two distinct approaches are possible. We can take the maps resulting from this first step and subject them to a transformation to meet such cartographic standards (this approach was adopted by Goodchild and Dubuc, 1987, and Goodchild and Wang, 1988), or we can constrain the first vector soils map subproblem to meet cartographic standards for soils maps.

In the case of soils, many authors have questioned whether presently accepted soil classification schemes used in mapping are meaningful at all, since, for example, some appear to contain self-contradictions leading to 'fuzzy' classes (Webster, 1968). Furthermore, one can ask to what extent spatial phenomena can be classified without taking into consideration spatial properties. Since many natural phenomena can be regarded as at least two-dimensional in this respect, such evaluation of measurement data ignores significant information.

4.6 Transformations

Four soil-related concepts are of interest. Firstly, there is the actual phenomenon in the 'real world'. Next, there are the two distinct models of space, which lead to raster soils images or vector soil maps. Finally, there are simulations of the soil phenomena, which are of interest both to confirm our understanding of the actual phenomenon, and also to model spatial variation in a controlled environment for the study of accuracy statistics.

Given these four concepts, 6 of the 12 possible transformations amongst them are of interest. These are the transformations which lead to: the construction of the raster soils image from real-world or simulated data; the construction of vector soils maps from real-world or simulated data; and the two transformations between the two types of map/images (Figure 4.1).

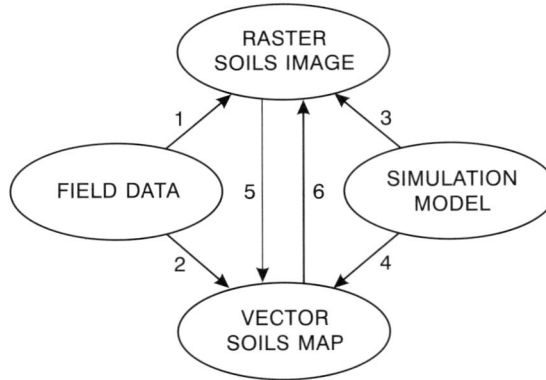

Figure 4.1   Six transformations of soil-related phenomena.

The transformation from field data to a raster soils image (1) is essentially an interpolation problem. One recent example of such a transformation was presented by Bregt, Bouma and Jellinek (1987). Data on three soil variables were determined from borings collected at 60 sites. These then were independently interpolated to the 500 grid points in a 20 by 25 grid, using Kriging. Next, each vector of three interpolated soil properties at a grid point was used as input to a deterministic soil classification equation, determining the most probable soil type at each grid cell. Finally, the resulting raster area-class image was vectorized (transformation 6) and compared (qualitatively) with a soils map determined by standard field methods, and tested against additional field data. They found that the result was of similar reliability to soil maps prepared by traditional methods. However, their map had obvious artefacts of the cells used, and did not have the 'look' of a traditional map.

The second transformation (2) is the way in which most soils maps often are produced in practice (Figure 4.1). Pedologists in the field often sketch the soil boundaries before taking samples (which in fact contradicts assumption 1 of the model presented above in the section 'A model of area-class data'); boundary locations are not directly based on soil properties, but rather are located at points where observable phenomena closely related to soils (for example, vegetation boundaries, breaks of slope, etc.) have sharp changes. Then, only after the boundaries have been determined, a soil sample from near the centre of each polygon is collected and analysed in order to determine the soil class of that polygon. It appears that attempts to automate this process in a direct way have not been published, and probably have not been made; rather, workers wanting to construct soil polygon maps from field data have performed transformation 1, followed by transformation 6 (Bregt, Bouma and Jellinek, 1987). An interesting topic for future research would be to take a 'knowledge-engineering' approach and interview professionals who make soils maps. One could then attempt to develop a system that would mimic the decision making of soil-mapping experts.

Similarly, in the case of simulation studies, no one seems to have attempted the direct simulation of vector soils maps (transformation 6); instead, raster soils images are first simulated (transformation 5) and then these are vectorized (transformation 4). The results of simulations involving raster-to-vector transformations have been viewed as somewhat unsuccessful (Goodchild and Dubuc, 1987; Goodchild and Wang, 1988) because the resulting boundary lines do not have the graphic character of the lines on traditionally prepared soils maps. 'It is likely that the boundaries produced by this simulation process are too irregular to be acceptable; they also show many isolated islands which are rare on real maps' (Goodchild and Dubuc, 1987: 169). Although they then '. . .suggest that these differences are the results of cartographic smoothings which take place during the drawing of choropleth boundaries' (Goodchild and Dubuc, 1987: 169), and thus would be artefacts of the drafting and symbolization process, they proceed to modify their statistical model to include spatial constraints through filtering.

Finally, the construction of a raster soils image from a vector soils map is usually a simple matter of vector to raster conversion (Peuquet, 1981b). It is more difficult to model probability surfaces given only a vector soils map as a starting point; that problem will not be addressed in this paper.

4.7 A Model for Generalization of Area-Class Boundaries

Our model for the generalization of boundaries on area-class maps is based on two assumptions. Firstly, we assume that, over sufficiently small distances, cartographic lines are smooth and can be adequately approximated by some family of parametric curves (splines, Bezier curves, etc.) with a relatively small number of parameters. Secondly, we assume a probabilistic epsilon band model of the position of the boundary line (Honeycutt, 1987). In this model as applied to soils maps, the probability that a sample point has the same soil type as that of the polygon within which it falls is some monotonic function of distance from the boundary (or boundaries) of that polygon. This function approaches one in the limit as the distance from any boundary goes to infinity; for a two-soils-type map, it would be 0.5 at the boundary. Whereas there is reason to suspect that the epsilon function may be asymmetric, and almost certainly varies with location along a boundary, a cumulative normal function with constant epsilon might be a good first approximation. The original epsilon model presented by Perkal (1966) and elaborated by Chrisman (1982a, 1982b) and Blakemore (1984), in which probability is 1.0 outside the epsilon bands and 0.5 within them, can be viewed as a discrete approximation to this continuous, symmetric epsilon model (Figure 4.2).

Given this conceptual framework and a vector model (digital or graphic) of a categorical coverage, one can define a probability surface for class membership by combining a parametric description of a cartographic boundary with a cumulative normal random deviate version of the epsilon band model (Figure 4.3). The parameterization of the line should ideally have associated with it a closed-form expression for

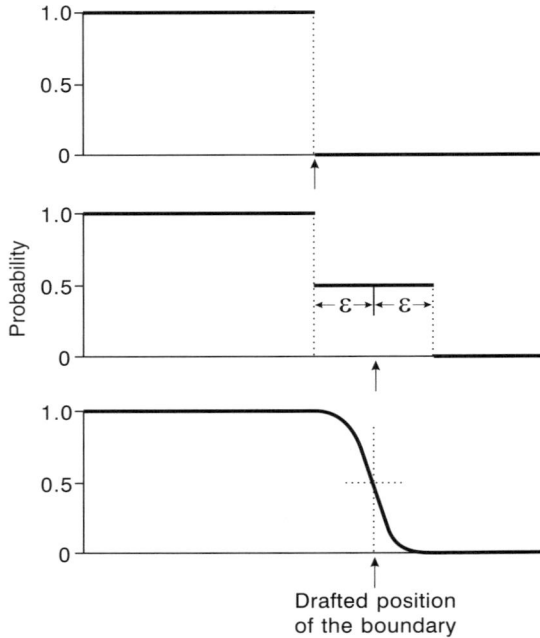

Figure 4.2 Three models of the probability of a point being a member of category 'A', as a function of position along a profile perpendicular to the boundary between a zone of category 'A' on the left and 'not-A' on the right; (top) implication of the vector model; (middle) discrete epsilon model of Chrisman (1982a, 1982b) and Blakemore (1984), after Perkal (1966); (bottom) probabilistic epsilon band model of Honeycutt (1987) and this paper.

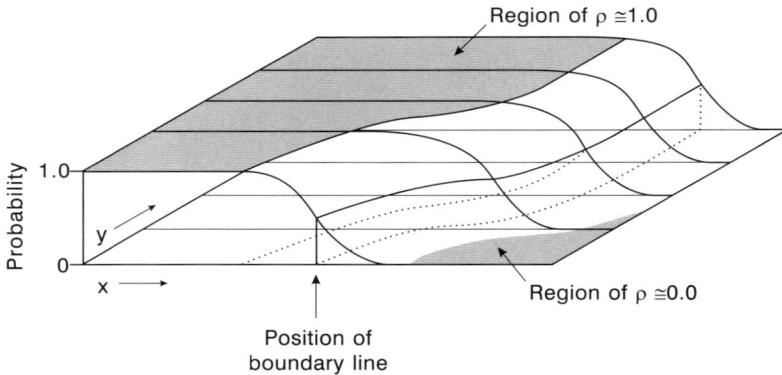

Figure 4.3 Probability surface for some class, based on a probabilistic epsilon band model.

distance from any point (x, y) to the line, given the line's parameters. Equations of this form can then be fitted to observed or simulated soils probability data using least-squares or other fitting techniques, with the results being the parameters of the line and the epsilon estimate. This approach can then be used to incorporate cartographic constraints (introduced as the constraints on the parameterization of the boundary line) into the simulation of area-class maps such as soils maps. A statistical form of cartographic generalization (see Brassel and Weibel, 1988, for a definition of statistical generalization) could be achieved through graphic constraints.

As a simple example, suppose that a square subregion of the plane is cut into two contiguous regions by a soils boundary, one in which soil type 'A' is most probable, and the other in which soil type 'B' is most likely. If we write an equation for the probability that soil type 'x' would be found at location (x, y) in terms of the parameters of the cartographic line and the parameter(s) of the monotonic function of certainty, then that function can be fitted to the probability surface using nonlinear least-squares. Some of the parameters of the best-fit surface relate to the epsilon-band model (in the normal version, there will just be the standard deviation of that normal distribution), and the others will be just a parametric description of the boundary line as it crosses the current square.

Because of the problems of global fitting to a complicated surface, a divide-and-conquer approach will probably be most practical. Quadtrees (Samet, 1984) provide a useful way to find boundaries in two-dimensional space (Mark, 1987), or to fit surfaces to spatial data (Martin, 1982; Leifer and Mark, 1987). Very briefly, a quadtree begins with a square region, and applies some test. Depending on the result of the test, the square may be retained as a data element, or may be split into four subquadrants. Then, the same test would be applied recursively to each subquadrant, and to their sub-quadrants, until no further subdivision occurs.

A recursive fitting procedure for finding area-class boundaries would proceed as follows:

Procedure FIT:

1.1 If the current quadrant contains just one cell, assign the cell's most probable soil type to the quadrant and return.

1.2 If the current quadrant contains more than one cell, then determine the most probable soil type for each cell in the current quadrant. Keep track of how many different soils types occur as most probable type for at least one cell in the quadrant.

    1.2.1 If the same type is most probable throughout the quadrant, then assign the current quadrant to that soil type and return.

    1.2.2 If exactly two soil types occur as 'most probable', then fit a probability surface (as described in the text) to determine the equation and the epsilon parameter(s).

        1.2.2.1 If the goodness-of-fit for the surface is adequate, store the resulting parameters and return;

        1.2.2.2 If the goodness-of-fit for the surface is inadequate, divide the current quadrant into four subquadrants, and apply FIT recursively to each of these subquadrants.

    1.2.3 If three or more soil types occur as 'most probable', divide the current quadrant into four subquadrants, and apply FIT recursively to each of these subquadrants.

Of course, lines found by this procedure would have discontinuities at patch boundaries; a post-processing phase would need to connect the equations of neighbouring patches and produce continuous boundary lines. There must also be a way to explicitly recognize nodes at which three boundary lines meet.

4.8 Discussion

At the end of the 1980s, cartographic generalization research is at a watershed. Procedures for the simplification, selection and reduction of cartographic line features have been developed, evaluated and perfected to the point where they can be valuable standard tools within automated systems. The challenge to cartographers is to go beyond geometric simplification. Automated methods for the generalization of cartographic and geographic phenomena are needed. For linear features, the recognition of feature types will form a first step in such generalizations (Buttenfield, 1989).

'Area-class' maps must be generalized and manipulated as whole phenomena (as 'surfaces'), especially if the resulting maps are to conform to cartographic traditions. Statistically, it may be appropriate to treat the attributes at each point (cell, pixel, resel) independently, ideally as a probability distribution across possible attributes. However, as described above, it has been found that when such raster images are vectorized, the resulting polygons do not have the 'look' of soils maps; to be acceptable as a component in a vector-based GIS, the information must be transformed under cartographic constraints.

This paper contains an approach to the construction of boundaries in 'area-class' maps. The approach proposed would allow graphic constraints on line character (introduced through the parameterization of the boundary) to be combined with statistical surface-fitting. Treating such data as probability surfaces should provide clear guidelines for future generalization and simplification research. Once the approach is implemented, extensive experimental work with real and simulated data will be needed to determine appropriate constraints on line parameters for maps of various purposes and scales. It is hoped that this general approach for combining phenomenon-dependent generalization procedures with cartographic constraints on the graphic product can and will be applied to other kinds of geographic data as well.

References

Ballard, D.H. (1981) Strip-trees: a hierarchical representation for curves. *Communications of the Association for Computing Machinery*, **25**, 310–321.

Blakemore, M. (1984) Generalization and error in spatial data bases. *Cartographica*, **21**(2/3), 131–139.

Brassel, K.E. and Weibel, R. (1988) A review and framework of automated map generalization. *International Journal of Geographical Information Systems*, **2**, 229–244.

Bregt, A.K., Bouma, J. and Jellinek, M. (1987) Comparison of thematic maps derived from a soil map and from Kriging of point data. *Geoderma*, **39**, 281–291.

Bunge, W. (1966) *Theoretical Geography*, CWK Gleerup, Lund.

Buttenfield, B.P. (1985) Treatment of the cartographic line. *Cartographica*, **22**(2), 1–26.

Buttenfield, B.P. (1989) Scale-dependence and self-similarity in cartographic lines. *Cartographica*, **26**(1), 79–100.

Burrough, P.A. (1986) *Principles of Geographical Information Systems for Land Resources Assessment*, Clarendon Press, Oxford.

Chrisman, N.R. (1982a) *Methods of Spatial Analysis Based on Error in Categorical Maps.* Unpublished PhD Thesis, University of Bristol.

Chrisman, N.R. (1982b) A theory of cartographic error and its measurement in digital data bases. *Proceedings, Fifth International Symposium on Computer-Assisted Cartography, Auto Carto 5* (ASPRS and ACSM, Falls Church, VA), pp. 159–168.

Csillag, F. (1987) A cartographer's approach to quantitative mapping of spatial variability from ground sampling to remote sensing of soils. *Proceedings, Eighth International Symposium on Computer-Assisted Cartography, Auto Carto 8*, Baltimore, MD (ASPRS and ACSM, Falls Church, VA), pp. 155–164.

Goodchild, M.F. and Dubuc, O. (1987) A model of error for choropleth maps, with applications to geographic information systems. *Proceedings, Eighth International Symposium on Computer-Assisted Cartography, Auto Carto 8*, Baltimore, MD (ASPRS and ACSM, Falls Church, VA), pp. 165–174.

Goodchild, M.F. and Wang, M.H. (1988) Modeling error in raster-based spatial data. *Proceedings, Third International Symposium on Spatial Data Handling*, Sydney, pp. 97–106.

Honeycutt, D.M. (1987) Epsilon bands based on probability. Paper presented at *Eighth International Symposium on Computer-Assisted Cartography, Auto Carto 8*, Baltimore, MD. Not published in proceedings. Unpublished manuscript distributed by the author.

Lakoff, G. (1987) *Women, Fire, and Dangerous Things: What Categories Reveal About the Mind*, University of Chicago Press, Chicago.

Leifer, L.A. and Mark, D.M. (1987) Recursive approximation of topographic data using quadtrees and orthogonal polynomials. *Proceedings, Eighth International Symposium on Computer- Assisted Cartography, Auto Carto 8*, Baltimore, MD (ASPRS and ACSM, Falls Church, VA), pp. 650–659.

Mark, D.M. (1987) Recursive algorithm for determination of proximal (Thiessen) polygons in any metric space. *Geographical Analysis*, **19**, 264–272.

Martin, J.J. (1982) Organization of geographical data with quad trees and least square approximation. *Proceedings, IEEE Conference on Pattern Recognition and Image Processing*, Las Vegas, pp. 458–463.

McMaster, R.B. (1986) A statistical analysis of mathematical measures for linear simplification. *The American Cartographer*, **13**(2), 103–117.

McMaster, R.B. (1987) Automated line generalization. *Cartographica*, **24**(2), 74–111.

National Committee for Digital Cartographic Data Standards (NCDCDS) (1988) The proposed standard for digital cartographic data. *The American Cartographer*, **15**(1), 9–140.

Pannekoek, A.J. (1962) Generalization of coastlines and contours. *International Yearbook of Cartography*, **2**, 55–74.

Perkal, J. (1966) An attempt at objective generalization (Trans. W. Jakowski from Julian Perkal, Proba obiektywnej generalizacji. Geodézia és Kartográfia, Tom VII, Zeszyt 2, 1958, 130–142). *Michigan Inter-University Community of Mathematical Geographers, Discussion Paper 10* (ed. J. Nystuen), University of Michigan, Ann Arbor.

Peucker, T.K. (1975) A theory of the cartographic line. *Proceedings, International Symposium on Computer-assisted Cartography, Auto Carto II* (US Department of Commerce Bureau of the Census and the ACSM), pp. 508–518.

Peuquet, D.J. (1981a) An examination of techniques for reformatting digital cartographic data; Part 1: The raster-to-vector process. *Cartographica*, **18**(1), 34–48.

Peuquet, D.J. (1981b) An examination of techniques for reformatting digital cartographic data; Part 2: The vector-to-raster process. *Cartographica*, **18**(3), 21–33.

Samet, H. (1984) Quadtrees and related hierarchical data structures. *Computing Surveys*, **16**, 187–260.

Strahler, A.N. and Strahler, A.H. (1984) *Elements of Physical Geography*, 3rd edn, John Wiley & Sons, Inc., New York.

Tobler, W. (1984) Application of image processing techniques to map processing. *Proceedings, International Symposium on Spatial Data Handling*, Zurich, pp. 140–144.

Tobler, W. (1988) Resolution, resampling, and all that, in *Building Databases for Global Science* (eds. H. Mounsey and R. Tomlinson), Taylor & Francis, London, pp. 129–137.

Webster, R. (1968) Fundamental objections to the 7th approximation. *Journal of Soil Science*, **19**, 354–366.

Weibel, R. (1987) An adaptive methodology for automated relief generalization. *Proceedings, Eighth International Symposium on Computer-Assisted Cartography, Auto Carto 8*, Baltimore, MD (ASPRS and ACSM, Falls Church, VA), pp. 42–49.

Weibel, R., Heller, M., Herzog, A. and Brassel, K. (1987) Approaches to digital surface modeling. Proceedings, *Primera Conferencia Latinoamericana sobre Informatica en Geografia*, San Jose, Costa Rica, pp. 143–163.

5

Reflection Essay: *The Nature of Boundaries on 'Area-Class' Maps*

David M. Mark

SUNY Distinguished Professor, Department of Geography, University of New York at Buffalo, USA

I am pleased to have this opportunity to write an essay to place the 1989 essay in context. This brief reflection is dedicated to my late co-author, Ferenc (Ferko) Csillag, who at the time of his premature death on 10 June 2005 was Chair of the Department of Geography at the University of Toronto at Mississauga, Canada.

The genesis of the article was loosely tied to the competition for the US National Centre for Geographic Information and Analysis (NCGIA). In 1988, I was a faculty member in the Geography department at the University at Buffalo (SUNY), recently promoted to the US rank of Full Professor; at the time, I was still teaching computer cartography and some physical geography. Mobility for researchers involved in university-based GIS and spatial analysis was high in 1988, as universities planning to submit NCGIA proposals to the US National Science Foundation (NSF) strengthened their rosters by hiring individuals in key areas. The University at Buffalo (SUNY) lost two such researchers to rival institutions, and funds were provided to allow us to hire visiting scholars to cover the teaching vacancies the following year. One of those 1988 Buffalo visitors was researcher Ferko Csillag from the Hungarian Academy of Sciences.

During his visit, Ferko and I had numerous academic discussions. Ferko was a soil scientist with knowledge of fieldwork and also of spatial statistics and geostatistics. My expertise was more in what we then called analytical cartography. Cartographic line generalization was a big topic (Douglas and Peucker, 1973, Chapter 2 of this volume), and I had long been concerned with how the nature of phenomena themselves should influence their digital and cartographic representations (Mark, 1979; see also Peuquet, 1984, Chapter 12 of this volume). Ferko wanted to co-author something while he was visiting, and eventually we decided to try to write something about how boundaries on area-class maps such as soils maps could be located and smoothed. The result was the *Cartographica* article (Chapter 4 of this volume). (We submitted the paper to *Cartographica* because, in 1989, it was perhaps the best North American cartography journal, and GIS journals were just gleams in the eyes of the publishers!)

In 1988 and 1989, the term ontology had not yet been extended to information systems (Gruber, 1993). But I'm sure that we would have used the term 'ontology' if we had been writing ten years later. Our title, 'On the nature of–', clearly signals an ontological bent. In order to figure out how to draw boundaries on area-class maps, it is necessary to understand the ontology of the phenomenon, namely soil types, and also the ontology of boundaries. The article also contains an interesting typology of geographical boundaries (Table 4.1). This was based on the character, dimensionality and source of the line, whether mathematical, by fiat or as a boundary, and also lines that represent linear geographic features. This typology would also now be considered to be an ontology, and hints at Barry Smith's (1995) distinction between fiat and bona fide boundaries. The article also pointed out the ontological distinction between choropleth maps and area-class maps, a point that was being missed in some GIS articles of the time. In true choropleth maps, the boundaries pre-exist independently of phenomena being mapped, and attributes are based on counts or densities that are dependent on where those boundaries fall. In area-class maps, phenomena exist as fields, and zone boundaries depend on the phenomena. The paper also briefly discussed the interaction between data resolution and taxonomic hierarchies for category maps.

Sadly, I wrote nothing else with Ferko. He returned to Hungary shortly after we wrote this paper, and by the time he returned to North America to take up a position at Syracuse University, my research had moved into cognitive and linguistic aspects of geographic information science (Mark and Frank, 1991) and Ferko was involved in fundamental research in geostatistics.

5.1 Impact

Despite the best of intentions, Ferko Csillag and I never implemented the method operationally in software. And as far as I know, neither did anyone else. Thus the article probably has had little impact on cartographic practice or the development GIS functionality. But the inability to deal with gradation is still an important deficiency of commercial GIS.

The article has, however, had a solid intellectual impact on the GIScience literature, as measured by citations. The ISI database shows 50 citations as of September 2009, and *Google Scholar* reports 101 citations as of January 2010. The paper appears to be Ferko's most frequently cited publication, and only eight other papers that I have written or co-authored have been cited more often. The article continues to be cited although I don't really know why – I suppose because it was an early reference to an important problem.

Gradation continues to be an important topic in GIScience, reasonably well understood in theory and in the literature, but methods to represent gradation of transition zones in spatial information are not readily available in off-the-shelf GIS software, despite being identified as a UCGIS Research Priority in 2003 (Kronenfeld, 2005). Perhaps the theory of boundaries on area-class maps has been developed, but practical implementations still remain a challenge.

With the birth of the NCGIA in 1988, I became co-leader of NCGIA's Research Initiative #2, 'Languages of Spatial Relations'. This confirmed my interest in cognitive science and artificial intelligence applied to geographical phenomena (Mark and Frank, 1991). This led to work on formalizing the meanings of spatial relationships as expressed in natural language (Mark and Egenhofer, 1994), then an ontology turn (Smith and Mark, 2001) and then the topic of how people conceptualize landscapes in different cultures (Mark and Turk, 2003).

Further Reading

There has been much research on uncertainty, gradation, and error in area-class maps since our paper was published. Five papers or chapters stand out as appropriate for further reading on the topic.

Chrisman, N. (1991) The error component in spatial data in *Geographical Information Systems: Principles and Applications, Volume* **1** (eds. D.J. Maguire, M.F. Goodchild and D.W. Rhind), Longman, Harlow, Essex, pp. 165–174 (Chrisman puts map 'error' in context, pointing out that so-called error is not always a mistake by data collectors, but may relate to an inherent mis-match between data models and reality.)

Goodchild, M.F., Sun, G. and Yang, S. (1992) Development and test of an error model for categorical data. *International Journal of Geographical Information Systems*, **6**(2), 87–104. (Significant in its use of simulation to explore the spatial structure of error in categorical maps.)

Worboys, M. (1998) Imprecision in finite resolution spatial data. *GeoInformatica*, **2**(3), 257–279. (This paper discusses relationships between resolution and imprecision in spatial information.)

Worboys, M.F. and Clementini, E. (2001) Integration of imperfect spatial information. *Journal of Visual Languages & Computing*, **12**(1), 61–80. (Worboys and Clementini discuss approaches for data integration when data inputs are imperfect.)

Bennett, B., Mallenby, D. and Third, A. (2008) An ontology for grounding vague geographic terms. *FOIS'08: Proceedings of the international conference on Formal Ontology in Information Systems*, pp. 105–116. (The paper provides a comprehensive review of ontological aspects of vagueness in spatial information.)

References

Douglas, D.H. and Peucker, T.K. (1973) Algorithms for the reduction of the number of points required to represent a digitized line or its caricature. *Cartographica*, **10**(2), 112–122. (Reproduced as Chapter 2 of this volume.)

Gruber, T.R. (1993) A translation approach to portable ontology specifications. *Knowledge Acquisition*, **5**(2), 199–220.

Kronenfeld, B. (2005) Gradation and map analysis in area-class maps, in *Spatial Information Theory, Lecture Notes in Computer Sciences No. 3693* (eds. A.G. Cohn and D.M. Mark), Springer-Verlag, Berlin, pp. 14–30.

Mark, D.M. (1979) Phenomenon-based data-structuring and digital terrain modelling. *Geo-processing*, **1**, 27–36.

Mark, D.M. and Csillag, F. (1989) The nature of boundaries on 'area-class' maps. *Cartographica*, **26**(1), 65–78. (Reproduced as Chapter 4 of this volume.)

Mark, D.M. and Egenhofer, M.J. (1994) Modeling spatial relations between lines and regions: Combining formal mathematical models and human subjects testing. *Cartography and Geographic Information Systems*, **21**(4), 195–212.

Mark, D.M. and Frank, A.U. (1991) *Cognitive and Linguistic Aspects of Geographic Space*, Kluwer Academic Publishers, Dordrecht.

Mark, D.M. and Turk, A.G. (2003) Landscape categories in Yindjibarndi: Ontology, environment and language, in *Spatial Information Theory: Foundations of Geographic Information Science, Lecture Notes in Computer Science No. 2825* (eds. W. Kuhn, M. Worboys and S. Timpf), Springer-Verlag, Berlin, pp. 31–49.

Peuquet, D.J. (1984) Conceptual framework and comparison of spatial data models. *Cartographica*, **21**(4), 66–113. (Reproduced as Chapter 12 of this volume.)

Smith, B. (1995) On drawing lines on a map, in *Spatial Information Theory. A Theoretical Basis for GIS, Lecture Notes in Computer Science No. 988* (eds. A.U. Frank and W. Kuhn), Springer-Verlag, Berlin, pp. 475–484.

Smith, B. and Mark, D.M. (2001) Geographic categories: An ontological investigation. *International Journal of Geographical Information Science*, **15**(7), 591–612.

6

Strategies for the Visualization of Geographic Time-Series Data

Mark Monmonier[1]

Abstract

Strategies for the visual display and analysis of geographic time-series data may be spatial or non-spatial, single view or multiple view, static or dynamic. Labels for place names or other geographic metaphors can describe symbols on aspatial time-series charts. Single-static-map strategies incorporate the temporal dimension through techniques ranging from complex point symbols, or temporal glyphs, to generalized trend-surface or flow-linkage maps focusing on movement. The multiple-static-maps strategy juxtaposes two or more maps for a simultaneous visual comparison of time units, whereas the single-dynamic-map strategy either presents maps in a temporal sequence or shows the evolution of a geographic pattern through a temporally sequenced accretion of symbols. In contrast, the multiple-dynamic-maps strategy provides programmed sequences of multiple views or allows the viewer to interact with maps and statistical diagrams representing different instants or periods. Electronic graphics systems have added time to the cartographer's list of visual variables. This paper addresses the graphic portrayal of geographic time-series data. It explores a variety of graphic strategies for the simultaneous symbolic representation of time and space, and summarizes these strategies in a conceptual framework of potential use to cartographers, geographers and graphic designers. These strategies range from statistical diagrams to maps to video animations to interactive graphics systems with which the analyst might freely manipulate time as a variable.

[1] Originally published: 1990, *Cartographica*, **27**(1), 30–45.

At the time of publication: Monmonier was Professor in the Department of Geography, Syracuse University, New York.

6.1 Graphic Representations in Time-Attribute Space

Time-series data traditionally have called for statistical diagrams like Figure 6.1, with time as the horizontal axis and a single variable as the vertical axis (Du Toit, Steyn and Stumpf, 1986: 265–274). A separate trend line represents each place, and each trend line requires a label identifying the place represented. The label might be a place name, a directional abbreviation such as 'NE', or some other geographic metaphor.

Munchie® consumption per capita

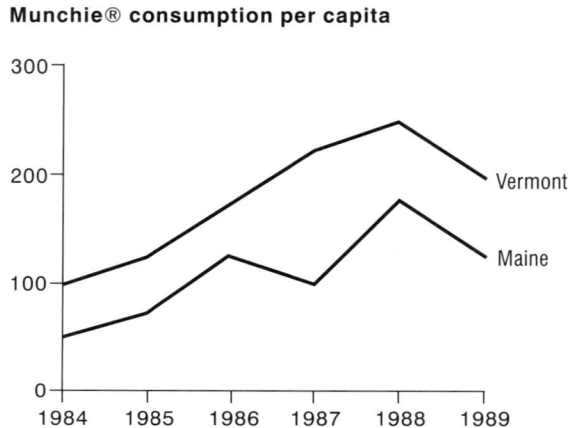

Figure 6.1   A typical time-series graph, with time scaled along the horizontal axis and the attribute scaled along the vertical axis. Two trend lines, both with a place-name label, illustrate geographic variation in an aspatial context.

A logarithmic scale for the vertical axis (Figure 6.2) is a useful modification that allows the slope of the trend line to portray relative rates of change. With an arithmetic scale the slopes portray only absolute change, not the rate of change. With Figure 6.1 the viewer should not compare slopes, whereas with Figure 6.2, they may validly interpret a steeper slope as representing a sharper rate of change than a gentler slope.

The graphic might also focus attention on time periods, as in Figure 6.3, rather than upon sample points on the temporal continuum. In this case the vertical axis might show absolute change or the rate of change.

When the data include many places, symbols and labels readily overload the time-series graphic (Figure 6.4). Assigning each place a unique line symbol might alleviate graphic congestion but, as in Figure 6.5, a wide variety of qualitative line symbols yields a complex key and makes the graph difficult to read. Even when the number of places is not ridiculously large, crisscrossing trend lines are visually complex and require frequent references to a legend.

Various strategies can help the analyst cope with a plethora of places. As in Figure 6.6, a mean or median might represent all places for each time slice. Adding separate trend lines to show temporal trends for key places, as in Figure 6.7, can focus the viewer's attention on important departures from the average trend and provide meaningful

Munchie® consumption per capita

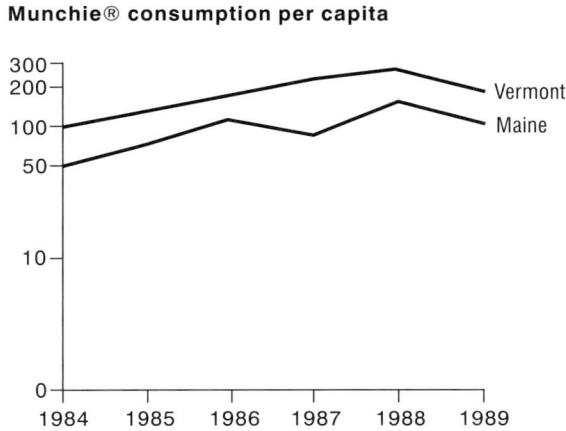

Figure 6.2 A logarithmic vertical scale replaces the arithmetic vertical scale of the diagram in Figure 6.1. This adjustment promotes comparison of the rate of change.

Munchie® consumption per capita

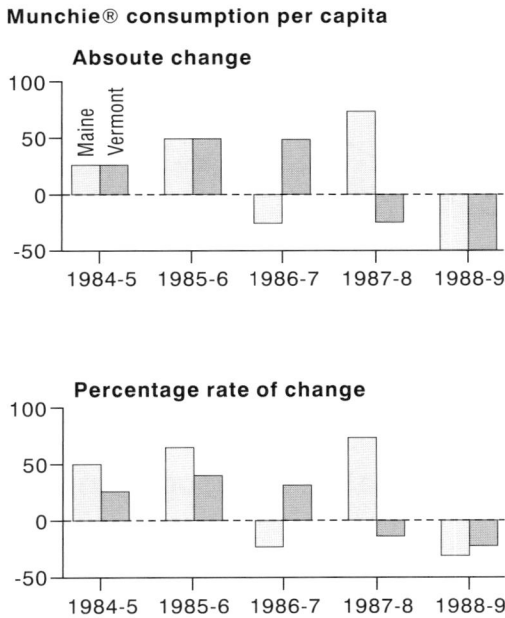

Figure 6.3 Time-series graphs can use bars, instead of lines, to focus on absolute change for time periods (above) and to focus on the rate of change (below). These graphs are based upon the data for Figures 6.1 and 6.2.

comparisons for selected places. Figure 6.8 illustrates how the addition of so-called *error bars*, representing a standard deviation or the interquartile range, can portray variation throughout the region for each time sample as well as temporal trends in regional variation.

Munchie® consumption per capita

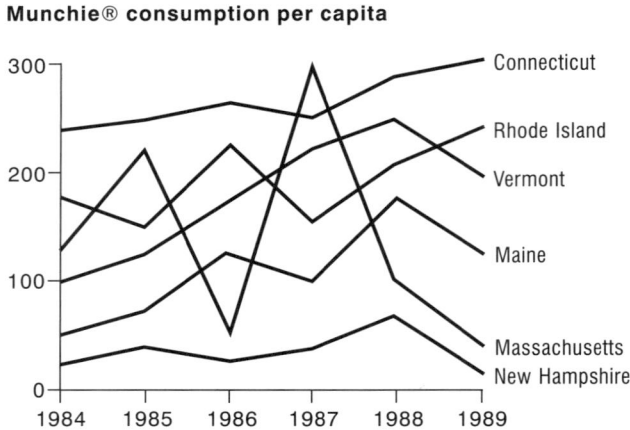

Figure 6.4   A time-series chart with many trend lines, each labelled with a place name.

Munchie® consumption per capita

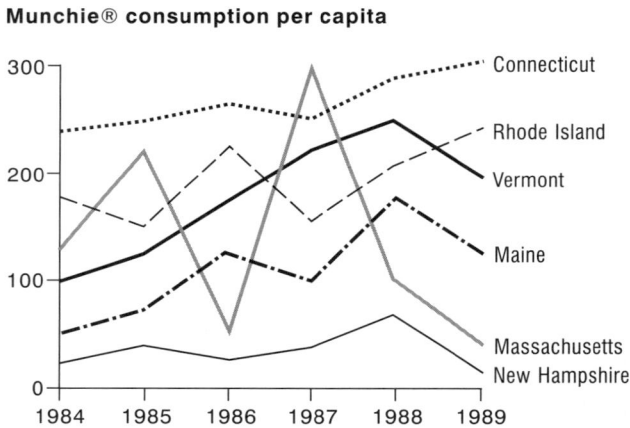

Figure 6.5   A time-series chart with many trend lines, each with a different patterned symbol.

6.2 Graphic Representation in Geographic Space

Thus far, place names and other geographic metaphors have provided the only link with the geographic space, and the phenomena have been univariate, not multivariate. For many applications, though, the analyst must cope with a set of places, each with its own data array (Figure 6.9), in which the rows, say, represent attributes and the columns represent instants or periods. To show relative location, the analyst might also treat these data as sets of maps, perhaps organized as in Figure 6.10, with one set for each attribute subdivided by time sample. Historical atlases based on quantitative data commonly take this form, with each attribute's set of maps arranged on a page or distributed over a sequence of adjacent pages.

   If the number of instants of time is small, if the number of attributes also is small, and if the number of places is not too large, then a cartographic cross-classification array

**New England States median
Munchie® consumption per capita**

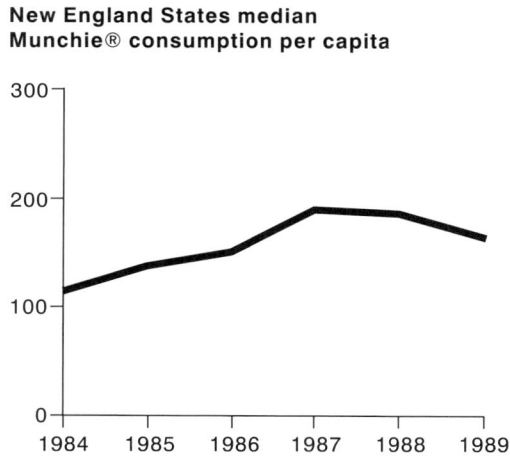

Figure 6.6 A time-series chart based upon the median value for all places for each recorded instant or period.

**New England States median
Munchie® consumption per capita**

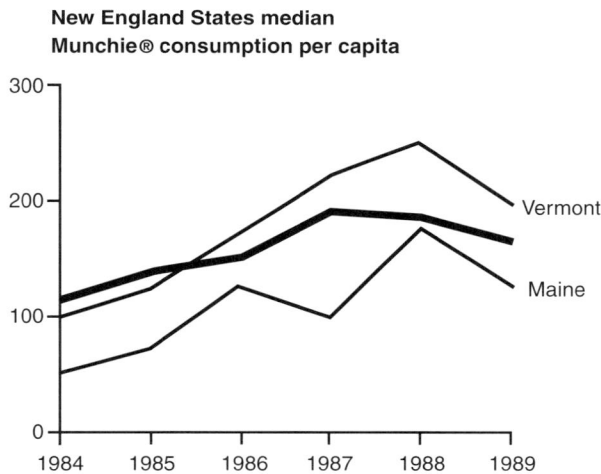

Figure 6.7 A time-series chart based on median values but including for comparison the trend lines for two significant places.

(Figure 6.11) might represent all the data in a single graphic (Monmonier, 1979). Additional columns might even be inserted for periods between time samples, or additional rows might portray rates of change. The eye can readily slew from map to map, and the analyst can examine spatial and temporal trends simultaneously and even infer cross-correlation between variables. Yet for most data sets, small display screens or small pages render this approach unsuitable.

For a single variable observed for several instants or periods, individual cartographic point symbols, or *glyphs,* might portray a separate, spatially positioned series for each place. Figure 6.12 illustrates a few typical temporal glyphs. Although tiny time-series

**New England States median
Munchie® consumption per capita**

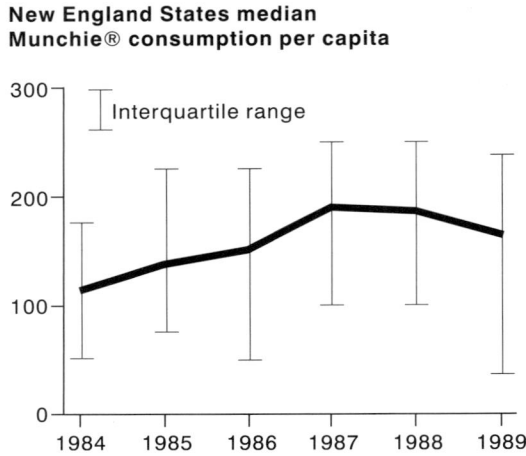

Figure 6.8 A time-series chart based on median values but including error bars to indicate each value's representativeness.

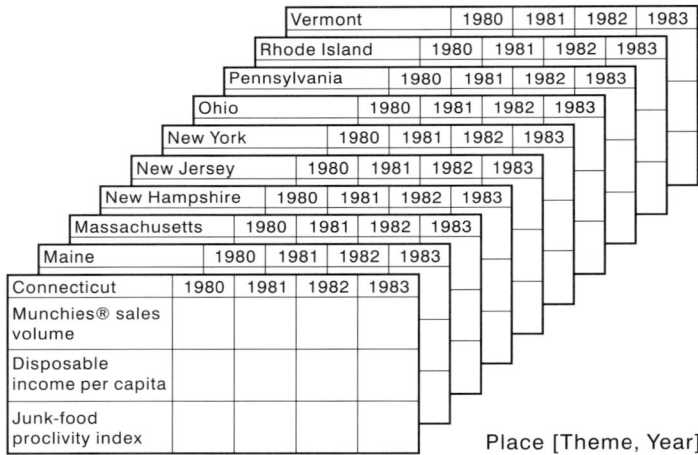

Figure 6.9 Data arrays, one for each place, with rows representing attributes and columns representing time periods.

line graphs might be too visually complex for a map with many places, small clock-face, calendar, or framed time-line symbols could be useful for some applications.

Another strategy is aggregation, perhaps the most severe example of which is the centre-of-population map used for decades by the US Bureau of the Census. As Figure 6.13 shows, for each census year, a single point symbol represents the centre of mass of the national population. By showing these symbols for successive censuses, the map demonstrates most dramatically the westward – and, more recently, the south-westward – movement of the nation's population.

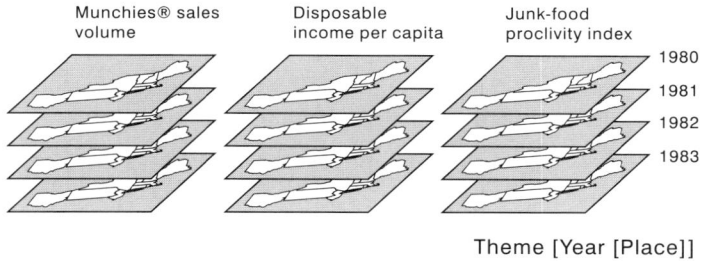

Figure 6.10 Stacks of maps grouped by attribute and within each attribute, by time.

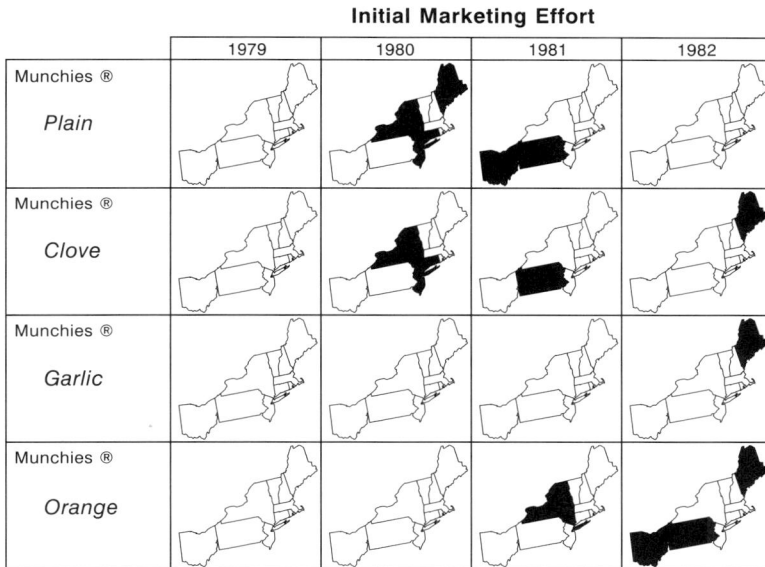

Figure 6.11 A cartographic cross-classification array, with rows representing attributes and columns representing time units.

Other single-map generalizations include displays based upon single dates for each areal unit. A good example of this strategy is the county-unit map showing census year with peak population. This type of map can shock naïve viewers unaware that many counties had more people fifty years ago than they do today.

Another spatial variable well suited to a single-map portrayal is the time of first settlement. Isochronic lines for a polynomial trend-surface (Chorley and Haggett, 1965) provide a concise generalization of major thrusts in the advance of the settlement frontier, as in Figure 6.14, a county-level example for New York State. Canonical trend surfaces (Monmonier, 1970) might be of use as well, to treat simultaneously the frontiers or innovation waves for several different ethnic groups or ideas.

Flow-linkage diagrams similar to Figure 6.15 offer a further generalization of advancing settlement. These directed links, which attempt to reveal principal avenues

Coffee time   Work time   Productivity   Munchies ® purchases

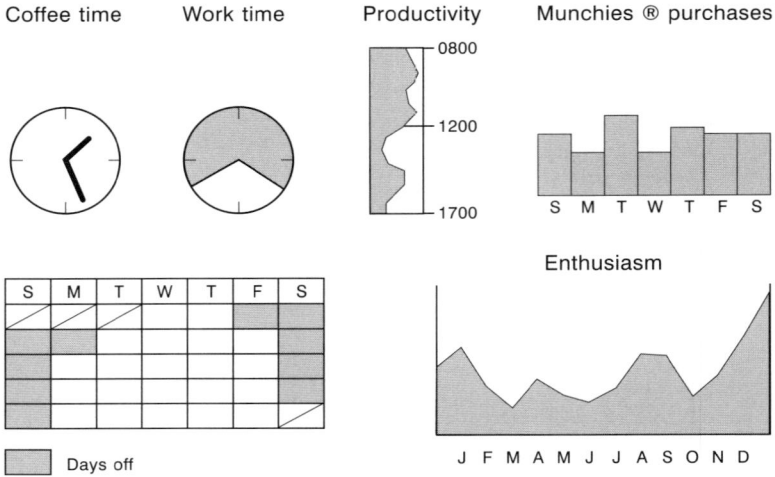

Days off

Enthusiasm

J F M A M J J A S O N D

Figure 6.12 Typical temporal glyphs: the clock face, the calendar and the framed time-line symbol.

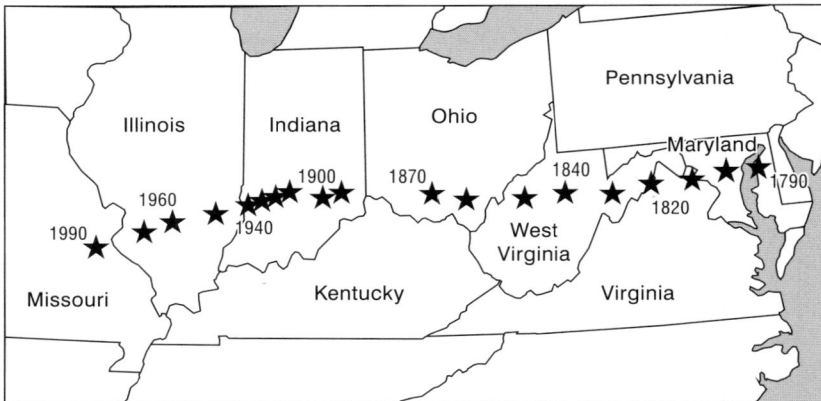

Figure 6.13 Map similar to the centre-of-population map used by the US Bureau of the Census to summarize the general westward shift of the US population since the first census in 1790. (Adapted from Statistical Abstract of the United States, 1984, 104th edition, page 7.).

of movement, focus the viewer's attention on corridors and direction, not on the extent of settlement at particular times (Monmonier, 1972).

Historical and cultural geographers frequently employ directional symbols to portray change over time. Arrow symbols are particularly useful in showing migration streams, the spatial diffusion of ideas, the migrations of tribes and refugees, and the advance and retreat of armies. Directional symbols can vary in size to represent relative magnitude, or vary in label, colour or pattern to represent a particular group or time period. Edward Tufte (1983: 40–41), in his widely acclaimed essay *The Visual Display of Quantitative Information* lavishly praises Charles Joseph Minard's use of

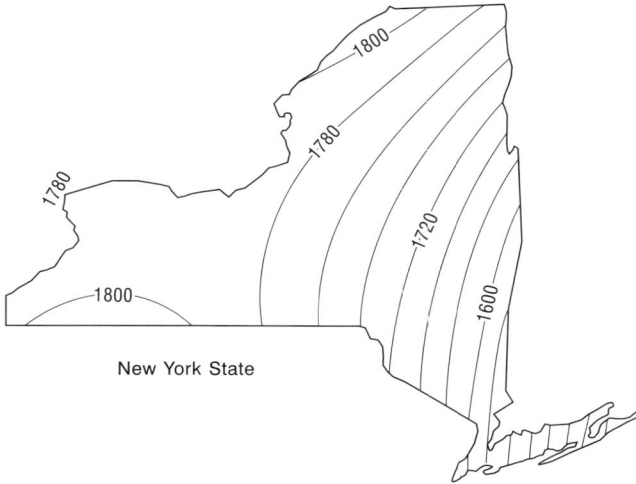

Figure 6.14 Quadratic polynomial trend surf ace showing general pattern of the time of first settlement of New York State counties.

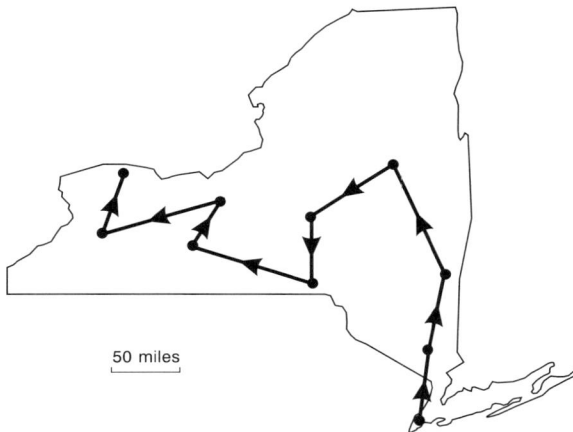

Figure 6.15 Flow-linkage trend diagram showing general pattern of the time of first settlement of New York State counties.

a variable-width flow-line symbol on a map showing the declining size of the Napoleon's army in its abortive Moscow campaign of 1812: 'It may well be the best statistical graphic ever drawn'.

As a cartographic genre, these maps might be termed *dance maps,* after the choreography diagrams (Figure 6.16) used to teach the spatial mechanics of ballroom dancing to generations of students. These maps cover a period marked by several events, each described by map symbols describing a transition from one place to another. Dance maps are one of the three most common spatial-temporal displays.

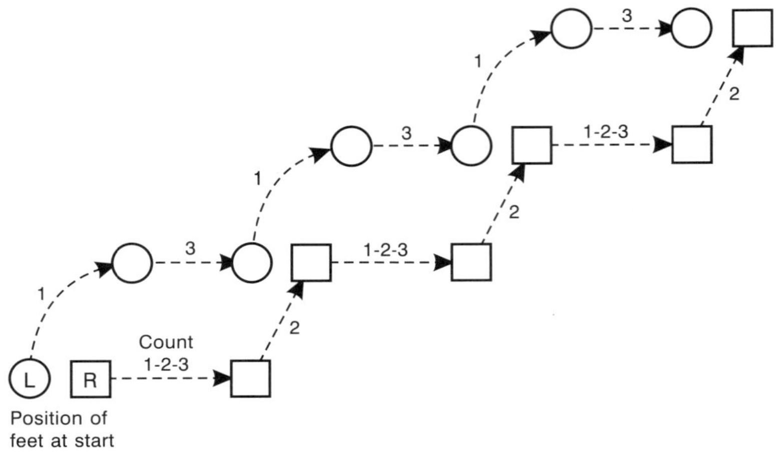

Figure 6.16 A prototypic dance map showing the woman's steps for the Hesitation Waltz (Source: Walker, 1914: 46).

The second type is the *chess map* (Figure 6.17), so called because a separate map presents a snapshot for a discrete instant or period. Chess maps are juxtaposed so that the user can compare the pattern at time 1 with the pattern at time 2. The visual focus here is on the status of the phenomenon at these particular times, not on change *per se*. The chess map strategy would include a pair of choropleth maps representing the same population trait for 1980 and 1990, say, or a set of point-symbol maps showing the distribution of military bases for 1945 and 1985.

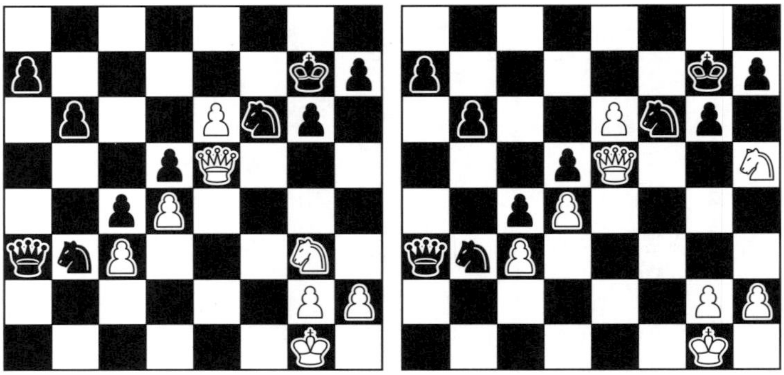

Figure 6.17 Chess maps are two or more geographic-space displays juxtaposed so that the viewer can compare spatial patterns for different times.

The third type (Figure 6.18) is called simply the *change map*. It refers to a single map on which symbols vary in value, size or some other appropriate visual variable to represent the direction, rate or absolute amount of change.

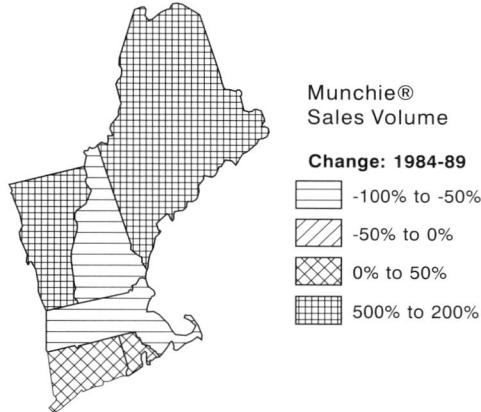

Figure 6.18 A change map showing rate of change by state.

Multivariate data compression techniques, such as principal components analysis (Johnston, 1978: 127–82) or canonical correlation (Davis, 1986: 607–15), might prove useful in reducing the number of variables that need to be displayed. As Palm and Caruso (1972) have aptly demonstrated, labelling factors is often problematic, and map titles such as 'Factor 1' or 'Socioeconomic Status' can be confusing. Indeed, multivariate statistical methods are seldom suitable if the audience does not understand the underlying statistical and geometric principles. Yet for the analyst well-grounded in both statistical and graphic analysis, multivariate methods – including classification techniques – can be highly effective for identifying redundant measures and for extracting summary maps from spatial-temporal data.

6.3 Hybrid Representations with Spatial and Time-Attribute Axes

No multivariate analysis should be attempted without a prefatory univariate analysis. Statisticians advocating *exploratory data analysis* call for graphing the frequency distributions of each and every variable. Statistician John Tukey (1977: 56) warned that 'We have not looked at our results until we have displayed them effectively'. When the data are geographic, though, we need to display them in both the attribute space familiar to the statistician (Figure 6.19, right) and the geographic space that provides the necessary sense of place and relative location (Figure 6.19, left).

Treating just a single variable measured at two different times calls for an array of graphics (Figure 6.20) in both attribute space and geographic space. As a minimum, this array would include separate maps portraying the spatial variance at each of the two times, a map of change or the rate of change, three separate univariate histograms or cumulative frequency diagrams showing frequencies for each time as well as for change, and three separate scatter plots portraying in attribute space the bivariate relationships

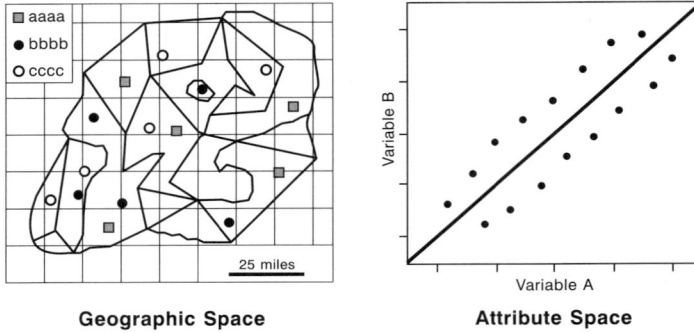

Geographic Space   **Attribute Space**

Figure 6.19   A comparison of attribute space and geographic space.

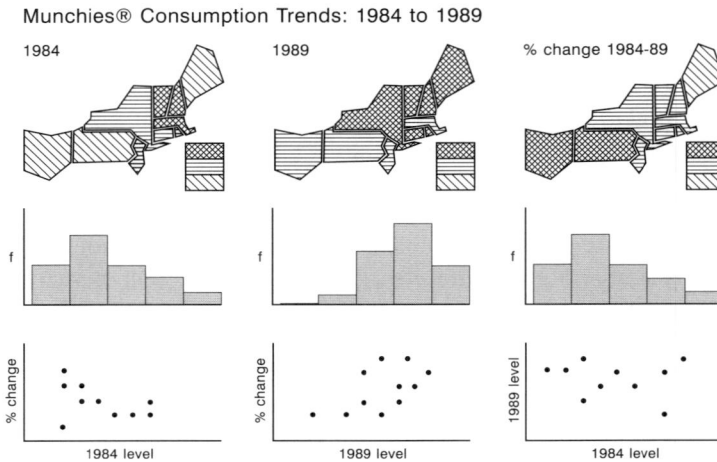

Figure 6.20   An array of graphics for exploring a single geographic attribute measured at two instants of time.

between the two sets of values and between each set of values and the rate or amount of change.

If the geographic space can be reduced to a single measure, such as distance from the equator or distance from the centre of town, a single hybrid graph might combine elements of both geographic and attribute space. The non-spatial axis might be the rate of change or even time itself. Graphic train schedules can employ this concept to reveal, at a glance, relative rates of travel between stations and the length of stop-overs at terminals. As Figure 6.21 describes, each train starts at the top of the graph and moves downward and toward the right. Fast trains have a steep descent, and slow trains a comparatively gentle descent. A horizontal terrace represents layover time at a station. A graph with many trains, some fast and others slow, can show differences throughout the day in frequency, the relative advantages of express service and trains that cover only a portion of the route (Tufte, 1983: 115–116). For commuter routes, with generally frequent service and many stops, a graphic train schedule providing a quick overview of train service might be more useful than a cumbersome numerical schedule consisting of lengthy tables and numerous footnotes.

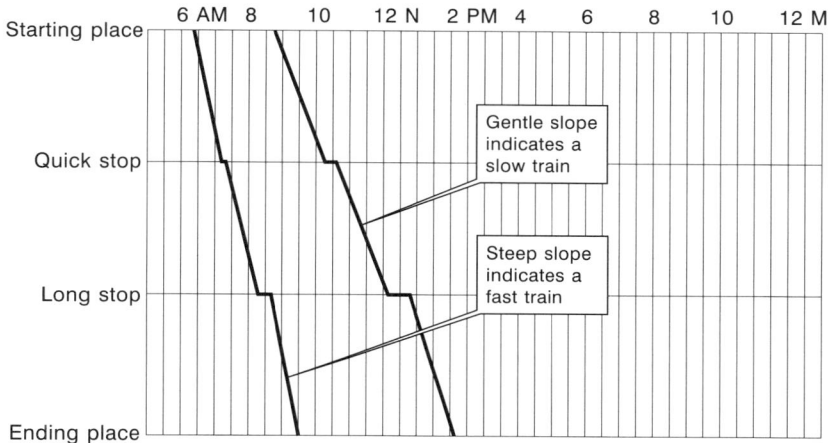

Figure 6.21   A graphic train schedule, with time scaled horizontally (from left to right) and distance shown vertically (from top to bottom).

6.4 Time as a Visual Variable

Computer graphics can elevate time to its proper place in graphic analysis. Graphics systems can do this in one of two ways. The first strategy is the interactive graphics system that allows the user to manipulate time freely, as with a *temporal scroll bar* (Figure 6.22), for moving between time periods. The viewer uses a mouse to point to the box in the scroll bar, and to change the time displayed on the map by 'dragging' this box to the left (to select an earlier time) or to the right (to select a later, more recent time). A discrete temporal scroll bar (Figure 6.22, above) might address time in years or decades, whereas a continuous scroll bar might provide a graphic scale and reference time more precisely, for example, in days, hours, minutes or even seconds. As Carter (1988) notes, the map viewer often needs to flip back and forth through a sequence of maps, and to move at his or her own pace. The second strategy is *animation graphics,* which provide a temporally ordered sequence of views, so that the map becomes a scale model in both space and time.

Over the past three decades a number of geographic cartographers have addressed the uses of animation techniques for dynamic maps (Berlyant, 1988; Moellering, 1980; Thrower, 1959, 1961; Tobler, 1970), and in the 1980s videotex and microcomputers provided dynamic sequencing for the categories displayed on choropleth maps (Slocum *et al.*, 1988; Taylor, 1982). Dynamic maps can range in complexity and sophistication from a simple temporal sequence of complete maps, as might be shown in sequence with

Discrete

Continuous

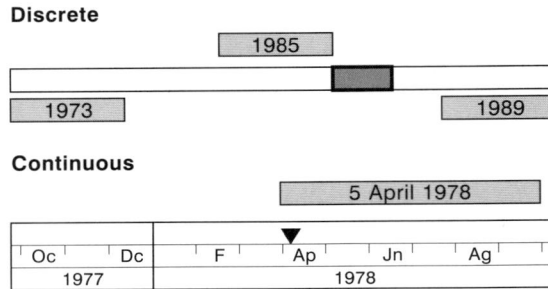

Figure 6.22   Temporal scroll bars can provide discrete (above) or relatively continuous (below) references to time.

a single slide projector, to dynamic symbols that move across the map in the manner of video games. Other intriguing animation effects are possible, of course, including dramatic fades or dissolves of multiple views, progressive zooms and rotating oblique views of statistical surfaces approached gradually, in the manner of an aeroplane circling an airport. Fading, fuzzy or blinking symbols might be particularly useful in dealing with data whose accuracy vary over time.

Another form of animation is the meaningful program or succession of views. In the 1970s statisticians addressed the problem of exploring graphically a database with a large number of variables with a technique called *projection pursuit* (Friedman and Tukey, 1974; Huber, 1985; Tukey and Tukey, 1981). Projection pursuit leads the analyst to one or more potentially significant scatter plots by selecting a small number of optimum two-dimensional 'interesting' views chosen for their degree of clustering or 'clottedness'. The *grand tour,* a more recent elaboration of the projection pursuit model, seeks an optimal sequence of such interesting views (Asimov, 1985; Buja and Asimov, 1986). Both highly promising techniques are still largely experimental and little used.

An obvious extension, of course, is the addition of an optional geographic space viewport. This approach, called 'atlas touring' and currently under development, integrates maps and statistical graphics through the use of *graphic scripts,* composed using basic sequences called *graphic phrases* (Monmonier, 1989b). A graphic phrase for the visual analysis of spatial-temporal data might, for example, partition the screen into four windows and generate an animated sequence of juxtaposed chess maps for pairs of individual years at the top of the display, a change map for the period in question at the lower right and a time-series statistical diagram for the entire period of analysis at the lower left. Other graphic phrases might explore the spatial trends on a particular map or the spatial correlation amongst two variables. A map author might use several such graphic phrases to develop a graphic script examining the spatial-temporal trends, geographic trends and spatial co-variation of an electronic atlas in the form of a large spatial-temporal data set. Monitoring the script-writing behaviour of map authors should provide data to support the development of a still more advanced system – a system able to generate automatically a meaningful sequence of graphics that serve as a guided tour of an electronic atlas.

These techniques beg the question of how best to define and measure the meaningfulness of a map or scatter plot or the interest it might hold for a viewer. As statisticians associate inherent interest with the clottedness of a point cloud, geographic cartographers might regard as interesting a map that resembles a known set of regions, demonstrates a straightforward spatial trend or otherwise exhibits a moderate to strong level of spatial autocorrelation, the geographic space equivalent of clottedness in attribute space. But for some distributions, a pattern that exhibits no discernable trend or lacks regional homogeneity might be more intriguing and noteworthy than one that meets such expectations.

High-interaction graphics avoid this issue by allowing the user to search the data for views he or she finds interesting. High-interaction graphics require a high-resolution display, a pointing device, a 'friendly' direct-manipulation graphics interface and a high-speed processor that allows the analyst to interact creatively with the display (Becker, Cleveland and Wilks, 1987). The software must be flexible and the response time minimal. Historians and geographers have demonstrated the pedagogic utility of interactive systems that allow the user to juxtapose maps and frequency diagrams and to classify distributions portrayed on choropleth maps (Miller and Modell, 1988).

Interactive systems for studying spatial-temporal data might incorporate *scatter plot brushing* (Figure 6.23, top) and a modification, the *geographic brush* (Figure 6.23, centre), with which the analyst highlights areas on the map and views them as well on the scatter plot matrix (Monmonier, 1989a). The viewer might use a scatter plot brush to select a number of places by enclosing, with a variable-size rectangular frame, the points that represent them in any of the nine scatter plots; the dots representing these places would be highlighted (darkened in this case) within the rectangular brush on the scatter plot in question and also highlighted in each of the other eight scatter plots as well as on the accompanying map. With a geographic brush, the analyst could select and highlight places by pointing to them on the map, by drawing a polygon around them, or by choosing one or more established regions, such as the Northeast, from a pull-down menu. For spatial-temporal data, addition of a *temporal brush* (Figure 6.23, bottom) is appropriate as well. The viewer might manipulate the time represented by the display by using the temporal scroll bar to call up the point clouds for some other time unit.

'Atlas touring' and geographic-temporal brushing can be complimentary. Data collected with an interactive system can be helpful in refining or tailoring definitions and measures of interest or meaningfulness. And a programmed series of views, which the analyst can interrupt at will, might serve as a useful introduction to the data – a graphic pump-primer, of sorts, to initiate analysis and encourage the generation and testing of hypotheses.

6.5 Concluding Remarks

Table 6.1 is a conceptual framework that summarizes the range of options for the graphic display of quantitative spatial-temporal data. It demonstrates that the principal tools available for the visualization of spatial-temporal data are:

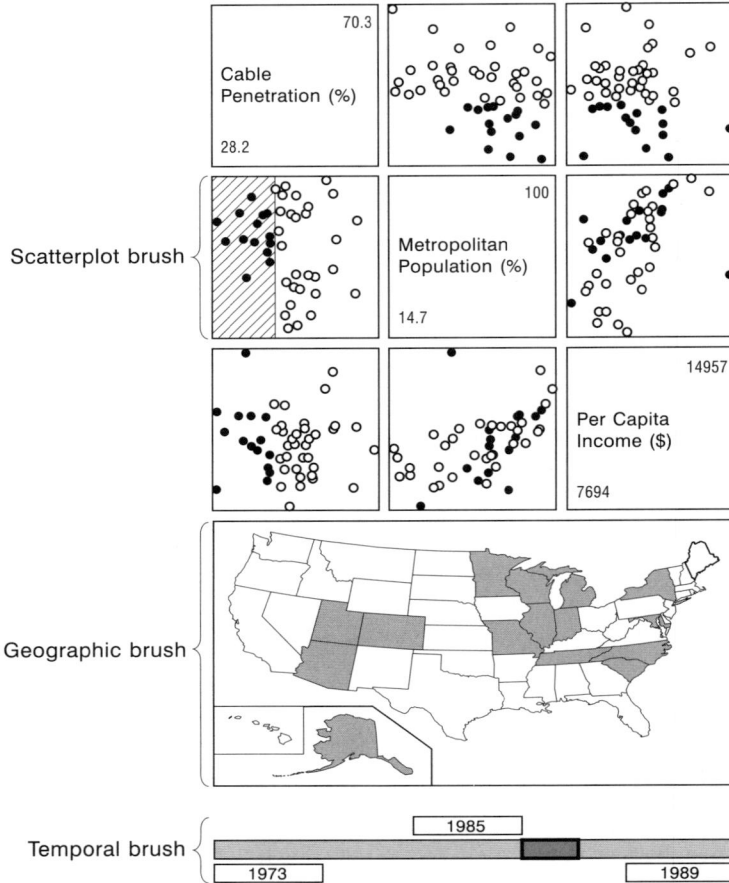

Figure 6.23 A scatterplot matrix with a scatterplot brush, a geographic brush and a temporal brush.

1. multiple views, either in sequence or simultaneously in the same window;

2. time scaling, as well as space scaling;

3. interaction with the data and the display; and

4. integration of maps and time-series graphics.

Cartographic research must turn away from its search for the single optimum map and begin to deal with sequences of maps and the need to integrate maps with statistical diagrams and text blocks containing definitions and other relevant information.

Table 6.1 A conceptual framework for the visualization of geographic time-series data

Time-series graphs with place names or other geographic metaphors

Single static maps

- Temporal symbols
- Temporal aggregation
- Focused measurements (e.g. peak-year maps)
- 'Dance Maps' (movement)
- 'Change Maps' (rates, absolute change)
- Generalized maps focusing on transition and or diffusion

Multiple static maps (and graphs)

- 'Chess Maps' (juxtaposition)
- Cartographic cross-classification arrays (with maps and statistical graphics)

Single dynamic maps

- Sequenced symbols (accretion)
- Temporal sequences of views
- Symbols suggesting motion (pulsating directional symbols)

Multiple dynamic maps (and graphs)

- High-interaction graphic analysis
  - Scatter plot brushing
  - Geographic brushing
  - Temporal brushing
- Programmed sequences of 'interesting' or 'meaningful' views
  - Authored video animation
  - 'Projection Pursuit' (not geographic)
  - 'Grand Tour' (not geographic)
  - 'Atlas Touring' (geographic)

References

Asimov, D. (1985) The grand tour: a tool for viewing multidimensional data. *SIAM Journal of Scientific and Statistical Computing*, **6**, 129–143.

Becker, R., Cleveland, W.S. and Wilks, A.R. (1987) Dynamic graphics for data analysis. *Statistical Science*, **2**, 355–395.

Berlyant, A.M. (1988) Geographic images and their properties. *Mapping Sciences and Remote Sensing*, **25**, 133–143.

Buja, A. and Asimov, D. (1986) Grand tour methods: an outline, in *Computer Science and Statistics: The Interface* (ed. D.M. Allen), Elsevier Science Publishers, New York, pp. 63–67.

Carter, J.R. (1988) The map viewing environment: a significant factor in cartographic design. *American Cartographer*, **15**, 379–385.

Chorley, R.J. and Haggett, P. (1965) Trend surface mapping in geographical research. *Transactions of the Institute of British Geographers*, **37**, 47–67.

Davis, J.C. (1986) *Statistics and Data Analysis in Geology*, 2nd edn, John Wiley and Sons, Inc., New York.

Du Toit, S.H.C., Steyn, A.G.W. and Stumpf, R.H. (1986) *Graphical Exploratory Data Analysis,* Springer-Verlag, New York.

Friedman, J.H. and Tukey, J.W. (1974) A projection pursuit algorithm for exploratory data analysis. *IEEE Transactions on Computers,* **C-23,** 881–890.

Huber, P.J. (1985) Projection pursuit. *Annals of Statistics,* **13,** 435–475.

Johnston, R.J. (1978) *Multivariate Statistical Analysis in Geography,* Longman, London.

Miller, D.W. and Modell, J. (1988) Teaching United States history with the Great American History Machine. *Historical Methods,* **21,** 121–134.

Moellering, H. (1980) The real-time animation of three-dimensional maps. *American Cartographer,* **7,** 67–75.

Monmonier, M. (1970) A spatially-controlled principal components analysis. *Geographical Analysis,* **2,** 192–195.

Monmonier, M. (1972) Flow-linkage construction for spatial trend recognition. *Geographical Analysis,* **4,** 392–406.

Monmonier, M. (1979) An alternative isomorphism for the mapping of correlation. *International Yearbook of Cartography,* **19,** 77–89.

Monmonier, M. (1989a) Geographic brushing: enhancing exploratory analysis of the scatterplot matrix. *Geographical Analysis,* **21,** 81–84.

Monmonier, M. (1989b) Graphic scripts for the sequenced visualization of geographic data. *Proceedings of GIS/LIS'89,* Orlando, Florida, 381–389.

Palm, R. and Caruso, D. (1972) Labelling in factorial ecology. *Annals of the Association of American Geographers,* **62,** 122–133.

Slocum, T.A., Egbert, S.L., Prante, M.C. and Robeson, S.H. (1988) Developing an information system for choropleth maps. *Proceedings, Third International Symposium on Spatial Data Handling,* Sydney, 293–305.

Taylor, D.R.F. (1982) The cartographic potential of Telidon. *Cartographica,* **19**(3/4), 18–30.

Thrower, N.J.W. (1959) Animated cartography. *Professional Geographer,* **11**(6), 9–12.

Thrower, N.J.W. (1961) Animated cartography in the United States. *International Yearbook of Cartography,* **1,** 20–30.

Tobler, W.R. (1970) A computer movie simulating urban growth in the Detroit region. *Economic Geography,* **46**(2), 234–240.

Tufte, E. (1983) *The Visual Display of Quantitative Information,* Graphics Press, Cheshire, CT.

Tukey, J.W. (1977) *Exploratory Data Analysis,* Addison-Wesley, Reading, MA.

Tukey, P.A. and Tukey, J.W. (1981) Preparation; pre-chosen sequences of views, *Interpreting Multivariate Data* (ed. V. Barnett), John Wiley and Sons Ltd, Chichester, UK, pp. 189–213.

Walker, C. (1914) *The Modern Dances: How to Dance Them,* Saul Brothers, Chicago.

7

Reflection Essay: *Strategies for the Visualization of Geographic Time-Series Data*

Mark Monmonier

Department of Geography, Syracuse University, USA

Where to start? There's much to say, and starting at the beginning will not help me reflect fluidly on everything that needs reflecting on. So I'll start in the middle and back up from there, eventually, in both directions.

7.1 'What's New Here?'

For me, the logical place to begin is with my mother's death, on 20 March 1989, the Monday of that year's annual meeting of the Association of American Geographers (AAG), held in Baltimore, where my parents lived and I had grown up. No other article or book I've written is so closely tied to family as *Strategies for the visualization of geographic times-series data* (Chapter 6). Mom's death is the first thing I thought of when I sat down to write.

I had driven down from Syracuse the afternoon before, 340 miles, Interstate all the way, leaving in late morning to spend a bit more time with wife Marge and daughter Jo, and planning to stay with Dad, just outside the city in Woodlawn, where he had lived since 1956, when my parents bought their first (and only) house. Several weeks earlier, Mom, whose terminal lung cancer had been diagnosed two years before, had been relocated to a nursing home because Dad could no longer care for her at home. (The hospice movement, which eleven years later let Dad die at home, in familiar surroundings with his cat, had not yet taken hold in Baltimore.) I took Dad up to see Mom, who

Classics in Cartography: Reflections on Influential Articles from Cartographica   Edited by Martin Dodge
© 2011 John Wiley & Sons, Ltd

was hooked to a machine and seemed very tired but coherent for a brief period – a tender visit but without much conversation. When the phone rang around 2 a.m., we knew immediately what had happened. Monday was a day of grieving, making funeral arrangements and (for me) a quick trip downtown, to hand over my slides and script to a graduate student, who read the paper at the AAG conference the following day at its appointed time.

Early the following week several colleagues reported that the paper had gone well, as had the entire conference session, 'Geographic information systems: graphic display and analysis of spatial-temporal data', which I had organized and planned to chair. No problems with the slides, and my doctoral student had been clear, precise and probably more poised than I might have been, even under less sad circumstances. The only snag occurred during the question period, when a senior faculty member at another university opined, 'I don't see anything new here'. He had a reputation for sandbagging graduate students and probably would not have phrased his query in quite the same way, if at all, had I been there. Though my stand-in didn't have a ready response, I did, and thinking the challenge unfair under the circumstances, I e-mailed an indignant rebuttal to a list of addresses lost in a succession of computer upgrades but most likely acquired when I helped Joel Morrison set up the AAG's Cartography Specialty Group of perhaps a hundred cartographers in academia and government. In an era when *spam* largely referred to a canned meat product high in sodium and saturated fat, a few people thought my unprecedented rejoinder a bit excessive, but most of those who replied agreed that a response was warranted. (That retaliations of this sort remain rare amongst academics no doubt reflects fears of mutually assured decon-struction.) Challenged in absentia about the paper's originality, I was determined to give its content a wider airing in print.

'What's new here?' is actually a great question, raised far too rarely at academic conferences, and surely appropriate in this essay. The innovation in the *Strategies* article was an intentionally broad conceptual framework that treated the map in the context of geographic space, attribute space and time. Although someone familiar with GIS terminology might think (incorrectly) that *geographic space* and *attribute space* are merely coordinate data and feature attributes, a key diagram in both the AAG presentation and the *Cartographica* paper to follow (Figure 6.19) built upon the mathematical notion of a *space* defined by one or more measurements, or coordinates, by contrasting the two-dimensional space defined by the eastings and northings of a plane coordinate system with the statistician's scatter-plot describing the relationship between variables A and B. Orthogonal axes frame both graphics, but the first focuses on geographic relationships while the second portrays a bivariate correlation. If variables A and B are measured for specific points or regions in geographic space, their maps can be juxtaposed to focus on possible similarities in spatial pattern. If A and B share a coherent spatial pattern – an underlying causal relationship with a third variable, perhaps – their geographic correlation might be as meaningful as their aspatial statistical relationship (weak or strong, positive or negative, linear or curvilinear). For example, in the American Great Plains a shared west-to-east trend in the average value of an acre of farmland and the proportion of farmers younger than 40 might reflect the lower reliability of precipitation in the western part of the region.

As other examples in the *Strategies* article demonstrate, the analysis or presentation could accommodate more than two variables as well as measurements for more than one instant or period of time. What's more, time could play a variety of roles: in a traditional time-series graph, which is merely an attribute space with time as the horizontal axis; in a centrographic time-series map like that used by the US Census Bureau to describe the south-westward shift of the national centre of population; or in a cartographic animation consisting of a temporally ordered sequence of geographic-space views. My focus on the display and analysis of relatively small, comparatively undemanding data sets thus differs from the conceptual framework presented six years earlier by Donna Peuquet (1984; Chapter 12 this volume), who focused on data structures appropriate for massive cartographic databases, the need for temporal sampling and the development of efficient compression/decompression strategies for coping with temporal data – impediments rendered less daunting by subsequent advances in electronic storage media. By contrast, my goal was to explore the broad range of ways for presenting or exploring spatial-temporal data, to evaluate their relative merits and complementarity, and to uncover perhaps some new or largely under-utilized strategies. Collectively, the four elements of the paper's title – strategies, visualization, geographic, time-series – summarize concisely my effort to be original.

7.2 Origins

Admittedly, my comparison of geographic and attribute spaces was hardly new. I had used (and perhaps even coined) the terms in a 1979 *International Yearbook of Cartography* paper that included a 'cartographic cross-classification array', in which the rows and columns represent categorical breakdowns for a pair of interval-scale attributes (Monmonier, 1979). The version in *Strategies* (Figure 6.11) is a structurally similar array of mini-maps, but with rows representing a nominal-scale attribute (flavour of Munchies, a fictitious snack with a blatantly generic name, which Frito-Lay nonetheless registered as a trademark in 2004) and columns representing time in years. As attribute-space frameworks for tiny geographic-space mappings, my arrays of maps are akin to the 'small multiples' promoted by Edward Tufte (1983: 170–175) in his acclaimed *The Visual Display of Quantitative Information*. In the dozen years before *Strategies* I had published several papers on geographic-space treatments of bivariate correlation – quirky if not clever black-and-white alternatives to the full-colour two-variable covariance maps promoted by the US Bureau of the Census in the 1970s (Olson, 1981; Monmonier, 1976, 1977, 1978a, 1978b, 1979, 1988). Most of my prototypes were pairs of juxtaposed choropleth maps, one for each variable, with the symbols or class breaks tweaked to underscore the variables' attribute-space correlation. *Cartographica* and its predecessor, *The Canadian Cartographer,* published many of these essays – Bernie Gutsell, the editor, had become a close friend and, like many cartographic journal editors of the day, he was chronically hungry for content – amongst the three most prominent English-language academic cartographic journals, *The Canadian Cartographer* had the smallest membership/subscriber base.

The most significant part of the conceptual framework in the *Strategies* article was the integration of maps and time with statistical graphics, which had begun to attract interest from statisticians and computer scientists as a result of proselytizing by John Tukey, author of *Exploratory Data Analysis* (1979) and arch advocate of looking carefully at data and letting them speak for themselves. In May 1988 I attended the Symposium on Visualization in Scientific Computing: Trends, Tools, Techniques, held at Tukey's academic home, Princeton University. For me the conference was catalytic: I had gone as far as I cared to go with cartographic generalization and class intervals for choropleth mapping, and dynamic graphics beckoned with a range of intriguing strategies that might usefully be harnessed to geographic data and map-making. At the Princeton symposium I learned of scatter-plot brushing, which combined the notions of multiple linked windows, the conditional brush and the scatter-plot matrix. Relationships amongst three or more variables can be explored by simultaneously viewing all possible scatter-plots organized in a scatter-plot matrix, as demonstrated in *Strategies* in the upper half of Figure 6.23. Individual variables are named in the diagonal cells, and each scatter-plot appears twice, with complementary orientations. A rectangle called a 'brush' is inserted in one cell, or 'window', of the array, and all data points inside the rectangle are highlighted, in this instance by filling the centres of the open-circle data points. Because the brush covers the full length of one of the cell's axes but only part of the other axis, it describes a condition of the form $x_{low} \leq x \leq x_{high}$; hence the term 'conditional brush'. Moving the brush along this axis changes the limits (x_{low}, x_{high}) that define the condition, which in turn determines which data points are highlighted. Because the cells of the scatter-plot array are linked, cases or places highlighted within the cell containing the brush are also highlighted in the other cells. The analyst explores the multivariate relationship interactively by sweeping the brush back and forth, or up and down, or by jumping it to another window. Engaged interaction with data was the leitmotiv of exploratory data analysis, which is still under-represented in the analytical toolbox of GIS software.

My contribution was adding a map, which could be used passively, merely to display the highlighted places, or actively, to turn them on or off, individually or even regionally. To identify with the visualization tool thereby enhanced, I labelled the map window a 'geographic brush' and discussed its potential in a short *Geographical Analysis* paper titled Geographic Brushing (Monmonier, 1989). Although the 3×3 scatter-plot matrix and its accompanying map were actually programmed to work dynamically, the file containing my Pascal source code disintegrated without backup when my computer crashed – how easy it is to be too absorbed with making a program work to periodically save the file under another name. Fortunately, I had saved the screen capture used to illustrate the paper. Because of the crash, at the time I prepared by the *Strategies* presentation and its subsequent article, I had no prototype software, which didn't deter me from using MacDraw to add the 'temporal brush' at the bottom of Figure 6.23 – not inappropriate because this was a conceptual paper, and theoretical graphics were sufficient to explain and promote a concept. That's what I like about graphic theory – you do not actually have to write and debug computer code.

Several months later, using a more flexible C compiler and a vigilant backup strategy, I reconstructed the geographic brushing code, and much more, for another

Princeton-inspired technique, which I named 'Atlas Touring' after an innovative technique in statistical graphics called 'The Grand Tour' (Asimov, 1985). Statisticians had recently recognized that a multivariate data set allowed many different two-dimensional views. Some were simple pair-wise scatter-plots, others were more mathematically complex views based on pairs of principal components, perhaps enhanced by rotation, and some of the latter were arguably more 'interesting' than others because the points were aligned or clustered. Because each two-dimensional view was, in a sense, a 'projection' of an m-dimensional space into two (or perhaps three) dimensions, statisticians had devised a technique called 'projection pursuit' to search for the most 'interesting' projections. The Grand Tour was an enhancement of projection pursuit that promised an optimally informative sequence of interesting views – what a concept to apply to a rich but complex electronic atlas! That I was off and running in this new direction when I wrote *Strategies* is apparent in my having listed '"Grand Tour" (not geographic)' and '"Atlas Touring" (not geographic)' in the last two rows of Table 6.1, under 'programmed sequences of "interesting" or "meaningful" views'.

Atlas Touring introduced the notion of a programmed sequence of maps and statistical graphs as a tool for exploring multivariate (including multi-temporal) geographic data. I called this tool a 'graphic script' because it could be acted out by whatever variables the analyst chose to explore. With the support of a seed grant from the New York State Center for Advanced Technology in Computer Application and Software Engineering (the CASE Centre, naturally), and shortly afterward a larger grant from the National Science Foundation (NSF), I developed two prototype graphic scripts, which extended the theatrical metaphor to include stage sets (standard screen layouts) and costumes (consistent colours for specific variables) as well as division into distinct 'acts', subdivided in turn into 'scenes' (Monmonier, 1992). 'Women in Politics' ran just under ten minutes, while 'The American Newspaper Industry, 1900–1990' ran slightly more than 19 minutes. Both scripts employed numerous elements of the conceptual framework introduced in *Strategies*. The 'Women' script, for instance, integrated *chess maps*, *change maps* and attribute-space scatter-plots, while 'Newspaper' included a dynamic centre-of-mass *dance map* that dramatised a northeast-to-southwest horse race between population and newspaper publishing. Both were what I called 'closed' scripts because, once started, they ran to completion without interruption. Without a voiceover, they provided a dynamic background for numerous talks on the future of scripting for customised cartographic presentations, including analytical pump-primers highlighting a data set's interesting views before turning users loose to explore the data freely.

As a research project with creative ambitions, Atlas Touring was perhaps too successful. When my NSF grant was due to expire, I applied for additional funding using a new proposal format called accomplishment-based renewal. Instead of pitching a project from whole cloth, the investigator was asked to recapitulate project goals, summarize progress, articulate goals and strategies for the next stage, and submit relevant reprints and copies of articles accepted but not yet in print. With a number of journal articles in print and several accepted for publication, I asked the NSF program officer for advice, and he encouraged me even though the new format had never been used before in the Geography and Regional Science Program. Ironically, my proposal received more supportive comments and stronger ratings than previous proposals that the NSF had funded, but the review panel rejected it because, by their thinking, the new format gave me an unfair advantage.

Though hardly amused by the panel's rejection, I saw the decision as a long-awaited signal. For several years I had been torn between developing innovative cartographic software and writing books and articles on map history and the societal impacts of mapping. The two activities were invigorating in different ways but also time-consuming and a growing source of stress – all-nighters writing computer code do not mix well with early-morning scribbling. Although I occasionally wonder about this and other the-hell-with-it decisions I've made, I'm convinced that toppling in response to a strong gust is a lot saner than trying to balance indefinitely on the cusp between attractive options.

Although I wound down my work on Atlas Touring, I didn't abandon the concepts laid out in the *Strategies* paper. In *Mapping It Out* (Monmonier, 1993), initiated before the NSF debacle, the chapter 'Mapping Movement, Change and Process' not only draws on *Strategies* but employs the same coarse MacDraw fill patterns and heavy typography typical of laser printers in the late 1980s and early 1990s. A few years later, the second edition of *How to Lie with Maps* (Monmonier, 1996) included a new chapter on 'Multimedia, Experiential Maps and Graphic Scripts'. Like most moderately productive academics, I believe firmly in vertical integration.

7.3 Assessing Impact

For a fuller sense of the impact of the *Strategies* article I explored the citations that account for its sixth position in *Cartographica*'s top-ten list (see Table 1.1). I looked firstly for a temporal trend in the citations (Figure 7.1), and noted, with moderate disappointment, that the frequency had dropped from six citations in 1997 to none in 2006, two in 2007, and only one in 2008. More troubling was the complete lack of citations for the first five years following the article's publication in 1990 – hardly surprising, though, insofar as *Cartographica*'s top-ten list was based solely on Elsevier's Scopus database, which is notorious for its neglect of pre-1996 papers (see the discussion of these issues in Chapter 1).

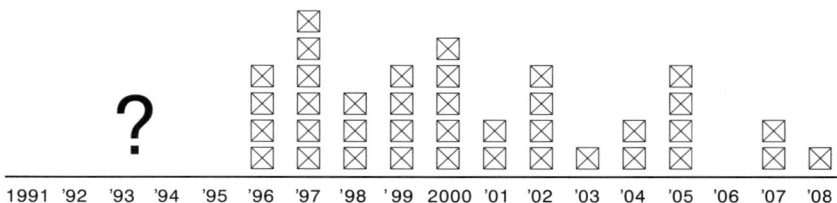

Figure 7.1   Citations reported by Scopus, by year, 1996–2008.

For a second opinion I turned to Scopus's principal competitor, Thompson ISIs Web of Knowledge, which reported 24 unique citations plus one self-citation, appropriately ignored. All but one of the 14 post-1995 citations had already been noted by Scopus, but the 10 pre-1996 citations were a useful supplement. That these two commercial databases yielded substantially different counts indicates that whether the journals in which one publishes are indexed – and if so, where – clearly affects the completeness of citation counts, especially for older works. Not surprisingly, my 1989 'Geographic Brushing' article, which registered 41 citations in Web of Knowledge, garnered only 32 mentions in Scopus.

I then turned to *Google Scholar*, a free service that tracks publications accessible online and includes a 'cited by' feature, which makes it a useful supplement to subscription databases like Scopus and Web of Knowledge. Google further filled in the pre-1996 period with 11 citations, only three of which had been found by Web of Knowledge. The post-1995 yield was substantially greater: 37 citations that had been found by neither Scopus nor Web of Knowledge, 10 that only Scopus had found, eight that both Scopus and Web of Knowledge had noted, and one reported only by Web of Knowledge. Although many of the papers reported by Google are grey literature – dissertations, theses, conference papers, workshop presentations and research reports – these citations confirm that *Strategies* has not fallen out of favour as greatly as its Scopus counts might suggest (Figure 7.2). Were I more vigilantly obsessive, I would construct a composite visualization-animation-interactive-exploratory index to put this trend in a broader context. Citation counts reflect not only the merit of an individual article but also the extent to which other researchers are exploring topics that make it worth citing. The counts for *Strategies* clearly reflect both the breadth of my conceptual framework and a sustained interest in geographic visualization, animation and temporal data.

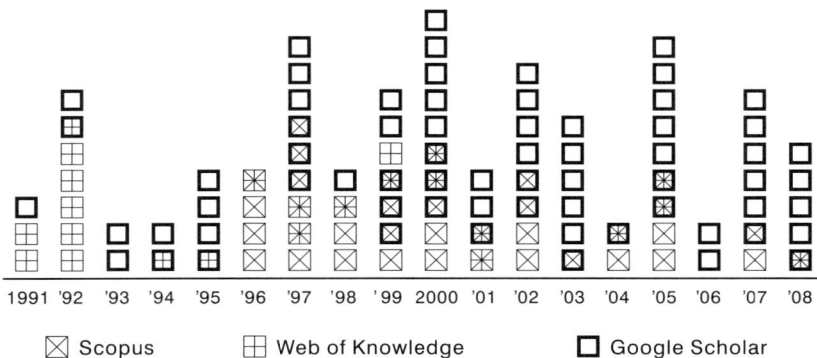

Figure 7.2   Citations reported by all three databases, by year, 1991–2008.

To better understand the meaning of these counts, I looked at the wording of the nine most recent citations in Scopus and the nine most recent citations reported only by Google. Eight were group citations, in which an author dutifully mentioned *Strategies* along with one to eight other articles to buttress assertions about 'ongoing challenges', underscore a dearth of studies on animation, exemplify the variety of 'tools [that] have been suggested' or reinforce similar points that apparently called for citations. Scopus accounts for six of these eight multi-article citations. The other ten articles cite *Strategies* alone as a useful (and possibly pioneering) source of additional information on: 'temporally focused' data (1); a 'tabular format' for time-series data (1); interactive linked displays (1); brushing (2); dynamic, interactive mapping (2); and the value of arrow symbols to represent movement or flows of various types (3). None of these citations strikes me as remotely controversial. The broad scope of my conceptual framework, which remains valid two decades later, no doubt made *Strategies* a conveniently credible source for researchers eager to demonstrate the shortcomings of existing approaches while offering new solutions. Whether or not my paper is a 'classic', it clearly found a niche in the canon of dynamic mapping.

Further Reading

DiBiase, D., MacEachren, A.M., Krygier, J.B. and Reeves, C. (1992) Animation and the role of map design in scientific visualization. *Cartography and Geographic Information Systems*, **19**, 201–214, 265–266. (A seminal exploration of the possibilities and promise of animated maps, broadly defined.)

Gregory, I.N. and Ell, P.S. (2007) *Historical GIS: Technologies, Methodologies and Scholarship*, Cambridge University Press, Cambridge. (A contemporary survey of scholarly applications of techniques for analysing and displaying temporal geographic data.)

Harrower, M. (2004) A look at the history and future of animated maps. *Cartographica*, **39**, 33–42. (An insightful examination of impediments to the development of animated maps.)

Peuquet, D.J. (1994) It's about time: A conceptual framework for the representation of temporal dynamics in geographic information systems. *Annals of the Association of American Geographers*, **84**, 441–461. (A conceptual basis for the efficient structuring of temporal data for electronic analysis and display.)

Vasiliev, I.R. (1997) Mapping time. *Cartographica, Monograph*, **49**, 1–51. (A philosophical examination of the treatment of time on maps and the broader implications for cartography.)

References

Asimov, D. (1985) The grand tour: a tool for viewing multidimensional data. *SIAM Journal of Scientific and Statistical Computing*, **6**, 129–143.

Monmonier, M. (1976) Modifying objective functions and constraints for maximizing visual correspondence of choropleth maps. *The Canadian Cartographer*, **13**, 21–34.

Monmonier, M. (1977) Regression-based scaling to facilitate the cross-correlation of graduated circle maps. *The Cartographic Journal*, **14**, 89–98.

Monmonier, M. (1978a) The significance and symbolization of trend direction. *The Canadian Cartographer*, **15**, 35–49.

Monmonier, M. (1978b) Modifications of the choropleth technique to communicate correlation. *International Yearbook of Cartography*, **18**, 143–158.

Monmonier, M. (1979) An alternative isomorphism for the mapping of correlation. *International Yearbook of Cartography*, **19**, 7–89.

Monmonier, M. (1988) Geographical representations in statistical graphics: a conceptual framework. American Statistical Association, 1988 Proceedings of the Section on Statistical Graphics, 1–10.

Monmonier, M. (1989) Geographic brushing: enhancing exploratory analysis of the scatterplot matrix. *Geographical Analysis*, **21**, 81–84.

Monmonier, M. (1992) Authoring graphic scripts: experiences and principles. *Cartography and Geographic Information Systems*, **19**, 247–260, 272.

Monmonier, M. (1993) *Mapping it Out: Expository Cartography for the Humanities and Social Sciences*, University of Chicago Press, Chicago.

Monmonier, M. (1996) *How to Lie with Maps*, 2nd edn, University of Chicago Press, Chicago.

Olson, J.M. (1981) Spectrally encoded two-variable maps. *Annals of the Association of American Geographers*, **71**, 259–276.

Peuquet, D.J. (1984) A conceptual framework and comparison of spatial data models. *Cartogaphica*, **21**(4), 66–113. (Reproduced as Chapter 12 this volume.)

Tufte, E.R. (1983) *The Visual Display of Quantitative Information*, Graphics Press, Cheshire, CT.

Tukey, J.W. (1979) *Exploratory Data Analysis*, Addison-Wesley, Reading, MA.

8

PPGIS in Community Development Planning: Framing the Organizational Context

Sarah Elwood and Rina Ghose[1]

Abstract

This article examines the local variability of public participation GIS (PPGIS) by urban community revitalization organizations, arguing that this variability is in part shaped by a variety of organizational factors. Existing research has shown PPGIS production to be highly context dependent, identifying an ever-growing set of key elements of this context, including a variety of locally available resources for GIS access and use as well as organizational capacities and characteristics. Contributing to current efforts to expand the conceptual basis of PPGIS research, this article argues that the conceptualization of organizational context must be expanded beyond internal capacities to include organizational networks with local actors, institutions, and resources; organizational knowledge and stability; and organization mission and priorities, all of which shape its activities and relationships, as well as the utility of available GIS resources. This broadened conception of organizational context enables a stronger explanation of the influencing role of organizations in PPGIS, as well as of local variability in PPGIS. These arguments are developed from comparative case study research with six Milwaukee, WI, community revitalization organizations engaged in PPGIS within a city-wide participatory planning initiative.

[1] Originally published: 2001, *Cartographica*, **38**(3/4), 19–33.

At the time of publication: Elwood was at the Department of Geography, DePaul University, Chicago. Ghose was at the Department of Geography, University of Wisconsin, Milwaukee.

8.1 Introduction

With the adoption of GIS by an ever-expanding range of users, public participation GIS (PPGIS) is a growing area of enquiry for GIS researchers examining access to geographic information technologies and their social and political impacts, PPGIS practice is eclectic, encompassing a multitude of GIS applications and user communities around the world. Key PPGIS applications include natural resource management and conservation efforts (Jordan, 2002; Kyem, 2002; Macnab, 2002; Meredith, Yetman and Frias, 2002), community-based planning and neighbourhood revitalization in US urban areas (Bosworth, Donovan and Couey, 2002; Casey and Pederson, 2002; Elwood, 2002; Ghose and Huxhold, 2002; Sawicki and Peterman, 2002; Ventura *et al.*, 2002) and activism organized at local, national, global and multiscalar levels (Sieber, 2002; Stonich, 2002; Tulloch, 2002). These diverse user groups and application types share a common commitment to grassroots community involvement in knowledge production through GIS analysis and in application of knowledge produced.

As a set of practices and as a research agenda, PPGIS has engaged scholars in a research capacity, but also as participants and facilitators of PPGIS initiatives. This involvement by scholars in participatory GIS initiatives has generated a rich body of case-based research that examines in detail: the barriers to PPGIS production faced by different communities in different places (ranging from hardware, software and data access to unequal power relations amongst participants); strategies that are likely to foster community-based GIS use and equal access to the technology and decision making processes that use it; and the empowering and marginalizing impacts of PPGIS activities. We define PPGIS production not only as the processes of acquisition and application of hardware, software and spatial data for GIS analysis, but also the social and political contexts in which GIS is being employed, which influence its use and impacts. These contexts might include organizational capacities with respect to technology; local political opportunities for citizen access to neighbourhood revitalization planning, which can affect the extent to which community organizations' GIS-based knowledge may be inserted into local decision making; or community histories and power relations affecting whose knowledge is or is not included. More recently, the PPGIS research agenda – a PPGIScience – highlights a need to develop conceptual frameworks to synthesize and conceptualize processes of PPGIS production and impacts. In this agenda, a number of topics have been identified for further study: ways of conceptualizing participation and democratic practice within the diverse processes in which PPGIS occurs; explaining in greater detail the processes through which PPGIS might empower and disempower within socially differentiated communities; and understanding the variable role of socio-geographic context in shaping PPGIS (Carver, 2001; Craig, Harris and Weiner, 2002; Elmes, 2001).

A critical issue in strengthening conceptualizations of PPGIS production involves building a better understanding of how this process is shaped by the organizational context in which it is situated; PPGIS is often mediated through organizations, typically non-profit or non-governmental organizations, grassroots activist groups, community and neighbourhood development organizations, and others. Frequently, such organizations are simultaneously sites of PPGIS production as well as institutions through which citizens are seeking greater participation and influence in public decision making. As we will show, GIS research that highlights the importance of organizational characteristics and capacities has focused primarily upon different types of organizations than those typically engaged in PPGIS. This research has tended to focus on GIS *implementation*, the processes of preparation, adoption and initial use of GIS, with an emphasis on the importance of internal actors and characteristics of an organization in shaping GIS implementation. We build on the work of scholars who have suggested that the explanatory frameworks coming out of such studies are not sufficient to explain the complex process of PPGIS production (Ramasubramanian, 1999; Sieber, 2000a, 2000b). We develop an explanatory framework of organizational factors shaping PPGIS production by citizen organizations engaged in community development, planning and revitalization in urban neighbourhoods.

This article will show that PPGIS production by community-based organizations is quite variable, even when these organizations are situated in the same local political context of urban planning and neighbourhood revitalization and have access to the same local support resources for PPGIS. We contend that this local variability can be explained by differences in organizational characteristics and factors, but that conceptualization of this organizational context must be extended to include more than internal characteristics and capacities of organizations. Thus, organizational context must be expanded to include intra- and inter-organizational networks and relationships that influence how an organization navigates the broader situation in which it is embedded. In this notion of organizational context, we include knowledge of staff and community residents about local support resources for PPGIS and neighbourhood participation in local decision making; networks of relationships with public and non-profit agencies or local government officials that may help an organization locate or gain access to PPGIS resources; stability of organization's funds, staff and goals; and the priorities and strategies that inform an organization's PPGIS production and the 'fit' of publicly available PPGIS resources with these priorities.

8.2 Organizational GIS Use: Differentiating Community-Based Organizations

Researchers studying GIS implementation in organizations were amongst those scholars who first emphasized the necessity of understanding GIS as a socially constructed technology (Campbell, 1991, 1994; Campbell and Masser, 1995; Obermeyer and Pinto, 1994). Specifically, these scholars noted the important role of institutional culture and practices in shaping GIS adoption, identifying both material and ideological factors. GIS implementation in organizations was shown to be contingent upon such factors as the openness of the organization's leadership and staff members to the new technology, as well as their level of flexibility in altering organizational procedures for information management, acquisition, sharing and dissemination to accommodate GIS use. In particular, this literature noted that implementation rests heavily on allocation of an appropriate level of financial and technological resources to the endeavour. Some scholars link material support for GIS implementation to ideological support, noting that organizational implementation of GIS is often fostered through the efforts of a single individual who acts as a 'champion' for the initiative (Campbell, 1991; Campbell and Masser, 1995; Nedović-Budić, 1998; Nedovic-Budic and Pinto, 2000). Most of the factors identified in the traditional organizational GIS literature have been developed without a great deal of reference to conditions and relationships beyond the organization. However, a few scholars show that public policies regarding data access and data sharing shape GIS implementation by organizations, since these policies strongly affect the quality of data organizations are able to obtain for their GIS use (Innes and Simpson, 1993; Nedovic-Budic and Pinto, 2000; Onsrud and Pinto, 1991). As well, Obermeyer and Pinto (1994) point out that in implementation of collaborative GIS endeavours, power dynamics between participating organizations will structure information access and information sharing, as well as problem definition and decision making in GIS.

Explanations in the organizational GIS literature concerning the role of institutional practices in shaping GIS implementation are a useful starting point, but they cannot fully conceptualize the organizational contexts that shape PPGIS production. Because the organizational factors identified in this literature were developed through a focus on implementation, they address only those material or ideological factors that directly facilitate or impede adoption of the technology, as well as organizational acceptance of the technology and its associated changes. In examining the longer-term processes of knowledge construction and application in GIS, a different conceptualization of relevant organizational factors must be considered. Research examining GIS access and use by community groups, social movement organizations and non-governmental or non-profit organizations (henceforth referred to as 'grassroots' organizations) similarly notes that the 'traditional' organizational GIS implementation literature is not uniformly useful in this context. Scholars studying grassroots organizations' use of GIS have identified a few similarities with traditional organizations in implementation. Ramasubramanian (1998, 1999) and Sieber (2000a, 2000b), for instance, have shown that upper management commitment to GIS use, organizational resource base, allocation of sufficient resources to GIS activities and presence of an organizational 'champion' affect GIS adoption and use in grassroots organizations. However, they and others note that the capacities and characteristics of grassroots organizations relative to the resources needed for GIS use create significant variation from GIS use in public and private for-profit organizations (Barndt, 2002; Elwood, 2002; Leitner *et al.*, 2000).

PPGIS research has begun to build a framework for understanding its organizational context by identifying key situational factors that shape grassroots organizations' use of GIS. Some of this research identifies organizational limitations inhibiting GIS access and use by grassroots and other participatory citizen organizations. These constraints include limited financial and time resources to devote to GIS; staff members without specialized training or experience in data access and management, computer use and GIS use; and high staff turnover, especially amongst those staff members who obtain specialized training in information technologies (Barndt, 1998; Elwood and Leitner, 1998). Other studies identify the local presence (or absence) of institutions and resources supporting PPGIS as a critical element of the context affecting grassroots organizations' use of GIS. Sawicki and Craig (1996) and Barndt and Craig (1994), amongst others, have noted the manner in which the local presence of a dense network of agencies orientated toward fostering public data provision, providing assistance in obtaining and using information technologies, and directly providing GIS services to grassroots organizations can affect the use and impacts of GIS use by these organizations. Sieber (2000b) suggests that not only is the *presence* of such support structures for PPGIS an important influencing factor, but so too are the 'resource substitution' strategies developed by grassroots organizations. Sieber (2000b) and Leitner and others (2000) note that grassroots organizations respond to constraints upon their GIS use through creative and multifaceted strategies to draw upon available support resources.

Political roles and relationships of grassroots organizations (at multiple scales) have further been identified as affecting GIS use by these groups. Sieber (2000b), for instance, argues that an important variable shaping GIS use by grassroots groups is their particular role in society. As advocacy groups, she contends, grassroots organizations have different knowledge production strategies, political mandates and constituent communities than a public agency might have, leading to different kinds of knowledge production through GIS. Other scholars have noted that these organizations generally occupy a position of lesser social and political power (Barndt, 1998; Harris and Weiner, 1998). This marginal position alters their access to GIS, the knowledge they might produce and the manner in which this knowledge is employed in efforts to leverage greater social and political influence and access a greater voice in decision making (Craig, Harris and Weiner, 2002; Leitner *et al.*, 2002; Sieber, 2002). In the specific instance of PPGIS in an urban planning and revitalization context, organizational relationships with the local government institutions are identified as a key factor affecting both GIS production and the impacts and application of the knowledge produced. Barndt (1998), Leitner and others (2000) and Tulloch (2002) have noted that the openness of state actors to sharing the data and support resources necessary for using GIS and to granting citizen organizations meaningful opportunities to insert their knowledge into decision making processes shapes both how community organizations are able to use GIS and how this use might inform their activities.

Together these various accounts of GIS implementation and production by different types of organizations suggest that two types of factors may significantly impact PPGIS production in community development and urban revitalization: (1) organizational characteristics and capacities and (2) elements of the local support structures for GIS and spatial data access, as well as for citizen involvement in local governance. However, since existing research has tended to focus on one area or the other, it is not entirely clear how these two sets of influencing factors come together to affect PPGIS production. That is, how is the process of PPGIS production affected by the intersection of organizational capacities and characteristics with local context of public data access, available resources to support community-based GIS use and local level support for citizen participation in local governance? How can variability in PPGIS practice amongst organizations in the same local context be more fully understood as stemming not only from differences in internal organizational capacities but from a broader set of organizational relationships, opportunities, constraints, and choices that shape the social production of GIS and application of GIS-based knowledge to organizational activities?

Examining multiple organizations engaged in PPGIS production in the same locality reveals that not all organizations can or do engage in the same way with local opportunities for data sharing, GIS access and citizen involvement in urban governance.

Different types of organizational knowledge and experience, networks of relationships between organizations, organizational stability and organizational mission differentiate the manner in which a community group engages with the local contexts of GIS and spatial data access and use, as well as opportunities for application of these analyses within local planning processes. These propositions are developed from a comparative research project carried out with six community organizations in Milwaukee, WI (profiled in Table 8.1), that have been active in PPGIS and in a city-wide participatory planning initiative. The case study organizations were selected from amongst the 17 organizations that coordinated and implemented Milwaukee's Neighbourhood Strategic Planning (NSP) programme at the neighbourhood level because these groups had completed their NSP plans and used GIS and spatial analysis in the strategic planning process. Project findings are drawn from analysis of intensive interviews conducted with organization staff, local government officials from divisions overseeing citizen participation in planning and digital information and technology services, and staff from non-governmental organizations supporting PPGIS in Milwaukee. Additionally, we analysed strategic planning and community revitalization documents obtained from participating organizations and local government offices.[2]

8.3 Local Infrastructure Supporting PPGIS and Citizen Participation

Milwaukee is a useful case through which to examine PPGIS production by community revitalization organizations because of its strong traditions in urban GIS analysis and local government efforts to facilitate community-based GIS use. Milwaukee's urban spatial data infrastructure is well developed, supported by the city of Milwaukee's own GIS efforts, which began as early as 1975 (Huxhold, 1991). The city has developed several initiatives to foster citizen participation in urban governance and, in a recent neighbourhood-based strategic planning process, has established formal structures to assist community groups in gaining access to and using spatial data and GIS-based analysis. As a result, a large number of community-based organizations in Milwaukee have been engaged in PPGIS since the early 1990s (Barndt, 1998; Ghose, 2001; Ghose and Huxhold, 2001; Ghose and Huxhold, 2002). However, this generally supportive context for citizen participation and PPGIS development is not experienced in the same way by all of Milwaukee's community-based organizations, fostering vastly different PPGIS capacities. We begin with a more detailed account of public participation opportunities in local governance in Milwaukee and of the support infrastructures for GIS use and digital data access, continuing with a discussion of the differential engagements of the case study organizations with the local opportunity structures for PPGIS.

Citizen involvement in neighbourhood revitalization planning, public data access and community-based GIS use is strongly institutionalized in Milwaukee. Additionally, the city has made efforts to integrate community-based GIS analysis into neighbour-

---

[2] Analysis of these data is also informed by the authors' previous research in Milwaukee related to GIS provision and participatory planning (Ghose and Huxhold, 2001; Ghose, 2001).

hood revitalization planning through its NSP. Begun in 1995, the programme was intended to foster citizen involvement in distribution of Community Development Block Grant (CDBG) funds[3] in the city and in planning specific revitalization projects to which these funds would be applied. In each of the 17 planning areas in the city (roughly corresponding to government- and resident-identified delineations of existing neighbourhoods), a single community organization facilitated citizen surveying, resident focus groups and community meetings to identify strengths, weaknesses, opportunities and threats in the area, as well as formulating a revitalization plan to be implemented with CDBG funds.

GIS-based spatial analysis is strongly institutionalized within the NSP. Under programme guidelines, strategic plans produced by community organizations are required to include neighbourhood data and GIS-generated thematic maps showing demographic, economic, crime and housing conditions (Barndt, 2001; Martin, 2001). Fulfilment of this requirement has been facilitated through a dense network of public and private supporting institutions. Government agencies within the city of Milwaukee and the Milwaukee Community Block Grant Administration and institutions such as the University of Wisconsin – Milwaukee (UWM) have collaborated to provide access to various types of data, analysis and GIS (Ghose and Huxhold, 2001). A key relationship supporting PPGIS in the NSP process has been the city's contract with the GIS and spatial analysis division of a local technical assistance organization for non-profits, the Data Center programme of the NonProfit Center, to assist community organizations in data access and spatial analysis for their NSP activities. The Data Center thus provides free access to data sets, GIS, analysis and maps for the 17 community organizations engaged in NSP planning and implementation.[4]

In addition to providing assistance for GIS use by these organizations, the city administration has designed strategies to broaden public access to digital spatial data. Municipal property data are available on-line, together with a mapping interface called Map Milwaukee that enables users to query and map a wide range of city data with many different scales and resolutions. Another recently developed online mapping application is the city's COMPASS project – an initiative funded through the National Institute of Justice that enables mapping and analysis of property, community safely, crime and public health data throughout the city.[5] The city of Milwaukee has also made a great deal of this digital information available to UWM, recognizing the university as another key provider of spatial data, GIS and GIS-based analyses to local community organizations as well as a source of planning advice and support for neighbourhood revitalization efforts.[6]

---

[3] CDBG funds arc an annual allocation of US federal funds made to cities, counties and states. Cities generally determine how the funds are allocated.

[4] This is only one of many projects through which the Data Center supports GIS and spatial analysis activities of community organizations in Milwaukee. While free services are provided only for NSP implementation organizations, similar services are available to other non-profit groups in the city at a very low cost.

[5] Map Milwaukee is available at www.mapmilwaukee.com. COMPASS is available at www.milwaukee.gov/compass/.

[6] See Ghose and Huxhold (2001) for a description and discussion of Milwaukee's network of community–university partnerships that have fostered PPGIS and neighbourhood revitalization planning efforts.

Table 8.1   Case study organizations, their activities and their staffing levels

---

WAICO-YMCA

• Housing and economic development
• Community organizing: crime and safety, community building
• Extensive university/community partnership activity
• Family and youth services
• 8–10 staff members[a]

Lisbon Area Neighbourhood Development (LAND)

• Housing and economic development
• Job training
• Community organizing: crime and safety, community building
• Public health improvement
• 8–10 staff members

Metcalfe Park Community Association (MPCA)

• Housing and economic development
• Community organizing: crime and safety, community building
• 1–2 staff members

Sherman Park Community Organization (SPCO)

• Housing and economic development
• Community organizing: crime and safety, community building
• Public school improvement
• 4–6 staff members

Northwest Side Community Development Corporation (NWSCDC)

• Housing and economic development
• Community organizing: crime and safety, community building
• Online organizing: crime and nuisance reporting, neighbourhood discussion network
• Workforce development
• 10–12 staff members

Harambee Ombudsman Project, Inc.

• Housing and economic development
• Community organizing: crime and safety, community building
• Child/youth services: after-school and summer activities, job fair, community service
• 6–8 staff members

---

[a]Community organizations experience a great deal of change in staff capacity, usually because of fluctuations in funding. To give some sense of staff capacity across the case study organizations, Table 8.1 provides staffing estimates based on levels in July 2001.

Whereas grassroots organizations in Milwaukee have relatively open access to a wide range of community-based spatial data and the potential to gain access to GIS technology at little or no cost, the actual PPGIS efforts of these organizations vary considerably. In the following sections we will describe and offer an explanation for this variability, drawing on the PPGIS efforts of six Milwaukee organizations. Each of these organizations, profiled in Table 8.1, has a place-based mission of fostering revitalization and improved quality of life in a particular Milwaukee neighbourhood. Though specific opportunities and challenges vary considerably amongst the target neighbourhoods of these organizations, all have experienced post-industrial decline and disinvestment, with accompanying concerns about job loss, declining housing conditions, ageing physical infrastructure, crime and safety and public health.

8.4 Variable PPGIS Production

The PPGIS efforts of these six NSP implementation organizations in Milwaukee show tremendous variation with respect to the richness of data and knowledge incorporated for analysis, the nature of their relationship with PPGIS supporting institutions in the city and the level of integration of GIS-based knowledge across a multifaceted range of organizational activities. The most extensive and integrated PPGIS processes of the six were developed by Northwest Side Community Development Corporation (NWSCDC), Walnut Area Improvement Community Organization (WAICO-YMCA), Lisbon Area Neighbourhood Development (LAND) and Harambee Ombudsman Project; they incorporate a wide range of spatial data and analyses gathered from multiple sources and applied in multiple organizational endeavours. PPGIS production of the other two organizations, Metcalfe Park Residents Association (MPRA) and Sherman Park Community Organization (SPCO), is characterized by constrained data sets and a limited range of applications of GIS-based knowledge to community revitalization priorities.

The case study organizations with more extensive and integrated PPGIS activities tend to have sought access to a much wider range of spatial data to include in their maps and analyses and are more likely to maintain at least some spatial data in their own offices, rather than only seeking data maintained by external actors. WAICO-YMCA, for instance, maintains an extensive in-house spatial database that is used frequently in community organizing, housing improvement and economic development activities. The data include both public governmental data and local experiential knowledge provided by community residents; WAICO-YMCA staff members are frequent users of the online Map Milwaukee site to create maps. LAND and Harambee similarly rely on Map Milwaukee's online GIS resources, but both also have a more extensive in-house database capacity than WAICO-YMCA, relying on locally designed software called Community Expert.[7] Both organizations used Community Expert to create and manage their own databases for community organizing activities, and both have been using these databases since the mid-1990s. Though Harambee has become less active in using Community Expert to manage its spatial databases, LAND has acquired ArcView GIS to enable analysis and mapping of data in its community database. NWSCDC has the most extensive in-house data resources of the six case study organizations, with a database that incorporates information acquired from the city of Milwaukee as well as community-based information.

---

[7] See Ghose and Huxhold (2001) for a more detailed account of the university–community collaborations through which Community Expert was designed.

NWSCDC acquires local community data for its mapping and spatial analysis in a variety of ways, ranging from door-knocking by its community organizing staff to what the staff refer to as 'cyber-organizing'. The organization's Web site (www.nwscdc.org) hosts an e-mail-based community discussion list, as well as an application that enables residents to submit the location of a problem or concern in the neighbourhood, along with comments. These data are incorporated into the organization's database, and residents' comments, together with maps and digital photos, are used to direct and lobby for city services and intervention. The organization runs a computer donation and redistribution programme to try to increase residents' access to these digital forums.

The more extensive and varied data resources of these four organizations can be attributed in part to the active role the organizations have assumed as knowledge producers. They employ a range of strategies and resources for acquiring and developing spatial data, as exemplified in this explanation by a NWSCDC staff member:

> [Our only resource] isn't just either at City Hall or in the Police Department, or at UWM, it's like we can make some of our own stuff . . . our own technology and our own ability to be able to connect people to one another . . . today, with the kind of technology that we have available to us, it's GIS, it's crime mapping, we have the city's COMPASS program, we can analyse, we can do trend analysis . . . (NWSCDC, 2001)[8]

WAICO-YMCA, Harambee and LAND all pursue similar strategies of locating and using multiple types of data from a range of sources.

Those organizations with more robust PPGIS processes also show greater diversity in application of the resulting knowledge, data and maps across a variety of organizational activities and community concerns. For instance, all four have used the spatial analyses created for their NSP plans in other activities, particularly to inform their implementation of revitalization strategies. LAND continues to use the spatial analyses of its NSP plan in this manner, as explained by a staff member:

> the implementation process is actually sitting down to talk about addressing [neighbourhood issues]. Okay, let's pull those maps back out. Okay, we appear to have a concentration of one [issue] here, so why don't we concentrate our efforts over here? (LAND, 2001)

These organizations further utilize the knowledge produced in their PPGIS efforts to garner funding support outside of the NSP programme. Describing Harambee's ongoing use of the data and maps generated as part of their NSP plan, a consultant to the organization explained,

---

[8] In keeping with our agreement with interviewees and with Institutional Review Board policies, we will identify interview material only by the agency with which the interviewee was affiliated.

I think that the strong agencies, I think Harambee, figured out that, hey, we [have] to do this anyway ... Banks and everyone starts to expect it at some point. Most potential investors, and banks and even the national chains that have rediscovered the central city – they'd like to see a plan ... [for] getting neighbourhood development programs, attracting new businesses, you have to have a neighbourhood plan, that gives us a snapshot of the neighbourhood. (Sanders, 2001)

Harambee also uses the NSP-generated maps and statistics in seeking greater access to local government assistance programmes for neighbourhoods, as in its 2001 application for inclusion in the city's Targeted Investment Neighbourhood programme.

In contrast, the other case study organizations show more limited or narrowly focused PPGIS initiatives. These endeavours are characterized by more limited data access, heavy reliance on external assistance and application of PPGIS-produced maps and analysis to a relatively narrow scope of community priorities. The Sherman Park and Metcalfe Park organizations, for instance, maintain very limited digital data in-house compared to the other organizations, or the data are not in an accessible form. Sherman Park obtained and began to use Community Expert in the mid-1990s, but outdated data and failed hardware have resulted in the computer (and the data it contains) being used, in the words of one organizer, 'as a doorstop!' (SPCO, 2001). Metcalfe Park, with a single computer in its organizational offices, maintains most spatial data in hard copy, and a good deal of this has gone missing during office moves to new locations. In both cases, organizational capacity to systematically obtain and incorporate local knowledge into their spatial data holdings is similarly limited. In describing the organization's difficulties in sustaining GIS efforts, a staff member commented, 'staff turnovers [and] salary issues are concerns in using GIS and data ... [I] wish we had time to do more training, workshops and Seminars' (MPRA, 2000).

These organizations are not completely blocked from PPGIS production. Through their involvement in the NSP process, and their collaborations with local GIS and spatial analysis assistance providers, they are engaged in acquisition and use of maps and spatial data. However, this process is different from those of other organizations. All of the case study organizations rely on assistance received through such collaborations, but organizations with more limited PPGIS production tend not to work collaboratively and actively to produce information with the assistance providers, instead simply receiving spatial analysis products in a 'client–provider' type of interaction. Staff members from both Sherman Park and Metcalfe Park describe this kind of interaction with the Data Center, frequently a phone call or visit with Data Center staff just to request a map. Further, the maps, data and knowledge produced in the PPGIS efforts of these organizations tend to be integrated into organizational activities only in limited ways, sometimes only for the purpose of satisfying the NSP programme's requirement that the neighbourhood plans include these forms of information. One organizer explained that the group only briefly consulted the GIS-generated maps and statistical analyses and that these materials were incorporated in the strategic plan 'because it was required, no question. I wouldn't have put too many of those in there myself' (SPCO, 2001).

While, in theory, all of the Milwaukee organizations have access to similar support resources for GIS use, spatial data access and strategic planning, these two groups have not exploited available resources to the same degree as the others, for a variety of reasons. As we will discuss in more detail below, these groups experienced instability in funding or staffing that limit their PPGIS and strategic planning endeavours. Sherman Park, as noted above, has intentionally limited its involvement with the GIS mandate of the NSP programme, feeling that it could retain greater organizational independence by doing so and by pursuing revitalization strategies outside of the priorities being advanced by the NSP. Level of knowledge about available GIS and data resources also influences PPGIS production by these two groups. Thus, within the PPGIS efforts of Milwaukee community organizations, experiences vary widely. Whereas all of the organizations appear to have similar opportunities available to them, it is clear that they navigate these PPGIS opportunities in quite different ways.

8.5 Accounting for Variability: The Organizational Contexts of PPGIS Production

The variable forms of PPGIS co-existing in Milwaukee do not result from a simple concentration of skills and resources in some organizations relative to others. Rather, a number of organizational factors differentiate the ways in which organizations navigate local PPGIS opportunity structures. These organizational factors, outlined in Table 8.2, include: knowledge vested in staff or residents concerning local support resources for PPGIS and community organizations; broader networks of relationships that inform and extend organizational PPGIS capacities; organizational stability; and organizational priorities, strategies and status.

Table 8.2 Organizational context of PPGIS production

Organizational knowledge and experience

• Staff and resident knowledge of local support resources for PPGIS (funds, data, hardware, software)
• Knowledge/experience of locally successful resource acquisition strategies
• Knowledge/experience of locally successful political strategies

Network of collaborative relationships

• Relationships with public and private support institutions for community organizations and PPGIS
• Formal collaboration with resource institutions
• Informal personal relationships with local political actors, public agency staff, community organizers

Organizational stability

• Duration of leadership/low rates of staff turnover
• Consistency of organizational mission and goals
• Consistency of funding support

Organizational priorities, strategies, status

• 'Fit' between organizational priorities/strategies and those of local government priorities
• 'Fit' between (GIS/data needs for these strategies and publicly available GIS resources

8.5.1 Organizational Knowledge and Experience

Organizational knowledge and experience have a tremendous influence on the way an organization is able (and chooses) to negotiate systems of participatory planning and GIS support – together shaping PPGIS production. Previous research has demonstrated the importance of staff members with technical training in use and maintenance of hardware, software and digital data to PPGIS development (Leitner *et al.*, 2000; Sieber, 1997, 2000a, 2000b). However, we argue that organizational knowledge and experience relevant to PPGIS production also includes knowledge about sources of assistance for PPGIS endeavours and for community revitalization organizations. As well, organizational knowledge about, and experience with, locally effective strategies for enhancing political influence and resource acquisition play a significant role in production of PPGIS.

Organizational knowledge about sources of assistance is essential to PPGIS production. Of the six case study organizations, three knew about and utilized the Data Center to obtain GIS-based maps and analysis prior to the inception of the NSP programme and its contract for the Data Center's services. The PPGIS efforts of these organizations have also been more diverse in their sources of information and applications, extending beyond the usual analyses produced for a community's NSP plan. Organizational knowledge of support resources is not limited to GIS support but also includes funding resources. The organizations in our study show a great deal of variability in their knowledge of and ability to access multiple forms of funding, which, of course, informs their capacity to obtain and use spatial data and GIS. The Metcalfe Park, Sherman Park and Harambee organizations have less diverse forms of funding, as well as a more limited range of organizational activities. In contrast, LAND's expansive PPGIS and other activities are supported by public funding as well as by:

> *foundations, corporations, looking at our own way in which we can increase our own income ... we're reaching out to traditional foundations, corporations for philanthropic support. We're also looking at ... entrepreneurialism, saying how can LAND generate income for itself? How docs LAND begin to look at [funding sources] creatively? (LAND, 2001)*

NWSCDC has benefited similarly from staff members' knowledge of alternative and additional funding sources. For instance, NWSCDC staff members know about additional funding sources that might vastly expand their spatial analysis and technology capacity. At the time of writing, the organization was a finalist for a grant from the US Department of Commerce's Technology Opportunity Program with a proposal to use information technologies in housing development (NWSCDC, 2001). Within organizations that have relied more on NSP funds as a primary source of support, PPGIS and other activities have been significantly retrenched as funding provided through NSP has declined (Barndt, 2001).

Organizational knowledge important to PPGIS production also includes a strategic understanding of how to navigate structures of local governance to the greatest benefit of the organization, the community and its PPGIS process. We found that the organizations with the most complex and sustained PPGIS efforts were also those with the most detailed knowledge of local political opportunity structures for community participation. For instance, not all organizations in Milwaukee were aware that a city funded programme provided limited funding for organizations to hire a consultant with strategic planning or community development expertise to help prepare of their strategic plans (Sanders, 2001). NWSCDC, WAICO-YMCA, LAND and Harambee all knew of this option and took advantage of it – producing plans that, based on our examination of them, proved to be significantly more developed with respect both to articulation of goals and strategies and to interpretation and integration of maps and spatial analysis to inform the plan. Harambee staff members further recognized the process of NSP as a strategic opportunity to expand organizational knowledge, as evident in the words of this organizer:

> *I think Harambee is probably one of the most successful with NSP because we've used it as a vehicle to identify concerns ... [also], when neighbourhood planning first came, I remember saying to the group that [we should) use this as an opportunity to learn how to do planning. (Harambee, 2001)*

Together, these forms of organizational knowledge foster PPGIS production in terms of gaining access to and using GIS, but also in interpreting and applying the resulting knowledge to inform organization activities.

8.5.2 Network of Collaborative Relationships

Whereas organizational knowledge of locally available resources and opportunities fosters PPGIS production for community organizations, so, too, does creating a dense network of collaborative relationships. An organization may have knowledge about possible resources and forms of assistance for PPGIS and citizen participation in planning, but a strong network of relationships with individuals and agencies helps in gaining access to these resources. Personal and professional acquaintances between community organization staff members and staff in these public and private institutions enhance an organization's capacity to take advantage of local opportunity structures supporting PPGIS, as do formal collaborations between a community organization and other institutions. These networks of relationships sometimes grow out of a staff member's long-term experience in a single community, as with the leader of NWSCDC, who grew up in the community and has worked in community organizations in the area for more than 20 years. They may also emerge from a community organizer's experience working in community development across several different communities. This is true of WAICO-YMCAs leader, who has worked in multiple neighbourhoods in Milwaukee, in the process developing connections with many supporting agencies, key individuals within government agencies, stakeholder organizations and staff members of private philanthropic organizations.

These dense networks of relationships with individuals and institutions that support spatial analysis and technology use in urban improvement efforts can significantly enhance an organization's PPGIS capacity. The NWSCDC, WAICO-YMCA and LAND, with the most complex and diverse PPGIS initiatives, have relied heavily upon their collaborative relationships with university researchers and other local and non-local relationships. NWSCDC, for instance, has collaborated with UWM faculty and student researchers in multiple projects, including an extensive GIS-based analysis of housing, economic, crime and demographic conditions, as well as on background research and consultation in preparing its NSP plan. Describing some of the benefits of this relationship, and the results of the collaboration, one of the university partners commented,

> *My early work with the university and with Northwest Side CDC [showed that] Northwest Side was really able to take full advantage of the planning process, because it had the university making its planning process the topic of one of our [urban planning] studio workshops. . .. [T]he idea was to develop a plan that could serve as a prototype for the neighbourhood . . . But what happened is that Northwest Side demonstrated all the benefits of being able to Lake advantage of this process.* (Sanders, 2001)

WAICO-YMCA has similar collaborative relationships with UWM, but also has a dense array of partnerships with private and public agencies that have enabled it to access further support for PPGIS endeavours. For example, WAICO-YMCA worked with both UWM and the Wisconsin Housing and Economic Development Authority to create GIS-based maps and analysis informing neighbourhood redevelopment activities. The collaboration produced a multiscalar GIS-based neighbourhood indicator study assessing indicators of neighbourhood quality of life in comparison to other city neighbourhoods (Ghose and Huxhold, 2002). The results of the study were further applied to WAICOs efforts to leverage assistance from the city of Milwaukee – particularly the creation of the first Tax Incremental Financing (TIF) district in a residential area.

As for all of the elements of organizational context discussed in this section, there is a mutually reinforcing relationship between these networks of relationships and organizational knowledge. Organizational knowledge about local opportunity structures for PPGIS and citizen participation may guide an organization's efforts to foster relationships with individuals and institutions capable of offering assistance or other resources. At the same time, such networks foster organizational knowledge of possible sources of assistance, as well as of successful community revitalization strategies within the local political context. Networks of relationships between people, organizations, and communities have been shown to be particularly important to the effectiveness and sustainability of community-based organizations. In their study of grassroots groups active in community-based workforce development and employment training across the United States, Harrison and Weiss (1998) found that the most effective workforce development organizations were those that prioritized forming and sustaining these networks; the authors predict that the importance of networks in this context might be transferable to other kinds of community-based organizations. In the case of community-based organizations active in PPGIS production, this certainly seems to be the case. PPGIS production requires assembling a complex set of resources, including (but not limited to) data, knowledge, training, hardware and software. Organizations with dense social networks have a wider range of potential support in their efforts to assemble these resources than organizations with more limited networks.

8.5.3 Organization Stability

The stability of an organization, understood as duration of leadership, low rates of staff turnover, consistency in organizational mission and consistency of funding support, further affects the ways in which an organization engages with local participation and GIS support structures. Organizational stability is important because it influences the areas of organizational knowledge and network formation described above, shaping the extent to which these resources are retained and sustained in an organization. The PPGIS initiatives of the MPRA in the mid-1990s, for instance, were significantly hampered by internal struggles over power and mission that resulted in leadership changes and staff turnover. The organization's funding was highly variable, as well, as the new leader attempted to stabilize the organization and establish relationships with community stakeholders in addition to fundraising activities. Absence of sustained leadership also disrupted this organization's university collaboration partnership, which was supporting its PPGIS endeavours (Ghose, 2001). Similar disruptions have characterized the PPGIS and strategic planning initiatives of the Sherman Park Community Organization. Queried about her knowledge of SPCO's PPGIS efforts and NSP activities, one organizer said,

> *I can only guess, because it was two executive directors earlier. It [involved] lots of other people who aren't even here now on staff, lots of board members that have changed. (SPCO, 2001)*

In contrast, staff members at NWSCDC contend that their multifaceted PPGIS efforts have been strongly informed by the organization's stability with respect to funding levels and staff involvement. LAND and WAICO-YMCA have experienced similar stability, with primary staff members who have worked in their organizations for several years and diverse funding sources that have remained constant. Organizations such as Metcalfe Park, much more dependent on NSP programme funds, have found their capacity significantly limited by the sharp reduction in funds allocated to NSP since 2001.

8.5.4 Organizational Priorities, Strategies and Status

The extent to which community revitalization organizations do or do not engage with locally available public support structures for spatial data access and GIS use in strategic planning is further shaped by each organization's revitalization priorities and strategies, and by the relationship between its dominant approaches to revitalization and those of local government. Organizations whose mission, priorities and strategies for revitalization more closely match the goals and strategies of local government revitalization programmes are more likely to find that public data and GIS resources effectively support their activities. In contrast, organizations that pursue alternative goals and strategies may find that public support resources for PPGIS production do not enable access to the types of data and analysis that would best support their activities. As well, these organizations may choose not to make use of available resources, feeling that PPGIS production that relies on local government data, or that is undertaken in the context of revitalization initiatives with very different priorities, may threaten the independence of their alternative approaches.[9]

The Milwaukee case study organizations exemplify both of these scenarios. NWSCDC, for instance, engaged in a wide array of industrial and commercial property improvement and seeking to promote home ownership, found that these existing priorities and programmes fit quite well with local government priorities emphasized in the NSP programme. Data and maps it acquired through the Data Center's NSP contract were useful in preparing its plan for this programme but were also more broadly applicable within the organization's activities, perhaps because of this similarity between their organizational activities and those encouraged and supported by NSP. In contrast, a Sherman Park organizer noted that she had little use for the maps and analysis produced for their organization's NSP plan and that she included them only because they were required. She spoke of the difficulty of inserting alternative analyses and priorities into local government programmes such as NSP, noting that 'what the city wanted (in terms of spatial analysis and revitalization strategies) was what they were used to getting' (SPCO, 2001). Organizations whose priorities differ from the revitalization agenda of the local state are less likely to find public data and GIS opportunities that support these efforts.

In assessing how PPGIS production may be shaped by community revitalization, organizations' strategies and priorities, and the relationship between these and local government priorities, two trends emerge between different types of organizations. In Milwaukee, as in other US cities, a variety of grassroots groups are engaged in community revitalization. However, scholars and activists generally identify these organizations as one of two different types: community development corporations (CDCs) and community-based organizations (CBOs). CBOs tend to have a stronger orientation toward community organizing activities, greater reliance on public funding and a lesser focus on large capital investment projects. CDCs tend to have a strong economic and housing development focus, stronger links to the private philanthropic sector and the corporate community for funding and technical support, and more involvement as direct sponsors of large capital investment projects (Gitell and Wilder, 1999; Stoecker, 1997; Stoecker and Vakil, 2000).[10]

---

[9] Sieber (2000a) and Elwood and Leitner (2003) show that grassroots groups may resist negotiating directly with government officials and participating in governmental programmes because these interactions may require shifting the terms of their debate toward differing priorities and values advanced in these official structures.

[10] Any division between activities pursued by CDCs and those pursued by CBOs is not absolute. In practice, many organizations that identify as CDCs also do some community organizing, and many CBOs have become active in projects such as housing development and commercial revitalizalion.

Frequently, there is a mutually reinforcing relationship between CDCs' goals and strategies for community revitalization and those of dominant local government paradigms that have emerged in US cities over the past decade. In many urban areas, capital investment and home ownership have become dominant strategies for neighbourhood revitalization, frequently implemented through local government partnerships with organizations like CDCs. In a climate of community planning and revitalization that is characterised by competition amongst organizations for a declining pool of local state funds, CDCs have a distinct advantage. As Lake and Newman (2002) note, declining local government funding for urban revitalization puts growing pressure upon non-profit organizations to 'leverage' their own funds for revitalization. Amongst our case study organizations, those that position themselves primarily as CDCs are notably more successful in doing so than those that identify primarily as CBOs. The attitudes of city officials also work to the advantage of the CDCs, which, in our interviews, were uniformly characterised by these officials as more 'effective' and 'efficient' (DCD, 2001). Not only do local government practices prioritise these CDCs, so, too, do private funding agencies' practices. A staff member of a local non-profit development organization offered an example of support available only to CDCs from the Milwaukee Partnership for Community Development:

> [A] group of local funders come together and they provide . . . annual operating grants to a small select group of CDCs. And in return these CDCs make a contract to go through an assessment process and 10 come up with a plan to improve their organizational effectiveness. (LISC, 2001)

This staff member went on to describe the different status of organizations in the City's NSP process:

> [Y]ou had one level of competition between the groups that were in control of the [NSP] process and had the inside track in terms of getting the contracts and that sort of thing. And . . . the groups that had kind of a predictable source of revenue suddenly were competing against other groups who may or may not have been as politically well connected . . . (LISC, 2001)

The advantaged groups to which he referred were all neighbourhood-based CDCs, such as NWSCDC and LAND, that had been selected as NSP implementation groups. The economic and political advantages of CDCs relative to CBOs foster organizational capacities that enhance the ability of CDCs to sustain and effectively utilize PPGIS initiatives, including expanded networks of collaborative relationships, greater organizational knowledge of potential resources and greater organizational stability. Notably, three of the case study organizations with the most complex PPGIS applications are those that act primarily as CDCs: NWSCDC, WAICO-YMCA and LAND. The PPGIS capacities of the organizations identified as CBOs – Metcalfe Park, Harambee and Sherman Park – are much more narrowly focused.

Whereas both types of organizations are operating within the same supportive local context for PPGIS, the CDCs have engaged with this context to create more extensive and integrated PPGIS initiatives. These variations are not wholly due to differences in capacity between CDCs and CBOs with respect to their financial resources or organizational size. It is not simply that the organizations that have not engaged extensively in PPGIS in Milwaukee are not able to do so. Rather, while their PPGIS production is certainly shaped in part by their levels of funding and staffing, it is also affected by the relationship between their organizational priorities and the priorities of local government, as well as by the manner in which these government priorities are woven into local support resources for PPGIS. For an organization with distinctly differing strategies, such PPGIS support resources may not be a particularly useful or attractive opportunity to advance their own agenda.

8.6 Conclusions and Future Directions

Grassroots organizations are tremendously adaptable in their pursuit of PPGIS. Differential organizational characteristics, relationships and priorities help explain some of this variation in PPGIS production by community development and revitalization organizations. Organizational knowledge, stability, relationships and status all affect the manner in which organizations gain access to sources of GIS or strategic planning assistance (or choose alternative paths) and consolidate the financial, data and technological resources necessary to sustain robust PPGIS activities. These four organizational factors are closely linked. For instance, a dense network of organizational relationships is likely to foster greater organizational knowledge, and organizational stability provides the opportunity to form dense networks of relationships. These organizational factors must be seen not just as facilitating or impeding the use of GIS but, rather, as playing a role throughout PPGIS production. They affect access to data and technology, as well as the application of GIS-based knowledge to revitalization efforts and the manner in which this application of knowledge might alter an organization's participation and influence in local government decision making. The framework developed in this article provides a way of conceptualizing the role of organizational capacities and characteristics in this complex process. The case study organizations with the most developed PPGIS initiatives are those that possess broad knowledge of available support resources, dense networks of relationships with other organizations and local political actors, levels of stability that enhance organizational knowledge and ability to cultivate available resources, and revitalization priorities and strategies that can be advanced through PPGIS resources available in the locality. Organizations more limited in their engagement in PPGIS are those with, for instance, less knowledge about potential resources or organizational instability in funding, staffing or mission that impedes a variety of efforts, PPGIS included. In other instances, an organization may have priorities and strategies that differ from those of local government and may, as a result, have either chosen not to engage with governmentally structured PPGIS resources in order to retain more independent focus on their own initiatives or found that available resources do not fit their GIS, data and application needs.

By framing the organizational context of PPGIS production as encompassing multiple elements of organizational stability, relationships and networks with local political actors, public agencies and PPGIS support resources; and the relationship between organizational missions and goals being pursued by local government, this article attempts to further explain how and why organizations matter in PPGIS production. Extending our understanding of organizational context beyond the confines of an organization itself to include its positioning within the broader situation of PPGIS production enables a more detailed assessment of how groups engage with locally available resources, as well as a stronger explanation of the local variability of PPGIS. Further, the conceptualization of organizational context developed in this article may suggest a broadened range of strategies for enhancing PPGIS production. Moving beyond strategies that focus on providing technology and training, it may be useful to consider strategies directed at organizational stabilization, network building or diversification of the range of revitalization priorities and approaches that can be supported by local PPGIS resources.

Organizational context is, of course, only part of a complete theorization of PPGIS production, and a number of avenues for further conceptual development remain. PPGIS is at heart participatory, so it will continue to be important to conceptualize different forms of participation evident in and fostered through various PPGIS processes. Differing forms of participation, in GIS use as well as in application of the knowledge produced, potentially play an important role in further differentiating PPGIS production in various application types and organizations. As well, it is critical to further examine the role of local political context in differentiating PPGIS production from place to place – to develop a more expanded theorization that is able to explain PPGES production across different organizational contexts, but also across variable local political contexts, PPGIS remains a product of multiple contextual factors, and further theorization of the importance of local political context may significantly build upon the ideas developed here.

Acknowledgements

This research was carried out with funding support provided by Illinois State University and DePaul University. We are particularly grateful to Renée Sieber and two anonymous reviewers for their constructive reading of an earlier draft and their suggestions for revision.

References

Barndt, M. (1998) Public participation GIS: Barriers to implementation. *Cartography and Geographic Information Systems*, **25**(2), 105–112.

Barndt, M. (2001) Interview by authors, Milwaukee, WI, 23 July.

Barndt, M. (2002) A model for evaluating public participation GIS, in *Community Participation and Geographic Information Systems* (eds. W.J. Craig, T.M. Harris and D. Weiner), Taylor and Francis, New York, pp. 346–356.

Barndt, M. and Craig, W. (1994) Data providers empower community GIS efforts. *GIS World*, **7**, 49–51.

Bosworth, M., Donovan, J. and Couey, P. (2002) Portland Metro's dream for public involvement, in *Community Participation and Geographic Information Systems* (eds. W.J. Craig, T.M. Harris and D. Weiner), Taylor and Francis, New York, pp. 125–136.

Campbell, H. (1991) Organizational issues in managing geographic information, in *Handling Geographic Information* (eds. I. Masser and M. Blakemore), Longman, London, pp. 259–282.

Campbell, H. (1994) How effective are GIS in practice? A case study of British local government. *International Journal of Geographic Information Systems*, **8**, 309–325.

Campbell, H. and Masser, I. (1995) *GIS and Organizations: How Effective Are GIS in Practice?* Taylor and Francis, London.

Carver, S. (2001) Participation and geographical information: a position paper. *Workshop on Access to Geographic Information and Participatory Approaches Using Geographic Information*, Spoleto, Italy.

Casey, L. and Pederson, T. (2002) Mapping Philadelphia's neighborhoods, in *Community Participation and Geographic Information Systems* (eds. W.J. Craig, T.M., Harris and D. Weiner), Taylor and Francis, New York, pp. 65–76.

Craig, W.J., Harris, T.M. and Weiner, D. (2002) Conclusion, in *Community Participation and Geographic Information Systems* (eds. W.J. Craig, T.M. Harris and D. Weiner), Taylor and Francis, New York, pp. 367–372.

Department of City Development (DCD), Milwaukee (2001) Staff member interview by authors. Milwaukee, WI, 26 July.

Elmes, G. (2001) Responses to papers on access and public participation using geographic information. *Workshop on Access to Geographic Information and Participatory Approaches Using Geographic Information*, Spoleto, Italy.

Elwood, S. (2002) The impacts of GIS use for neighbourhood revitalization, in *Community Participation and Geographic Information Systems* (eds. W.J. Craig, T.M. Harris and D. Weiner), Taylor and Francis, New York, pp. 77–88.

Elwood, S. and Leitner, H. (1998) GIS and community-based planning: Exploring the diversity of neighborhood perspectives and needs. *Cartography and Geographic Information Systems*, **25**(2), 77–88.

Elwood, S. and Leitner, H. (2003) Community-based planning and GIS: Aligning neighborhood organizations with state priorities? *Journal of Urban Affairs*, **25**(2), 139–157.

Ghose, R. (2001) Use of information technology for community empowerment: Transforming geographic information systems into community information systems. *Transactions in GIS*, **5**(2), 141–163.

Ghose, R. and Huxhold, W. (2001) The role of local contextual factors in building public participation GIS: The Milwaukee experience. *Cartography and Geographic Information Systems*, **28**(3), 195–208.

Ghose, R. and Huxhold, W. (2002) Role of multi-scalar GIS-based indicators studies in formulating neighborhood planning policy. *URISA Journal*, **14**(2), 3–16.

Gitell, R. and Wilder, M. (1999) Community development corporations: Critical factors that influence success. *Journal of Urban Affairs*, **21**, 541–562.

Harambee Ombudsman Project (2001) Staff member interview by authors. Milwaukee, WI, 27 July.

Harris, T. and Weiner, D. (1998) Empowerment, marginalization, and community-oriented GIS. *Cartography and Geographic Information Systems*, **25**(2), 67–76.

Harrison, B. and Weiss, M. (1998) *Workforce Development Networks: Community-Based Organizations and Regional Alliances*, Sage, Thousand Oaks, CA.

Huxhold, W. (1991) *An Introduction to Urban Geographic Information Systems*, Oxford University Press, London.

Innes, J. and Simpson, D. (1993) Implementing GIS for planning: Lessons from the history of technological innovation. *Journal of the American Planning Association*, **59**, 230–236.

Jordan, G. (2002) GIS for community forestry user groups in Nepal: Putting people before technology, in *Community Participation and Geographic Information Systems* (eds. W.J. Craig, T.M. Harris and D. Weiner), Taylor and Francis, New York, pp. 232–245.

Kyem, P. (2002) Promoting local community participation in forest management through a PPGIS application in Southern Ghana, in *Community Participation and Geographic Information Systems* (eds. W.J. Craig, T.M. Harris and D. Weiner), Taylor and Francis, New York, pp. 218–230.

Lake, R.W. and Newman, K. (2002) Differential citizenship in the shadow state. *Geojournal*, **58**(2/3), 109–120.

Leitner, H., Elwood, S., Sheppard, E. *et al.* (2000) Modes of GIS provision and their appropriateness for neighborhood organizations: Examples from Minneapolis and St. Paul, Minnesota. *URISA Journal*, **12**(4), 43–56.

Leitner, H., McMaster, R., Elwood, S. *et al.* (2002) Models for making GIS available to community organizations: Dimensions of difference and appropriateness, in *Community Participation and Geographic Information Systems* (eds W.J. Craig, T.M. Harris and D. Weiner), Taylor and Francis, New York, pp. 37–52.

Lisbon Area Neighbourhood Development (LAND) (2001) Staff member interview by authors. Milwaukee, WI, 24 July.

Local Initiatives Support Corporation (LISC) (2001) Staff member interview by authors. Milwaukee, WI, 26 July.

Macnab, P. (2002) There must be a catch: Participatory GIS in a Newfoundland fishing community, in *Community Participation and Geographic Information Systems* (eds. W.J. Craig, T.M. Harris and D. Weiner), Taylor and Francis, New York, pp. 173–191.

Martin, M. (2001) Personal interview by authors, July 2001.

Meredith, T., Yetman, G. and Frias, G. (2002) Mexican and Canadian case studies of community-based spatial information management for biodiversity conservation, in *Community Participation and Geographic Information Systems* (eds. W.J. Craig, T.M. Harris and D. Weiner), Taylor and Francis, New York, pp. 205–217.

Metcalfe Park Residents Association (MPRA) (2000) Staff member telephone interview by R. Ghose, 6 October.

Nedović-Budić, Z. (1998) The impact of GIS technology. *Environment and Planning B: Planning and Design*, **25**, 681–692.

Nedovic-Budic, Z. and Pinto, J. (2000) Information sharing in an inter-organizational GIS Environment. *Environment and Planning B: Planning and Design*, **27**, 455–474.

Northwest Side Community Development Corporation (NWSCDC) (2001) Staff member interview by authors. Milwaukee, WI, 27 July.

Obermeyer, N. and Pinto, J. (1994) *Managing Geographic Information Systems*, Guilford Press, New York.

Onsrud, H. and Pinto, J. (1991) Diffusion of geographic information innovations. *International Journal of Geographic Information Systems*, **5**, 447–467.

Ramasubramanian, L. (1998) *Knowledge Production and Use in Community-Based Organizations: Examining the Impacts and Influence of Information Technologies*. Unpublished PhD Thesis University of Wisconsin – Milwaukee, WI.

Ramasubramanian, L. (1999) GIS Implementation in developing countries: Learning from organizational theory and reflective practice. *Transactions in GIS*, **3**(4), 359–380.

Sanders, W. (2001) Interview by authors. Milwaukee, WI, 25 July.

Sawicki, D. and Craig, W. (1996) The democratization of data: Bridging the gap for community groups. *Journal of the American Planning Association*, **62**, 512–523.

Sawicki, D. and Peterman, D. (2002) Surveying the extent of PPGIS practice in the United States, in *Community Participation and Geographic Information Systems* (eds. W.J. Craig, T.M. Harris and D. Weiner), Taylor and Francis, New York, pp. 17–36.

Sherman Park Community Organization (SPCO) (2001) Staff member interview by authors. Milwaukee, WI, 27 July.

Sieber, R. (1997) *Computers in the Grassroots: Environmentalists, Geographic Information Systems, and Public Policy*. Unpublished PhD Thesis, Rutgers University, New Brunswick, NJ.

Sieber, R. (2000a) Confronting the opposition: The social construction of geographical information systems in social movements. *International Journal of Geographic Information Systems*, **14**, 775–793.

Sieber, R. (2000b) GIS implementation in the grassroots. *URISA Journal*, **12**(1), 15–51.

Sieber, R. (2002) Geographic information systems in the environmental movement, in *Community Participation and Geographic Information Systems* (eds. W.J. Craig, T.M. Harris and D. Weiner), Taylor and Francis, New York, pp. 153–172.

Stoecker, R. (1997) The CDC model of urban redevelopment: A critique and an alternative. *Journal of Urban Affairs*, **19**(1), 1–22.

Stoecker, R. and Vakil, A. (2000) States, cultures, and community organizing: Two tales of two neighborhoods. *Journal of Urban Affairs*, **22**, 439–458.

Stonich, S. (2002) Information technologies, PPGIS, and advocacy: Globalization of resistance to industrial shrimp farming, in *Community Participation and Geographic Information Systems* (eds. W.J. Craig, T.M. Harris and D. Weiner), Taylor and Francis, New York, pp. 259–269.

Tulloch, D. (2002) Environmental NGOs and community access to technology as a force for change, in *Community Participation and Geographic Information Systems* (eds. W.J. Craig, T.M. Harris and D. Weiner), Taylor and Francis, New York, pp. 192–204.

Ventura, S., Niemann, B., Sutphin, T. and Chenowith, R. (2002) GIS-enhanced land-use planning, in *Community Participation and Geographic Information Systems* (eds. W.J. Craig, T.M. Harris and D. Weiner), Taylor and Francis, New York, pp. 113–124.

Walnut Area Improvement Community Organization (WAICO-YMCA) (2001) Staff member interview by authors. Milwaukee, WI, 24 July.

9

Reflection Essay: PPGIS in Community Development Planning

Sarah Elwood[1] and Rina Ghose[2]

[1] *Department of Geography, University of Washington, USA*
[2] *Department of Geography, University of Wisconsin – Milwaukee, USA*

Surrounded as we are now by Google Maps, the geospatial Web and so-called citizen cartographers, it may seem strange that, not so long ago, researchers were trying to understand the struggles of grassroots groups to gain access to and use of geographic information technologies and digital spatial data. Within what was then referred to as 'GIS and Society research', our 2001 paper in *Cartographica* was part of two closely related research engagements occurring at the time. On the heels of fierce debates in the early 1990s about the societal and disciplinary impacts of GIS, human geographers and GIScience scholars began studying the social and political implications of GIS, particularly its impacts for marginalized individuals, institutions and social groups (Aitken and Michel, 1995; Harris *et al.*, 1995; Sheppard, 1995).

In parallel, others were beginning to study (and facilitate) the role of GIS in a range of activities argued to be 'participatory' in some capacity, including collaborative planning, community organizing, community mapping, participatory action research and community resource management (Craig and Elwood, 1998; Elwood and Leitner, 1998; Harris and Weiner, 1998; Sieber, 2000; Ghose, 2001; Ghose and Huxhold, 2001). The terms 'public participation GIS' (Nyerges, Barndt and Brooks, 1996; Obermeyer, 1998) and 'participatory GIS' research (Rambaldi *et al.*, 2004) emerged early on to describe these interweavings of GIS and participatory practice. These labels were also used to name the growing body of research studying the modes of participation, power and

Classics in Cartography: Reflections on Influential Articles from Cartographica   Edited by Martin Dodge
© 2011 John Wiley & Sons, Ltd

knowledge politics being advanced through these engagements with spatial technologies and cartographies.[1]

The paper reproduced here emerged from research we conducted together in Milwaukee, Wisconsin, in 2001. We met and identified common interests in participatory GIS and urban revitalization at a 1999 meeting of GIS and Society researchers. Milwaukee was a clear choice for a case study exploring the range of factors that shape the sustainability of GIS efforts by community-based organizations, as it had the most robust range of such activities of any city in the United States at the time.

In the following pages we consider the position of our paper within these past and present practices and research agendas. We also reflect on the paper's role in our own thinking and the ways in which it has shaped the work we have gone on to do in the ensuing years. Our account is one genealogy of how concepts, practices and disciplinary norms are made and re-made within the churn of emerging research agendas, our own and those of others. Scholarly writings are waypoints along paths *from* somewhere and *to* somewhere in our collective knowledge making efforts, and here we describe what this terrain has looked like for us. In the ensuing years, we have built upon and re-worked the ideas in 'PPGIS in Community Development Planning' in different ways, so we offer our individual reflections in sequence here.

9.1 Reflections on Knowledge Making and Re-Making (By Sarah)

Early in my graduate studies, I participated in a seminar organized around an invited lecture series in feminist geography. Each Saturday morning (in a dreadful imposition on our guests!) we met to discuss the previous day's lecture, related readings and the visitor's ongoing research. By chance or design, our guests all opened by reflecting on how they had come to a particular piece of research or writing, and where they felt it fitted in their own thinking, in conversation with other scholars, in the evolving work of feminist geography. These narratives were profoundly helpful in my sense of how knowledge emerges from our personal and collective engagements in research and writing, and so I follow their example here.

New research agendas often seem to open with widely divergent (and often unsubstantiated) claims about their societal and disciplinary impacts, followed by exploratory research seeking to characterize the new phenomenon and theorize from this. The efforts of geographers to come to grips with the societal and disciplinary impacts of GIS certainly followed such a path. The 'GIS wars' of the early 1990s (Schuurman, 2000) gave rise to a host of new research areas that sought to examine

---

[1] In reading this literature, it is important to recognize that there is a great deal of variability in how the identifiers 'Public participation GIS' (PPGIS) and 'participatory GIS' (PGIS) are used. Some scholars use them interchangeably, while others use them to distinguish between initiatives that are situated in formal planning practice (PPGIS) and those that emerge from citizen-based activities (PGIS). For further discussion of these debates see: Rambaldi *et al.*, 2004; Sieber, 2006; Dunn, 2007. Here, we use 'P/PGIS' to refer to this body of work as a whole.

the ever-diversifying role of maps and spatial technologies in society, as well as their disciplinary and methodological possibilities: Critical cartography, critical GIS, feminist GIS, qualitative GIS and, of course, participatory GIS and public partici-pation GIS.

Our 2001 *Cartographica* paper entered these debates at a time when P/PGIS research was seeking to theorize across and amongst its first wave of exploratory research projects, as a means of conceptualizing some of the key structures and relationships shaping inclusion, exclusion, sustainability or empowerment in P/PGIS practice. Indeed, we wrote the piece for an edited journal issue organized around the notion of a 'PPGIScience', in which the editor of the collection urged us, and others, to theorize from the rich body of P/PGIS case studies that were being conducted (Sieber, 2001). Reflecting a focus at the time on studying institutions and organiza-tions as mediators of PPGIS practice, three papers in the collection dealt with institutional contexts. Kyem (2001) offered an account of the ways that inflexible institutions can stymie the transformative potential of participatory resource man-agement that uses GIS. Norheim (2001) theorized the role of institutional cultures, data sources, and validated methodologies and epistemologies upon the GIS practices of institutions, drawing evidence from a rich comparative case of parallel analyses carried out by the US Forest Service and an environmental non-governmental organization.

Our own focus on community-based organizations stemmed in part from the predominance of these institutions as PPGIS practitioners in the North American urban context. In this context, PPGIS was most often being practiced within collab-orative planning, community development and neighbourhood revitalization process-es, typically by non-governmental or quasi-governmental organizations (Barndt, 1998; Craig, Harris and Weiner, 2002; Ghose and Huxhold, 2002). In prior work (Elwood and Leitner, 1998; Ghose, 2001; Ghose and Huxhold, 2001), we had both observed stark differences in the access, application and sustainability of the grassroots GIS initiatives, and suspected that the material and socio-political situations of the participating organizations were important factors generating these differences. Existing concep-tualizations of GIS in organizations primarily focused on the processes and mechanisms of technology adoption and diffusion, with an empirical focus on public and private sector organizations (Obermeyer and Pinto, 1994; Campbell and Masser, 1995). These accounts did not speak to the social and political construction of spatial technologies in under-resourced voluntary organizations and community action initiatives. Thus, the research we conducted together relied upon Burawoy's (1998) extended case method, an inductive research design that facilitates development of theoretical propositions from ethnographic research, to 'extend' or bridge gaps in existing theorizations.

I appreciate the extended case method not just as a robust framework for building theory in qualitative research, but because it explicitly foregrounds the ways in which we are always extending and bridging from existing explanations to our own emergent understandings. My own thinking about the socio-political construction of GIS in activism, community development and participatory decision making is strongly informed by several of the original *Cartographica* papers reprinted in this collection. From Woods and Fels (1986; reproduced as Chapter 14) I drew ways of conceiving of maps as texts, as situated and contingent collections of symbols that are always necessarily political. Harley (1989; reproduced as Chapter 16) and Edney (1993; reproduced as Chapter 18) offered accounts of maps and cartography as dynamic praxis emerging from contingent cultural, social and technological relations. Sparke's work (1995; reproduced as Chapter 20) pushed me to look for resistant cartographies, and understand their emergence even within hegemonic knowledge systems and starkly unequal power relations. These pieces push us to think beyond cartography-as-defined-by-cartographers *and* offer ways of theorizing the new (and newly recognized) practices that are visible from this expanded perspective. For these reasons, I have found them invaluable for understanding the multidimensional politics and practices of P/PGIS. 'Classics', I would argue, do just this sort of work – they enable new openings and new directions in emergent research agendas.

While I am not entirely comfortable applying this language to our paper, other researchers' engagement with the piece suggests that it has had some broader resonance. Our attention to the range of roles and influences that institutions and organizations may have upon grassroots GIS, cartography and spatial politics has been taken up in other related work. In particular, our call to focus on how power and knowledge are negotiated within institutions *and* in the relationships of these institutions with others seems to circulate in other work. Kyem's (2001) account of participatory GIS practice in community forest management in Ghana, for example, uses this framing to illuminate the ways in which institutional priorities and practices can impede efforts to use PPGIS-based knowledge-making to rewrite inequalities in local forest management practices and relationships. Rattray (2006) uses some of our ideas about socio-political and institutional contexts to examine Web-based PPGIS initiatives, showing how institutional knowledge, skills and relationships play a strong role in shaping the spatial data, cartographic tools and representational options that sponsoring organizations can make available to their site users.

However, I also take as productive the cautions rendered in response to our work. Gilbert and Masucci (2006) for instance, cited our 2001 *Cartographica* paper as they warned against equating 'access' with 'empowerment' in P/PGIS research. This caution, and their related research, inspires my continued attention to multiple levels on which empowerment and marginalization are produced in purportedly participatory engagements with spatial data and geographic information technologies. Responding to our paper and others, Wilson (2005) offered a well-conceived call for a more carefully theorized account of politics and power in P/PGIS, another urging that I have tried to respond to in later work.

'*PPGIS in Community Development Planning*' has also pointed me along the way toward new questions. One of the productive joys of ethnographic work is its capacity to illuminate ideas that were not part of your initial research questions. In our Milwaukee fieldwork, Rina and I spent many hours listening to community organization staff describe their use of spatial data, maps and GIS, and the barriers they encountered in trying to gain access to and self-determination over these resources. In listening to these accounts, I became more and more interested in our informants' accounts of how they used spatial data and cartographic representations in the local politics of neighbourhood revitalization. This interest carried over into my next project, a long-term participatory research project with community organizations in Chicago. In one of my first papers from that project, I argued that grassroots GIS practices must be theorized outside of a binary of empowerment or disempowerment, articulating instead a flexible knowledge politics advanced through social data, maps and the priorities or positions that community level actors use them to illuminate (Elwood, 2006). '*PPGIS in Community Development Planning*' also piqued my interest in looking at spatial data, the governmental institutions that create, maintain and disseminate them, and the infrastructures through which these efforts are mediated. The community organizers we worked with in Milwaukee all reported difficulties in gaining consistent access to appropriate spatial data from local government. Many of them pointed to the scale, resolution and attribute schemes of the data themselves as problematic, as did my later research partners in Chicago. These ideas started me toward thinking about the role of local spatial data infrastructures in shaping grassroots GIS practice in some of my more recent work (Elwood, 2008).

As we noted at the outset of this essay, much is different about the technological, institutional and disciplinary contexts of grassroots cartographies today (see Crampton (2009) for a discussion of 'new spatial media' on the Web). In the world of [seemingly] free mapping and open APIs and interactive online cartographies, it would be easy to imagine that perhaps attention to the P/PGIS touchstones of access, empowerment and grassroots organizations' spatial politics will wane. Nonetheless, I hope that our paper is one of many from that period that facilitates future efforts to understand the new ways in which resources, relationships and knowledge are being negotiated in these new practices. In retrospect, I do wish that we had designed the study to focus beyond only 'quasi state' organizations doing community development, and perhaps written about their practices with spatial technologies a bit more broadly than the P/PGIS framing. These openings would, I think, make it easier to use ideas from our paper to theorize contemporary practices.

9.2 Plus Ça Change...(By Rina)

While Sarah's participatory GIS research began and continued in other cities, my own has been centred in Milwaukee since the early 1990s. This sustained ethnographic engagement with the efforts of non-profit community organizations to use spatial data and technologies has afforded me a rich longitudinal view of the emergence of P/PGIS research and practice and their current transformations. I reflect upon some of those shifts here.

My introduction to community organizing as a social movement to contest poverty and blight in inner-city neighbourhoods occurred in my graduate work, in particular in an applied GIS course where I grappled firsthand with some of the challenges of public participation GIS that continue to shape my work today. Based on the requests from the residents and community organizers of a severely impoverished inner city neighbour-hood, we worked together to create a GIS-based Community Information System that integrated their local knowledge with public data sets. Such a system, it was hoped, would enable them to be more effective in their community organizing and planning pursuits (Ghose, 2001). I was intrigued by how well the community organizers and residents understood the power of using spatial knowledge through maps to negotiate urban politics, despite not having any background in traditional cartography and GIS. We hoped that the collaboration would help ameliorate their limited access to digital spatial data, mapping technologies and experience using these resources in their neighbourhood revitalization efforts. These goals mirrored those of P/PGIS research and practice, which was then coalescing to focus on ways of providing opportunities for geographic information access by marginalized citizens, with the hope that some form of empowerment might follow (Craig, Harris and Weiner, 2002).

In today's world of Google Maps, Google Earth and other such Web services, citizens appear to have limitless opportunities to gain access to digital spatial data online, and to collect and distribute their own data. Open source spatial data initiatives such as OpenStreetMap (OSM) create data sets based on the volunteer efforts of citizens, and aim to produce base maps that are 'more accurate than anything in the market' (Shiels, 2009). Such 'volunteered geographic information' (VGI) (Goodchild, 2007) has the potential to be highly current, for, as one public official argued, 'we have new buildings and streets being built all the time and this notion of sneakers on the ground going around mapping everything means you get up-to-date data' (Shiels, 2009). Further, some of the early writing on VGI imagines that these citizen-produced data sets may be more freely available than spatial data created through more conventional public and private sector pathways. In theory at least, these developments over the past several years would seem to realize many of the goals of PPGIS practice – equitable access to digital spatial data and maps, as well as the means to produce and use these resources.

However, are these goals of P/PGIS likely to be realized by volunteer geographers who are gathering, mapping and disseminating community information through easy-to-use mapping technologies, such as Google Maps, and collaboratively pro-duced data sets, such as the ones emerging from OSM? The answer is rather complicated.

Echoing Crutcher and Zook (2009), I would argue that spatial knowledge production though Google Maps, Google Earth or OSM occurs within the same class and race inequities as conventional forms of GIS-based technologies and spatial data. My own recent research shows that the digital divide persists for citizens of impoverished inner city neighbourhoods, regardless of the purportedly greater accessibility of online mapping platforms (Ghose,). The hierarchies and inequalities of the political context in which community organizations operate remain largely unchanged. These groups continue to interact with powerful actors and institutions that question neighbourhood knowledge as mere anecdotal information, unless those are made legitimate by 'hard data' – which in this context tends to mean 'official' information gathered by the state. In contrast to the situation envisioned in the burgeoning literature on so-called 'neogeography' and VGI, community organizers and local residents continue to struggle to assert the validity of their local knowledge to powerful actors. Even while these new technologies and forms of spatial data may afford under-resourced groups a greater degree of access, they do nothing to close the gap with more powerful and better resourced actors, who have the option to rely on *both* costly commercial spatial technologies and open access resources.

More broadly, the citizen groups involved in inner city P/PGIS in the United States exist in a state of crisis that affects not only their process of participation and spatial knowledge production, but their very survival. The trends we observed in 2001 with respect to diminishing public funding for community organizations, accompanied by increasing responsibilities, has increased exponentially. Historically, community organizations have been largely supported through federal government funding, but in recent years, federal funding has been reduced significantly. Today, community organization are expected to be entrepreneurial, acquiring their own funds from private and philanthropic funders and promoting economic development through investments from the private sector. These groups are responsible for providing myriad services to inner city residents, with progressively more limited resources to do so. Resource and power imbalances make it difficult for community organizations to construct these changes, but recent research suggests that citizen groups have developed a wide range of creative strategies for doing so (Leitner, Peck and Sheppard, 2007).

Since our 2001 research in Milwaukee, I have observed community organizations intentionally developing ways of using spatial data and technologies to their own advantage amidst these increasing pressures (Ghose, 2003; Lin and Ghose, 2008). Investors, planners and philanthropic funders have come to expect greater 'professionalization' from community organizations and the use of spatial data and GIS is one way that Milwaukee organizations try to demonstrate this. In the increasingly entrepreneurial situation in which they find themselves, these organizations gather and map neighbourhood data to demonstrate their needs, support funding application or demonstrate neighbourhood assets as a means of luring investors. With funders increasingly emphasizing assessment of 'benchmarks' of progress, outcomes and impacts, Milwaukee groups often deploy spatial data and maps to respond to these requests. Thus, even as more powerful actors continue to foist their own ideas of 'best practices' upon community organizations, GIS remains an important resource in their ability to respond to these pressures.

Many of the 'organizational factors' discussed in our original *Cartographica* paper continue to play a critical role in shaping Milwaukee community groups' P/PGIS practices. My recent fieldwork suggests that in spite of the increasing 'user friendliness' of GIS software since our original research in 2001, many community organizations are still greatly challenged by the technical complexity of GIS. They continue to rely greatly on external GIS actors to provide them with customized spatial data and geographical analysis, and their ability to form critical 'networks of association' remains key to their success in gaining access to and using spatial technologies. This notion of networks of association as an important resource that community organizations rely upon in their GIS and neighbourhood revitalization efforts emerges from my later work in Milwaukee (Ghose, 2005, 2007). Building upon Leitner, Pavlik and Sheppard (2002) concepts of thematic and territorial networks, I found that, in the case of P/PGIS practice, some community organizations are able to create powerful networks of association with planning agencies, philanthropic foundations, investors and business communities – at the national and local scale (Ghose, 2007). These networks create opportunities for community organizations to contest the problems of inner city disinvestment and neoliberal urban policy, both materially and discursively, and GIS is an important element of these efforts (Ghose, 2007).

The long-term nature of my research engagement in Milwaukee has given me the opportunity to illustrate how P/PGIS politics and practices may evolve over time as actors and their relationships change. For example, one of the institutions we studied in 2001, the Sherman Park Community Association, had a limited network at the time and very restricted GIS access and use. More recently, this organization has built a strong collaboration with municipal and other government entities, forming an alliance called the Housing Coalition. Sherman Park later received funding and staff resources to gather very detailed building condition data of its neighbourhood, for later incorporation into the City's central property database (Ghose, 2007). The data gathered by the coalition enabled Sherman Park and the City to jointly pressure absentee landlords to improve blighted housing conditions. This situation illustrates how bottom up spatial knowledge can be used to improve neighbourhood quality of life, and also illustrates the persistent significance of network entities in sustaining GIS capacities for community organizations.

GIS-based spatial knowledge production by community groups is constantly negotiated in a dynamic and sometimes contradictory way, and I would argue that a range of intersecting contextual factors continue to be relevant. Organizational networks provide flexibilities and opportunities for substitution in the face of limited resources. Organizational stability and capacity affect network formation – larger and more stable community organizations are more likely to be engaged in collaborative governance networks. Organizational leadership continues to play a key role, shaping a group's capacity to understand and successfully negotiate the political contexts in which it is situated, tailoring their discursive and material GIS practices to support these efforts. The presence of staff members with technological skills can propel greater and more sustained use of spatial data and GIS. Finally, organizational mission remains as critical as ever, with these overarching goals framing the forms of spatial knowledge and politics that community organizations foster through their P/PGIS activities. Thus, while the assemblage of spatial technologies and geographic information that grassroots groups use in their community organizing and redevelopment work is in flux, the social, political and organizational contexts of these efforts remain highly relevant in shaping their sustainability and impacts.

9.3 Whither Citizen Cartographies, Whither P/PGIS?

It may seem strange that in a collection of 'Classics in Cartography', our original paper and the reflection we have offered here says precious little about cartography as such. Our focus now, as then, has been on the social and political construction of spatial technologies, spatial data and geographic knowledge, especially in the context of grassroots GIS practice. Yet critical cartography scholarship – particularly several of the pieces included in this collection – played a central role in enabling P/PGIS research and our own work within this tradition. Amongst other contributions, scholars whom we would identify today as 'critical cartographers' were amongst the first to call geographers' attention to the ways in which cartographies both constitute and are born of social and political relationships and identities, and to insist upon the necessity of studying cartographies by way of their processes, visual artefacts and constitutive outcomes. It is precisely this insistence on examining the politics and praxis of spatial knowledge-making and representation that inspired a diverse range of critical GIS research, including P/PGIS.

Yet for all this focus upon social and political practice and relationship, it also the case that we stand amidst distinct changes in the assemblages of technologies that citizens and community-based organizations leverage in their efforts to create and share digital spatial data and maps. For the future of P/PGIS research and practice, transformations in the collections of hardware, software and data that we define as a 'geographic information systems' are of particular importance. An interactive online map created and modified from the information contributed by many individuals is a sort of geographic information system, though not as traditionally defined over the past twenty or more years. Following Sheppard's (2006) assertion that GIS-as-we-know-it is being dramatically transformed through the advent of online spatial media, we would argue that so too there is likely to be transformation in PPGIS-as-we-know it. The citizen cartographies that have long been studied under the auspices of PPGIS are expanding beyond the GIS-based practices of the 1990s and early 2000s to include an ever-expanding range of Web-based and mobile citizen cartographies. As the range of spatial technologies that are used in grassroots spatial data production and geovisualization expands well beyond conventional GIS technologies, so too must P/PGIS scholarship expand its theorization to include these new phenomena and practices – perhaps well beyond its focus on conventional geographic information systems.

Further Reading

Cope, M. and Elwood, S. (2009) *Qualitative GIS: A Mixed Methods Approach*, Sage, London. (P/PGIS research and practice have often involved efforts to incorporate non-quantitative forms of evidence and modes of analysis and, as such, provided some early momentum toward intersections of qualitative research and GIS. This edited collection includes several chapters profiling research that draws upon P/PGIS theory and practice.)

Dunn, C. (2007) Participatory GIS: A people's GIS? *Progress in Human Geography*, **31**(5), 617–638. (This paper provides a thorough review and discussion of public participation/participatory GIS research.)

Haklay, M., Singleton, A. and Parker, C. (2008) WebMapping 2.0: The neogeography of the geoweb. *Geography Compass*, **2**(6), 2011–2039. (This paper provides a thoughtful review of emergent new 'more than GIS' spatial technologies and practices, including neogeography, the geospatial web and volunteered geographic information.)

Sheppard, E. (2006) Knowledge production through critical GIS: Genealogy and prospects. *Cartographica*, **40**(4), 5–21. (While the two fields are by no means synonymous, P/PGIS shares a common intellectual history with critical GIS research, as well as a common commitment to using GIS in socially transformative ways. This paper provides a review of the emergence of critical GIS, and poses key questions for its continued development.)

Tulloch, D. (2008) Is volunteered geographic information participation? *GeoJournal*, **72**(2/3), 161–171. (Within the nascent research agenda on volunteered geographic information, this paper offers a thoughtful reflection on its intersections with P/PGIS research and practice, providing a sense of how all elements around which P/PGIS has coalesced – the 'public', 'participation' and 'GIS' – must be re-considered in light of the geospatial web.)

References

Aitken, S. and Michel, M. (1995) Who contrives the 'real' in GIS? Geographic information, planning, and critical theory. *Cartography and Geographic Information Systems*, **22**(1), 17–29.

Barndt, M. (1998) Public participation GIS: Barriers to implementation. *Cartography and Geographic Information Systems*, **25**(2), 105–112.

Burawoy, M. (1998) The extended case method. *Sociological Theory*, **16**(1), 4–33.

Campbell, H. and Masser, I. (1995) *GIS and Organizations*, Taylor and Francis, London.

Craig, W. and Elwood, S. (1998) How and why community groups use maps and geographic information. *Cartography and Geographic Information Systems*, **25**(2), 95–104.

Craig, W., Harris, T. and Weiner, D. (2002) Introduction, in *Community Participation and Geographic Information Systems* (eds. W. Craig, T. Harris and D. Weiner), Taylor and Francis, London, pp. 1–16.

Crampton, J. (2009) Cartography: Maps 2.0? *Progress in Human Geography*, **33**(1), 91–100.

Crutcher, M. and Zook, M. (2009) Placemarks and waterlines: Racialized cyberscapes in post-Katrina Google Earth. *Geoforum*, **40**(4), 523–534.

Dunn, C. (2007) Participatory GIS: A people's GIS? *Progress in Human Geography*, **31**(5), 617–638.

Edney, M. (1993) Cartography without 'progress': Reinterpreting the nature and historical development of mapmaking. *Cartographica*, **30**(2/3), 54–68. (Reproduced as Chapter 18 of this volume.)

Elwood, S. (2006) Beyond cooptation or resistance: Urban spatial politics, community organizations, and GIS-based spatial narratives. *Annals of the Association of American Geographers*, **96**(2), 323–341.

Elwood, S. (2008) Grassroots groups as stakeholders in spatial data infrastructures: Challenges and opportunities for local data development and sharing. *International Journal of Geographic Information Science*, **22**(1), 71–90.

Elwood, S. and Leitner, H. (1998) GIS and community-based planning: Exploring the diversity of neighborhood perspectives and needs. *Cartography and Geographic Information Systems*, **25**(2), 77–88.

Ghose, R. (2001) Use of information technology for community empowerment: Transforming geographic information system into community information systems. *Transactions in GIS*, **5**(2), 141–163.

Ghose, R. (2003) Investigating Community Participation, Spatial Knowledge Production and GIS Use in Inner City Revitalization *Journal of Urban Technology*, **10**(1), 39–60.

Ghose, R. (2005) The complexities of citizen participation through collaborative governance. *Space and Policy*, **9**(1), 61–75.

Ghose, R. (2007) Politics of scale and networks of association in public participation GIS. *Environment and Planning A*, **39**, 1961–1980.

Ghose, R. and Huxhold, W. (2001) Role of local contextual factors in building public participation GIS: The Milwaukee Experience. *Cartography and Geographic Information Science*, **28**(3), 195–208.

Ghose, R. and Huxhold, W. (2002) Role of multi-scalar GIS-based indicators studies in formulating neighborhood planning policy. *URISA Journal*, **14**(2), 3–16.

Gilbert, M. and Masucci, M. (2006) Geographic perspectives on e-collaboration research. *International Journal of E-Collaboration*, **2**(1), 1–5.

Goodchild, M. (2007) Citizens as sensors: The world of volunteered geography. *GeoJournal*, **69**(4), 211–221.

Harley, J.B. (1989) Deconstructing the map. *Cartographica*, **26**(1), 1–20. (Reproduced as Chapter 16 of this volume.)

Harris, T. and Weiner, D. (1998) Empowerment, marginalization, and community-integrated GIS. *Cartography and Geographic Information Systems*, **25**(2), 67–76.

Harris, T., Weiner, D., Warner, T. and Levin, R. (1995) Pursuing social goals through participatory GIS: Redressing South Africa's historical political ecology, in *Ground Truth: The Social Implications of Geographic Information Systems* (ed. J. Pickles), Guilford Press, London, pp. 196–222.

Kyem, P. (2001) Power, participation and inflexible institutions: An examination of the challenges to community empowerment in participatory GIS applications. *Cartographica*, **38**(3/4), 5–17.

Leitner, H., Pavlik, C. and Sheppard, E. (2002) Networks, governance and the politics of scale: Inter-urban networks and the European Union, in *Geographies of Power: Placing Scale* (eds. A. Herod and M.W. Wright), Blackwell, Malden, MA, pp. 274–303.

Leitner, H., Peck, J. and Sheppard, E. (2007) *Contesting Neoliberalism: Urban Frontiers*, Guilford Press, New York.

Lin, W. and Ghose, R. (2008) Complexities in sustainable provision of GIS for urban grassroots organizations. *Cartographica*, **43**(1), 31–44.

Norheim, R. (2001) How institutional culture affects results: Comparing two old growth forest mapping projects. *Cartographica*, **38**(3/4), 35–52.

Nyerges, T.L., Barndt, M. and Brooks, K. (1996) Public Participation Geographic Information Systems. AutoCarto 13, ACSM/ASPRS 1997 Technical Papers, Seattle.

Obermeyer, N. (1998) The evolution of public participation GIS. *Cartography and Geographic Information Systems*, **25**(2), 65–66.

Obermeyer, N. and Pinto, J. (1994) *Managing Geographic Information Systems*, Guilford Press, London.

Rambaldi, G., McCall, M., Weiner, W. *et al.* (2004) Participatory GIS, www.iapad.org/participatory_gis.htm.

Rattray, N. (2006) A user-centered model for community-based web-GIS. *URISA Journal*, **18**(2), 25–34.

Schuurman, N. (2000) Trouble in the heartland: GIS and its critics in the 1990s. *Progress in Human Geography*, **24**(4), 564–590.

Sheppard, E. (1995) GIS and society: Towards a research agenda. *Cartography and Geographic Information Systems*, **22**(2), 5–16.

Sheppard, E. (2006) Knowledge production through critical GIS: Genealogy and prospects. *Cartographica*, **40**(4), 5–21.

Shiels, M. (2009) US city to start giant 'mapathon', *BBC News*, http://news.bbc.co.uk/2/hi/technology/8305924.stm.

Sieber, R. (2000) GIS implementation in the grassroots. *URISA Journal*, **12**(1), 15–51.

Sieber, R. (2001) Towards a PPGIScience? *Cartographica*, **38**(3/4), 1–4.

Sieber, R. (2006) Public participation geographic information systems: A literature review and framework. *Annals of the Association of American Geographers*, **96**(3), 491–507.

Sparke, M. (1995) Between demythologizing and deconstructing the map: Shawnadithit's New-Found-Land and the alienation of Canada. *Cartographica*, **32**(1), 1–21. (Reproduced as Chapter 20 of this volume.)

Wilson, M. (2005) *Implications for a Public Participation Geographic Information Science: Analyzing Trends in Research and Practice. Unpublished MA Thesis, Department of Geography, University of Washington, Seattle.*

Wood, D. and Fels, J. (1986) Design on signs/Myth and meaning in maps. *Cartographica*, **23**(3), 54–104. (Reproduced as Chapter 14 of this volume.)

SECTION TWO
ONTOLOGICAL UNDERSTANDING

10

Cartographic Communication and Geographic Understanding

Leonard Guelke[1]

Abstract

A communications model emphasizing the transfer of information from cartographer to map user by means of the map has been widely accepted in cartography. The adoption of this model has focussed map use research *on* symbols and their visual characteristics. The communications model is inadequate for a complete analysis of the map, because it is concerned more with information than meaning and understanding. A fundamental objective of cartography is to enhance a map user's understanding of reality by placing information in appropriate contexts. A criterion of meaning is needed in cartography as a basis for decisions on map content. Maps showing distributions of isolated data are often failures, because such data lack an appropriate context. In assessing the effectiveness of maps as transmitters of geographic information, comparisons with graphs, tables and words are needed. A knowledge of geography is essential for cartographers concerned with adding meaning to their maps.

In recent years cartographers have given increasing attention to the nature of the map and its potential as a medium of communication. Balchin and Coleman have proposed the term graphicacy for the spatio-visual skills involved in making and understanding maps and recommend that graphicacy be accorded a place alongside literacy, numeracy and articulacy (ability to communicate verbally) as a basic objective in general education (Balchin and Coleman, 1966; Balchin, 1976). This idea has gained wide-spread support amongst those cartographers who have sought to define the nature of map communication. A number of models have been put forward based upon the idea

[1] Originally published: 1976, *Cartographica*, **13**(2), 129–145.

At the time of publication: Guelke was Assistant Professor in the Department of Geography, University of Waterloo, Ontario.

that maps are elements in a communication process (Board, 1967; Kolacny, 1969; Muehrcke, 1972; Ratajski, 1973). These models differ as to detail but all of them are concerned with clarifying the process by which information is transferred from the cartographer to the map user by means of a map. The basic emphasis of contemporary cartography is summed up in the terms of reference of Commission V, Communication in Cartography, created by the General Assembly of the International Cartographic Association (Ratajski, 1974). The terms of reference are:

1. The elaboration of basic principles of map design.

2. The evaluation of both the effectiveness and efficiency of communication by means of maps with reference to the different groups of map users.

3. The theory of cartographic communication, that is the transmission of information by means of maps.

The above terms of reference clearly imply that cartography is an autonomous field of human communication with its own 'language' for the transmission of information.

There can be no doubt about the appeal of the communication model approach to cartography, yet this conception of cartography is not as complete a definition of the subject as it might first appear. The traditional view of maps as representations of reality or the earth's surface gives a clue to what is missing in the communications model; it contains no mention of what it is that cartographers are concerned to map. This omission, many would argue, is quite deliberate and underlines the idea of maps as transmitters of all kinds of spatial data ranging from maps of the brain to those of the planets. Yet this position ignores the long and close association between cartography and geography as well as the fact that many cartographers are still to be found in departments of geography. In few disciplines apart from geography are maps considered as basic research tools. This close relationship of geography and cartography would suggest that the nature of the earth's surface itself has something to do with the widespread use of maps in geography. In this paper it is argued that maps have traditionally been employed in geography and other disciplines as aids to understanding not merely as transmitters of information and that the communications model, by concentrating on information transfer, fails to include this basic aspect of map use. It is therefore concluded that the communications model is an inadequate conception of the nature of the map.

10.1 The Map as a Communications System

The best analysis of the map as a communications system is to be found in a recent paper by Robinson and Petchenik (1975), who have built on and extended earlier work on this theme. In looking at the application of the communications model to cartography, Robinson and Petchenik are at pains to point out that many of the concepts of the communications model, basically derived from electronic communications theory, cannot be taken over by cartographers without considerable modification. Before these modifications can be discussed, however, the basic elements of the communications model developed by Robinson and Petchenik need to be described.

At the heart of the communications model is the idea that maps (like written words) imply a reader who desires the information contained in them. Robinson and Petchenik (1975: 7–8) use the word 'percipient' to describe the map reader, the intended receiver of cartographic information. The word 'percipient' is preferred to map user or reader, which Robinson and Petchenik feel have connotations of specific and limited action. A map percipient in contrast to a viewer enhances his geographic understanding or spatial knowledge. The percipient is the receiver of the information prepared by the cartographer. The importance of the distinction between viewer and percipient can be illustrated with an example. A viewer of a map of Southern Ontario showing urban population (Figure 10.1) would simply see a pattern of different sized circles, whereas a percipient would enhance his understanding of the size, growth rate and distribution of urban places in the region. In other words, if the map correctly conveyed the information about the size, growth rate and distribution of towns in Southern Ontario to a percipient it would have fulfilled all the conditions demanded of it by the communications model.

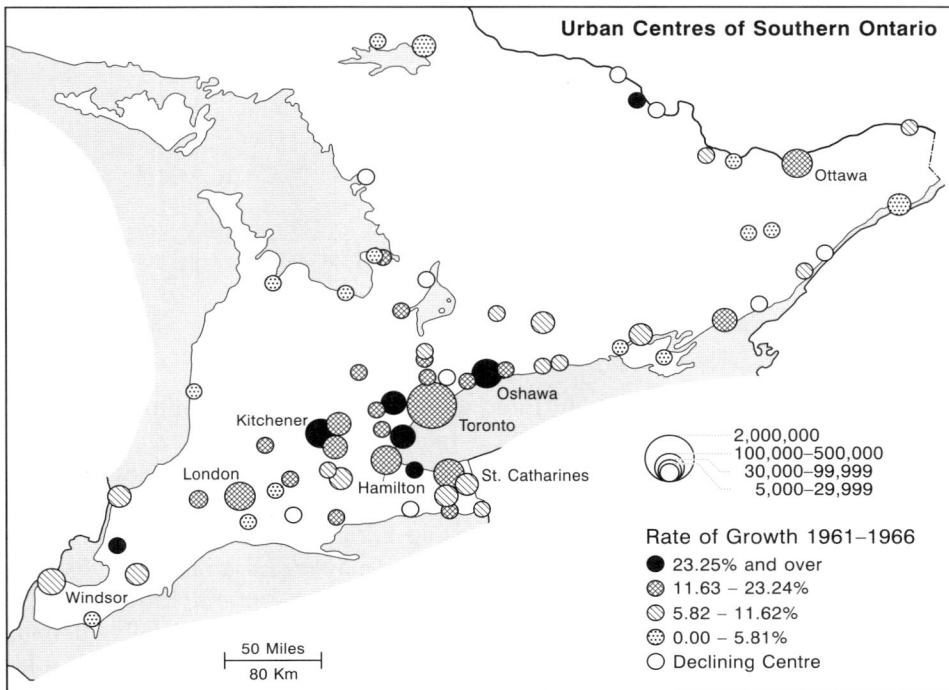

Figure 10.1 Urban Centres of Southern Ontario with 2000 or more inhabitants. (Adapted from Marshall, 1972: 68.)

A variety of models have been put forward to define the relationship between cartographer, map and percipient (Figure 10.2). All these models are based on electronic communications analogies. For example, a radio engineer is concerned with conveying a studio conversation to a group of listeners, and must concern himself with electronic coding of the message, its transmission and, finally, its decoding. Engineers call any unwanted interference with the clear transmission of a signal 'noise', and this term has been adopted by cartographers. But the use of the term 'noise' in cartography raises some problems. If by 'noise' one understands any eye distracting figures or shapes one would easily end up as Robinson and Petchenik (1975: 11) point out in the paradoxical situation in which an element on a map could be considered both noise and message. The problem is raised by Robinson and Petchenik (1975), but apart from suggesting that (by analogy) it would be reasonable to limit graphic noise 'to those delineations not necessary to the communication of the message', is left largely unresolved. The crucial question here is 'What kind of information or knowledge are cartographers interested in communicating?' Before considering this question some other aspects of the communications model need to be discussed.

Two Communication System Models

(a) The cartographic communication system with emphasis on conceptual aspects

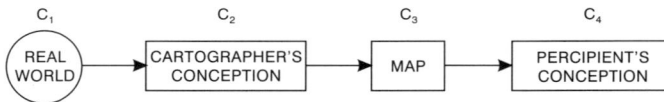

(b) A simple communication system

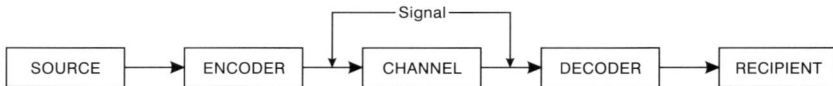

Figure 10.2   Communications Systems. (After Robinson and Petchenik, 1975: 9–10.)

The attempts by some to apply Information Theory to the communication of cartographic information receives critical treatment by Robinson and Petchenik (1975: 12–13). They point out that the assumption of linearity – or a sequential pattern of communication – in Information Theory is not a valid assumption when applied to the perception of maps. This indeed is telling criticism, but another even more important point they make is that while engineers may regard a piece of prose as a string of letters and spaces (i.e. information) what is, in fact, communicated is a set of *meanings*. Having established this basic point, Robinson and Petchenik fail to develop its implications, which are central to the whole question of the nature of cartographic communication. If cartographers are concerned to communicate meanings why do they concern themselves with analogies from electronics? It might be argued that carto-graphers are, in fact, involved in the transmission of information (along the lines of engineering) as well as meanings. If this position is accepted, the implication is that the communications system analogy is an inadequate model, because it fails to include a major aspect of cartographic communication.

It is clear from the preceding discussion that cartographers have been flexible in applying the model of the generalized communications system to their field, but few cartographers have questioned the utility of the model itself. Yet the model, because it is concerned with communication of information rather than meaning, fails to direct the attention of cartographers to the characteristics or content of that information. A fundamental question relating to the use of maps is not whether they are effective in transmitting information but whether the information communicated is meaningful. The adoption of the communications model of cartography has led cartographers to focus their research on map use on symbols in various combinations and their visual characteristics. For instance, research has recently been carried out on the effect contours of various characteristics (many, few, screened or unscreened) have on the legibility of the other information contained on a map (Cromie, 1976). This example raises the question posed above of whether a specific symbol should be regarded as noise (unwanted visual distraction) or message or both. In this case one might well argue that contours were both noise and message depending on the map user's purpose. The final decision about what one ought to include on a map must, therefore, be based upon a criterion of meaning which relates map content to the purpose of the user. If an element is likely to enhance the map user's understanding of reality it will be message not noise even if it is visually distracting. The concept of understanding employed here goes beyond the mere correct identification of symbols, that is, it implies more than the understanding achieved by Robinson and Petchenik's percipient. In the contour example it would imply that the map reader not only understood the symbolization but also comprehended the significance of the kind of terrain shown in relation to a specific purpose. For a hiker, a map with contours contains information which can be translated into meanings in terms of difficult and easy hiking terrain. Thus, in this example, the choice of whether to include contours or not and at what interval will depend upon how the cartographer interprets the requirements of the user of his maps, for this will determine whether the information he includes is potentially meaningful or not.

10.2 Maps and Meaning

It seems logical to argue that only information which is potentially meaningful to a map user ought to be included on a map. A criterion of meaning can, therefore, be considered a first principle of data selection and generalization for maps; this criterion immediately poses the question of just how a piece of information acquires meaning for a specific map user or group. This is a fundamental question in cartography and merits detailed examination. Maps show spatial patterns and spatial relationships yet a definition which emphasized the spatial or distributional aspect of the map would not provide a satisfactory basis for evaluating meaning. Consider the map showing the distribution of hotels in the downtown area of a hypothetical North American city (Figure 10.3).

Figure 10.3   The evolution of a map from spatial relationships to interrelationships in space.

This map we can assume contains accurate information about the location of each hotel within the area shown and the relationship of each hotel to every other one. In other words, we have a spatial representation of hotels. Yet this map on its own would be of limited value to potential tourists because the map user would be unable to give the information contained on the map a meaningful context (Figure 10.3a). The hotel locations become clearer as additional information is added to the map; as the rivers and coastline, the streets and, finally, points of interest are each added in turn the information on hotels acquires more meaning for the potential tourist (Figure 10.3).

The map user in looking at the final map can relate hotel locations to other features of interest to them. In other words, for the map user the meaning of a specific location is largely defined in terms of the location of related elements. In the example just cited, the location of a hotel acquires meaning in terms of its accessibility to other facilities. The value of this map derives not so much from the fact that it shows spatial relationships correctly, but that it shows the interrelationships amongst hotels and such items as streets and points of interests.

On the basis of the above example, it can be stated that the crucial feature of a map is not the information it contains *per se*, but whether that information is set in an appropriate context. The more appropriate the context of specific information to the needs or purpose of the user, the more meaningful it becomes. This point is central to the correct understanding of the nature of maps and merits further elucidation. Let us first consider general topographic maps. Such maps usually contain information on relief, rivers, swamps, roads, railways, roads, towns and a host of other physical and cultural features. In many countries these maps were first developed on a systematic basis by the military, and it will further the argument to enquire why topographic maps are considered so vital in military operations. If one imagines a war exercise involving the invasion of the territory shown (Figure 10.4), by a force from the north, how would a map be used? In using the map illustrated, commanders of both forces would plan their strategy in relation to the actual or potential positions of the enemy relative to the terrain. The terrain would be endowed with military meanings. Deep rivers and swamps would be obstacles, forests would be cover, high ground would have special strategic significance. A military commander once located on a map is able to relate his position to the total environment and thereby to comprehend his strategic situation in relation to the actual or potential dispositions of enemy forces. In the example under discussion, the invading force might plan: (1) to attack a position of the defending force located at Penetanguishene by a landing on the coast at Tiffin Basin (to avoid the enemy at Midland Bay); (2) to proceed along the east side of Wye Lake, skirting the swamp; (3) to move on to Wyebridge under cover of the forest, and (4) to attack Penetan-guishene from an unexpected direction. A map is essential for planning such a strategy because it shows interrelationships amongst the elements of the landscape. It shows the distribution of forests in relation to high ground and swamps, the distribution of swamps in relation to rivers and coasts, high ground and forests, or briefly, each element is shown in relationship to every other element included on the map. A series of maps of individual items would together include the same amount of information but would not serve the purpose of a military commander who needs to know how elements are interrelated if he is to endow them with their correct strategic significance or meaning.

If maps are essential in military operations, they are no less essential in other activities in which an understanding of interrelationships in space is important. In all kinds of planning and environmental work such understanding is central to the making of good decisions. In these cases various elements on the topographic map are endowed with new meanings in relation to the purpose of the users. If information not given on a general topographic map is required it can be added but its significance or meaning will ultimately rest on its relationship to other elements in the landscape. Thus a researcher

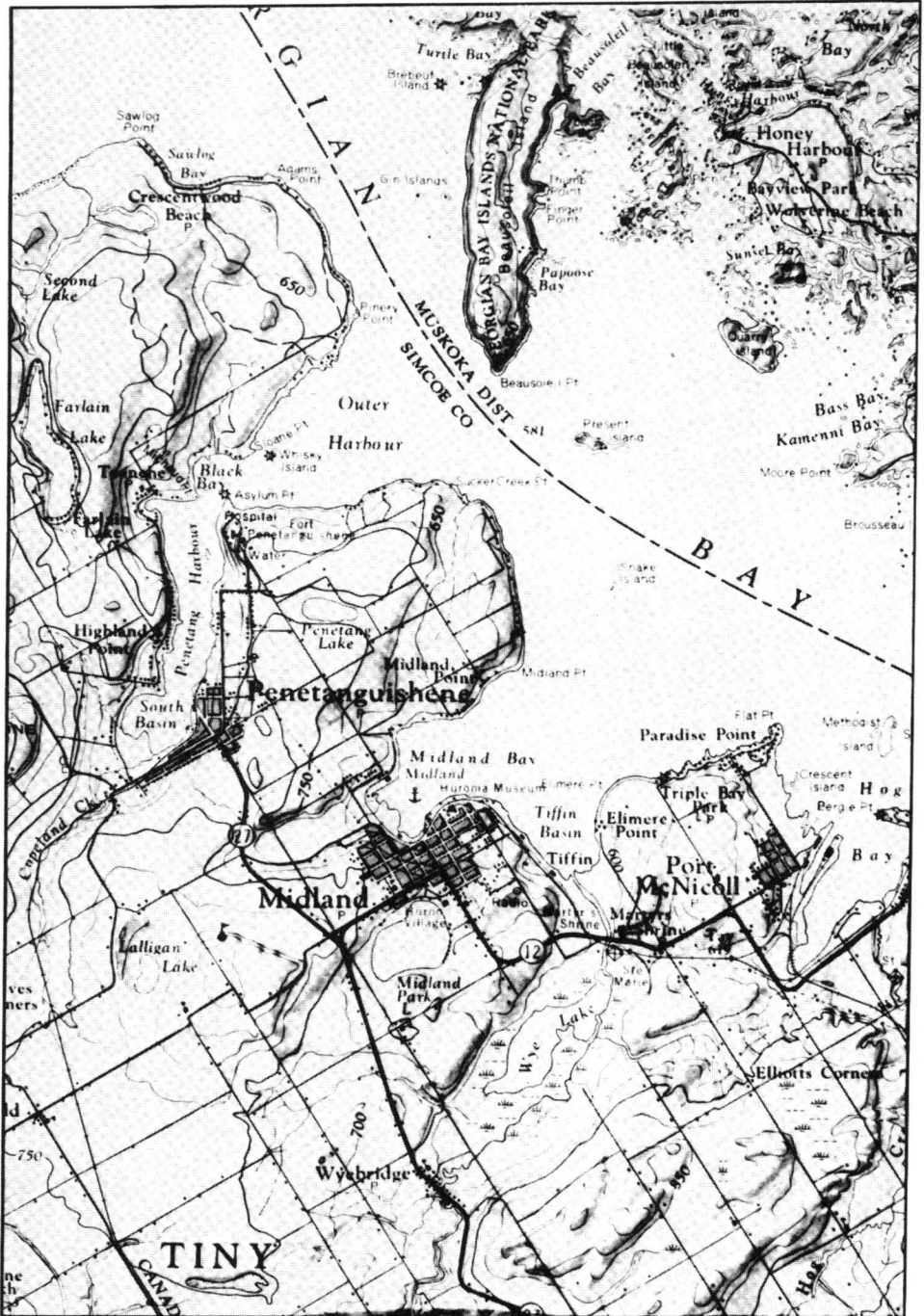

Figure 10.4   A topographic map of the Penetanguishene/Midland area at 1:125 000. (A section of the Orillia sheet of the 1:125 000 series of the National Topographic System.)

investigating air pollution might add factories and prevailing winds to a topographic map with a view to understanding the extent and distribution of polluted air. The human significance of the polluted areas would, in turn, be related to how these areas were used. In planning, an analysis of the adequacy of recreational facilities of an urban area could scarcely be carried out without maps of existing and potential recreational sites, but housing and demographic information would be needed to give the recreation data meaning.

To summarize this basic argument, a map is invaluable in many situations because it enables a user to see interrelationships amongst phenomena and to comprehend the meaning or significance of particular elements in relation to the whole; that is, their geographical contexts. The notion that the communications process in cartography involves the cartographer's conception of reality which is transmitted to a user via a map is an inadequate model when considering topographic maps because in many situations the map user's conception of reality is more precise than that of the cartographer. The cartographer who prepares general topographic maps is basically concerned with the symbolization and generalization of physical and cultural features on the landscape. In preparing his maps, the cartographer has some general conception of the kind of person who is likely to use the map and selects his information accordingly; however, the elements of the map only acquire their meaning in relation to the needs or purpose of a user. A soldier, hiker, ecologist and planner might all use the same map, but each of them would endow the information contained in it with different meanings. The value of the general topographic map lies in the fact that the cartographer does not commit himself to a specific conception of reality, but attempts to provide a general representation of the earth's surface.

The model of cartographic communication which includes the cartographer's conception of reality as an integral element is more appropriate to statistical or thematic maps. Such maps are, in principle, no different from topographic maps, but their narrower focus makes them less open to widely different uses and interpretations. A good thematic map is designed to illustrate a specific relationship. In this case the cartographer creates a map to bring out the meaning of specific information; it is basically a single purpose map. To the extent that the map user understands the purpose and comprehends the relationships illustrated, to that extent can the map be considered successful. In the foregoing discussion it is assumed that a good thematic map is concerned with conveying meaning or understanding to a potential user, not simply information. Just as in a topographic map, meaning is dependent on context; in other words, to endow a phenomenon with specific meaning implies that it be given a context that reveals this meaning. These points may be illustrated with examples.

Dr John Snow's map of cholera deaths in an area of London for the year 1855 is an example of a good thematic map (Figure 10.5). On this map, Dr Snow has shown deaths from cholera, streets and water pumps. The relationship of cholera deaths to the Broad Street water pump is immediately apparent. (It is assumed that the reader of Snow's map would be aware that cholera is a water-borne disease and that pumps within walking distance were the main source of water for many of London's inhabitants in the nineteenth century). The value of this map derives from the juxtaposition of data on cholera deaths with data on water pump accessibility. In this way, Dr Snow has given

Figure 10.5 Dot map illustrating deaths from cholera in London, by Dr John Snow, 1855. (After a reconstructed map in Thrower, 1972: 96.)

meaning to the cholera deaths by showing a relationship between the deaths and a particular water pump. A map which included nothing but the information on cholera deaths would convey factual locational information to a map reader, but lacking a context the information would be without much meaning. Snow's map is also an excellent example of economy, as no information not relevant to an understanding of the relationship between water pumps and cholera deaths has been included.

10.3 Maps, Figures or Words?

Examples of thematic maps lacking adequate contexts are the single topic maps illustrated (Figures 10.6 and 10.7). In the map showing the average sale price per house for Connecticut, the data exist in isolation (Figure 10.6). It would be impossible for a reader lacking special knowledge of Connecticut to make sense of the pattern (information) presented because the data lack a meaningful context. This map, which fails a test based on a criterion of meaning, might be considered successful in terms of

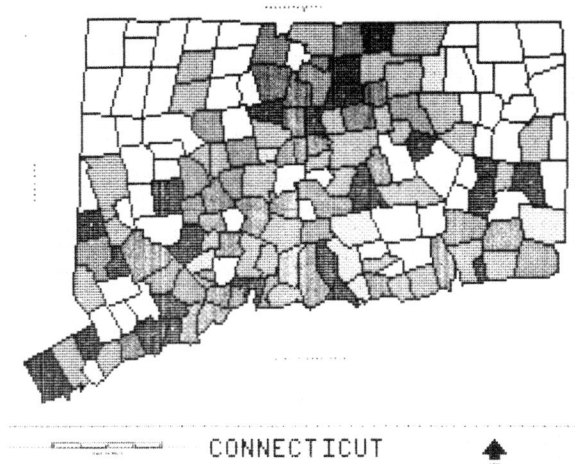

Figure 10.6 Choropleth Map of Connecticut showing average price per house by towns, prepared by Laboratory of Computer Graphics, Harvard University. (Source: Robinson and Sale, 1969: 148.)

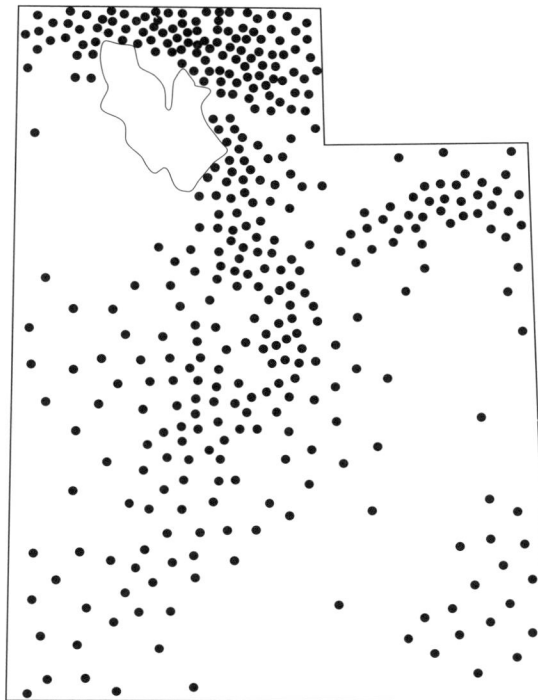

Each dot represents 5,000 acres of cropland

Figure 10.7 Crop Acreage in Utah. (After Jenks, 1976: 13.)

the information communications model, because on the latter test all that is required of a map is that it transmits accurately locational information to the map user. The map of crop acreage in Utah illustrates the same point (Figure 10.7). For Jenks, working within a communications framework, the map is a success (Jenks, 1976). Yet for anyone unfamiliar with Utah, the pattern presented would have little meaning. Recent research on map learning has shown that different types of information must be presented simultaneously for effective integration (Shimron, 1975). The Utah map could easily be improved with the addition of an isohyet and relief shading. The map reader would then recognize the importance of irrigation and be able to relate the crop acreage pattern to low lying flat areas near lakes and rivers.

The possibility of placing data in inappropriate or misleading contexts is a real danger. Just as the meaning of statistics can be radically altered by judicious manipulation of their context, so too can the meaning of map data be changed according to the contextual information included. It is, therefore, essential that a cartographer have a broad background in a wide range of subjects to enable him to select appropriate contextual information against which the data mapped can be more readily comprehended. This ability implies that he has a knowledge of factors behind distributional patterns and some understanding of their causal interconnections. The recent trend to see cartography as an emerging discipline is not without danger if such a development involves seeing cartographers as specialists in graphic communication. An emphasis on more effective ways of portraying data graphically is valuable but only if such developments do not take place at the expense of the cartographer's background in geography and other disciplines with an interest in understanding the distribution of phenomena on the earth's surface.

That cartographic information is spatial in character goes without saying, but this does not mean that maps are the only or best way of communicating such information. In other words, because something can be mapped – and anything with a spatial dimension can be mapped – it does not follow that it needs to be mapped to be communicated. The point is that there are several ways to communicate spatial information – words, figures, graphs and maps or some combination of all or some of these. For example, a map scale can, as all cartographers know, be given verbally, graphically or numerically (Figure 10.8). Each method of communication contains essentially the same information, and the preferred medium will depend on the kind of map user and his purpose. A first and crucial task in cartography is, therefore, to define areas where maps are preferable or superior for map users of given interests and education levels to other forms of communication.

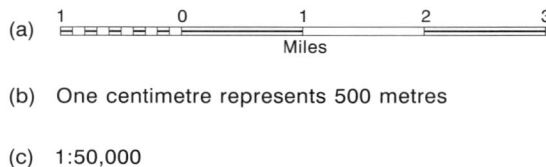

(a)   `1 ▭▭▭▭▭ 0 ▭▭▭▭ 1 ▭▭▭ 2 ▭▭▭ 3`
                    Miles

(b)   One centimetre represents 500 metres

(c)   1:50,000

Figure 10.8   Three ways of expressing the scale of a map.

If one assumes that maps are an important means of communication without demonstrating their value, cartographers could well spend effort and time perfecting a graphic language where other more effective methods of communicating information or knowledge exist. This problem is recognized by Jenks, who explored various methods of transmitting information on crop acreage (Jenks, 1976: 12–14). Amongst the examples he used were: (1) county map with separate table; (2) figures in county boundaries; (3) dot map with county boundaries; (4) dot map without county boundaries. The tests he gave, however, are inconclusive. For instance, the dot map without boundaries gives one the best visual impression of distribution, but just how much information is transferred in this manner? The table and figures in county boundaries appear ineffectual, but would a table arranged geographically (not alphabetically) with a small reference map be equally poor? For example, if one wished to show the distribution by provinces of average income in Canada an arrangement similar to the one shown would likely be more effective than a complete cartographic portrayal (Figure 10.9). In a table arranged geographically, one can often achieve enough locational information without having to sacrifice the clarity and precision of figures. In a complete cartographic presentation of income for Canada, precision is lost and the translation of area patterns into figures becomes necessary (Figure 10.10). Moreover, an

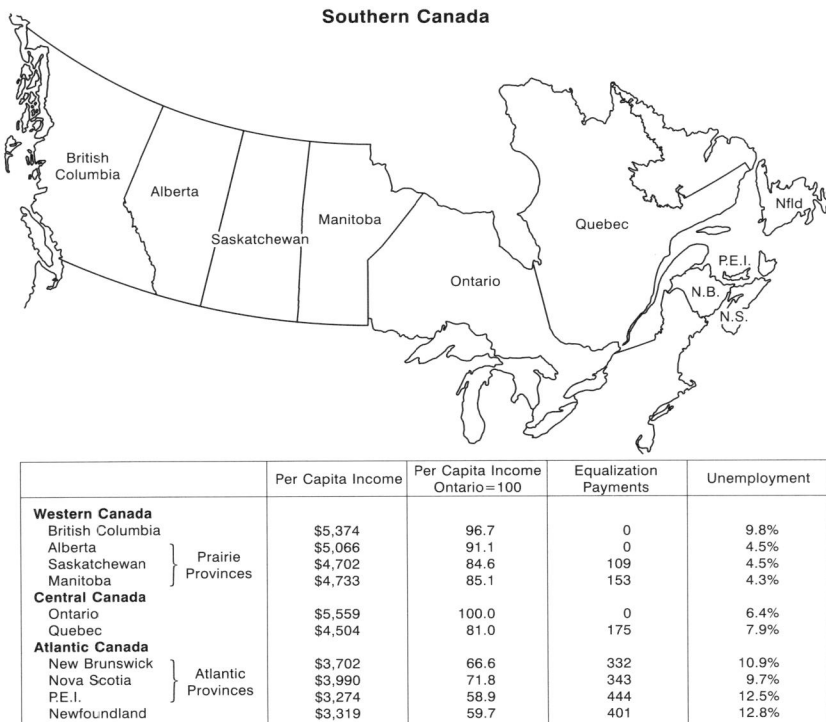

	Per Capita Income	Per Capita Income Ontario=100	Equalization Payments	Unemployment
Western Canada				
British Columbia	$5,374	96.7	0	9.8%
Alberta	$5,066	91.1	0	4.5%
Saskatchewan	$4,702	84.6	109	4.5%
Manitoba	$4,733	85.1	153	4.3%
Central Canada				
Ontario	$5,559	100.0	0	6.4%
Quebec	$4,504	81.0	175	7.9%
Atlantic Canada				
New Brunswick	$3,702	66.6	332	10.9%
Nova Scotia	$3,990	71.8	343	9.7%
P.E.I.	$3,274	58.9	444	12.5%
Newfoundland	$3,319	59.7	401	12.8%

(Alberta, Saskatchewan, Manitoba grouped as "Prairie Provinces"; New Brunswick, Nova Scotia, P.E.I. grouped as "Atlantic Provinces")

Figure 10.9 Southern Canada showing the combination of cartographic and statistical treatment of geographic information.

Southern Canada: Per Capita Income

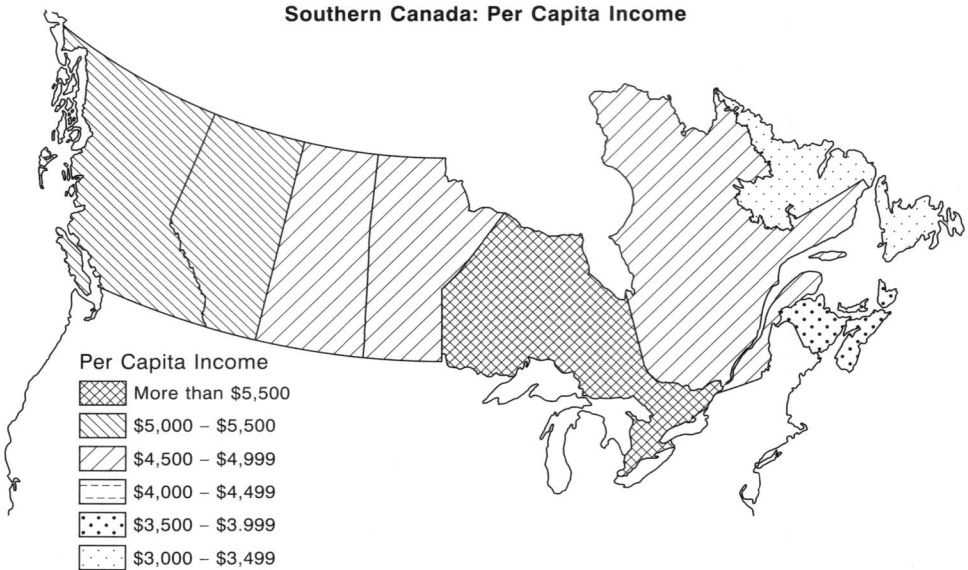

Per Capita Income
- More than $5,500
- $5,000 – $5,500
- $4,500 – $4,999
- $4,000 – $4,499
- $3,500 – $3.999
- $3,000 – $3,499

Figure 10.10   Cartographic treatment of geographic information.

impression of provincial regional equality is more likely to be obtained from a map in which each province is entirely covered with a certain kind of shading than it is from a set of figures. Obviously, much research needs to be done on a variety of topics before any firm statements about the effectiveness of maps over figures or words or combinations of these in transmitting information to specific map users can be made.

10.4 Conclusions

Some modern ideas on cartographic communication are based on models of information transfer derived from electronics. Although cartographers recognize the limitations of such models, few people have recognized them for what they often are, namely, a barrier to the correct understanding of maps. A good map is not simply concerned with transmitting information but with enhancing the map user's understanding of reality. Such understanding is possible only when the meaning of information is considered by the cartographer. Meaning, however, is not a category that falls within the domain of standard electronic communications models or Information Theory. If one adopts a criterion of meaning in assessing the value of maps one, at the same time, recognizes the frequent inadequacy of information-type models in cartography.

The goal of enhancing a map user's understanding of reality cannot be achieved by cartographers ignorant of the phenomena they map. Before a phenomenon can be mapped effectively it must first be understood by the cartographer. Hence, it is most appropriate that cartographers are generally found in the discipline of geography – a discipline concerned with understanding interrelationships on the earth's surface.

On the view expounded here, cartography is seen as a subfield of geography rather than a graphic art, although, of course, principles from any field which can improve the quality of maps will be used by cartographers. The main point is that the basic criterion of map making should be a criterion of meaning. Thus a cartographer would normally apply a criterion of meaning rather than a visual one in deciding whether, say, contours were required on a specific map or not.

Thematic maps differ from general topographic maps only in their narrower focus, not in their essential nature. All good maps are concerned with enhancing the map user's understanding of the subject matter of the map. A central concern of cartography ought to be in delimiting those areas of communication where maps arc most effective for enhancing understanding. In many cases, figures, graphs or even the written word are more effective media for the transmission of certain kinds of spatial information. In particular, it seems that map users have great difficulty in integrating map information lacking an appropriate context. However, this question has not yet been adequately researched, and it should be a priority. There would be little point in developing a cartographic language in areas where tables, graphs or verbal description are generally superior as communication media.

The emphasis on meaning as the primary principle of map making does not imply that work on the perception of map symbols is not valuable. The quality of cartographic work can obviously be enhanced if mapmakers have a sound understanding of the principles of perception and design, and research in this area is to be encouraged. The communications model of cartography has been criticised not for what it includes but for what it leaves out. A good cartographer should be a good symbolizer, but he/she should also have a good understanding of the phenomena he/she maps, as well as an idea of the characteristics of users for whom his/her maps are intended.

References

Balchin, W.G.V. (1976) Graphicacy. *The American Cartographer*, **3**, 33–38.

Balchin, W.G.V. and Coleman, A.M. (1966) Graphicacy should be the fourth ace in the pack. *The Cartographer*, **3**, 22–28. (Reprinted from The Times Educational Supplement, Friday, 5 November 1965.)

Board, C. (1967) Maps as models, in *Models in Geography* (eds R.J. Chorley and P. Haggett), Methuen and Co., London, pp. 671–725.

Cromie, B.W. (1976) Contour design and the topographic map user. Paper presented at the *First Annual Meeting of the Canadian Cartographic Association*, Queen's University, Kingston, May 1976.

Jenks, G.F. (1976) Contemporary statistical maps – evidence of spatial and graphic ignorance. *The American Cartographer*, **3**, 11–19.

Kolacny, A. (1969) Cartographic information – a fundamental term in modern cartography. *The Cartographic Journal*, **6**, 47–49.

Marshall, J.U. (1972) The urban network, in *Ontario* (ed. L.R. Gentilcore), University of Toronto Press, Toronto.

Muehrcke, P.C. (1972) *Research in Thematic Cartography*. Resource Paper No. 19, Commission on College Geography, Association of American Geographers, Washington, DC.

Ratajski, L. (1973) The research structure of theoretical cartography. *International Yearbook of Cartography*, **13**, 217–228.

Ratajski, L. (1974) Commission v of the ICA: The Task It Faces. *International Yearbook of Cartography*, **14**, 140–144.

Robinson, A.H. and Petchenik, B.B. (1975) The map as a communication system. *The Cartographic Journal*, **12**, 7–15.

Robinson, A.H. and Sale, R.D. (1969) *Elements of Cartography*, 3rd edn, John Wiley & Sons, Inc., New York.

Shimron, J. (1975) *On Learning Maps*, Center for Human Information Processing, University of California, La Jolla, CA.

Thrower, N.J. (1972) *Maps and Man*, Prentice Hall, Englewood Cliffs, NJ.

11

Reflection Essay: *Cartographic Communication and Geographic Understanding*

Mordechai (Muki) Haklay[1] and Catherine Emma (Kate) Jones[2]

[1] *University College London, UK*
[2] *University of Portsmouth, UK*

11.1 Introduction

At first sight Leonard Guelke's *Cartographica* article *Cartographic Communication and Geographic Understanding* is somewhat different from other 'classics' reproduced in this volume. It has not received much attention since its publication in 1976 in the form of references and citations, although it did provoke some lively discussion when it first appeared. Despite this lack of overt 'classic' stature, it clearly deserves a place in this collection and should serve as an example of the benefits gained from returning to articles written many years ago. In this case, they can illuminate current understanding of issues in cartography and geography. Moreover, the insights from Guelke's 1976 article lie not only in the development of ideas in cartography from an historical perspective, but more significantly on the way scholars understand contemporary cartographical practices.

Guelke's *Cartographica* article must be read and understood within the context of the development of concepts about cartographic communication, of which Board (1967), Koláčcný (1969), Robinson and Petchenik (1977) are oft cited. Guelke's paper is offering an antidote to the perception that the quantitative revolution in geography was the universally accepted path towards a 'new' geography and that resistance to it only evolved later. As you 'zoom out' from this paper and look at other writings of Guelke at the time, his critique of the positivist approach in geography (Guelke, 1971) or early behavioural geography (Bunting and Guelke, 1979) can be seen as part of his attempt to offer an alternative idealist framework to human geography (Guelke, 1974). The debate that he was engaged in has some resemblance to similar debates in the 1990s around the role of GIS and related technologies within geography (Schuurman, 2000).

Specifically, the paper provides a fairly early critique of the cartographic communication model, which had been suggested by Board (1967) and Koláčcný (1969), following concepts from the field of cybernetics and from Claude Shannon's information theory (Poore and Chrisman, 2006). With this context in mind, in this reflection essay we discuss four aspects relevant to contemporary readers.

Firstly, focusing on the core argument of Guelke's article we consider the role of the theory of cartographic communication and its development since its emergence in the early 1970s. As we shall see, Guelke's critique is a precursor to the analysis Poore and Chrisman (2006) provide some thirty years later. The communication model represents one of the clearest impacts of the quantitative revolution on cartographic scholarship and an attempt to establish map design on scientific grounds, led by Arthur Robinson and others (Kitchin, Perkins and Dodge, 2009; Montello, 2002). While the reductionist version of cartographic communication theory is not in vogue as it was in the 1970s, it remains a distinct theme within the cartographic literature and teaching – with current emphasis on spatial cognition and the understanding and sense making of information that the map carries.

Secondly, in the context of this collection of essays, Guelke's *Cartographica* article provides a reference point to technological aspects of cartography, with an implicit assumption that maps are static and that the role of the cartographer is to produce them as a final, immutable artefact that is then 'consumed' by map users. The changes in the media of creation and delivery, alongside the introduction of interactive and animated maps, have altered the scope of some of Guelke's original arguments although, as we shall see, many of the arguments are still applicable in some regards. Thirdly, the paper provides an early example of theoretical engagement with the technical and scientific practices of cartography, which in later years developed into critical cartography and critical GIS (Kitchin, Perkins and Dodge, 2009; Schuurman, 1999). By understanding past debates and intra-disciplinary tensions it might help us make sense of current scholarly discussions about the meanings of mapping.

Fourthly and finally, the paper also highlights the longevity of some of the most fundamental problems with the use of map representations and interpretation of geographical information. These problems include the process of knowledge construction, understanding of how maps are being created and used and how to bridge the conceptual gap between professional (be it trained cartographers or skilled GIS technicians), domain or topic experts and the final map users. The paper concludes with a discussion of these wider problems.

11.2 The Evolution of the Concept of Cartographic Communication

Whilst it is not our aim to present an exhaustive review of the development of cartography in this reflection essay, it is necessary to place Guelke's viewpoint within the context of contemporary writings. Therefore, it is crucial to consider the theory of cartographic communication, which was of significant importance to cartographic scholarship since the 1950s.

Communication is and always has been central to all human endeavours. Since the meteoric rise of telecommunication from the 1950s onwards, communication has become faster, easier and crucially cheaper for many. With this, the media of communication has also changed significantly, particularly with the development and widespread deployment of digital communication and the emergence of worldwide diffusion of the Internet in the last two decades. Within this framework, maps are most often understood as a form of visual communication used to encode geographic meaning and relationships. Since the 1950s there has been considerable effort to understand cartography as medium of visual communication.

Within the academic disciplines of cartography, and latterly in so-called GIScience, researchers and theoreticians have considered in detail how to conceptualize carto-graphic communication in order to understand the process and system of visual communication that is unique to map representations. These efforts led to a set of rules that improve the construction of maps and, by corollary, progress their communicative effectiveness. With this in mind, the 1950s agenda of the then newly established International Cartographic Commission (ICC; now the ICA) was driven by the processes underlying map design and how they are used and understood. From this period an intellectual alliance emerged between cartographic scholarship and the field of information theory. Many cartographic scholars were keen to embed 'real' scientific practice within the discipline and looked to information theory as a mechanism for describing and understanding cartographic communication in a rigorous and provable way.

Information theory was deemed relevant to cartography because, with its origins in electrical engineering, it used a series of complex mathematical equations to summarize the capacity of a given communication media to transmit a message accurately and effectively. It summarized the transmission of electronic signals, which of course represented our verbal language (Robinson and Petchenik, 1977). It then seemed natural to apply this systems thinking to cartographic research. In its most simple form, a communication system has three components: (1) an encoder (the cartographer), (2) a method of transmission (the map) and (3) a decoder (the map user). This can be understood simply as a message which is transmitted and received. Applied to cartog-raphy though, the process is much more complicated (Poore and Chrisman, 2006). The visual messages presented via a map representation are the result of complex and subjective processes which include selection, classification, generalization as well as interpretation. A process further complicated by the cartographer's decisions on such issues as the underlying projection and scale of the map. In essence, a map is a compromise, representing decisions taken by the cartographer which aim to fit the underlying purpose of the map.

The idea of cartographic communication as a system based on inputs, flows and outputs can be traced in the literature back to the late 1960s (Board, 1967; Koláčcný, 1969). At this time the map is compared to a linear communication system where the message is encoded, transmitted and received. For example, this can best demonstrated in Board (1967: 673) where he states the truism that maps are vehicles for information flow, although some cartographic designs will be more effective vehicles than others. In the same vein, Koláčcný (1969: 47) states that cartographic representations are more successful when the production and consumption of the map is clearly defined and communicated to all. In other words, the purpose of a map should inform the construction of the map. In these communications models information is transferred from the cartographer to the end user via the map representations. The map has two roles as derived from the map production and its consumption. Acting both as the *medium*, which is the method of information transference, and as itself - the *message*. As Poore and Chrisman (2006) noted, this is a rather muddled view which confuses the different meanings of information as both an immutable object of which no semantic meaning is assigned, and seeing information as part of the hierarchy of data, information and knowledge, where the semantic meaning is crucial.

In a slightly different vein, Muehrcke (1972) considered cartographic communication as a process system representing the most important relationships and transformations. Within his logic, real-world data are first selected and then undergo a series of transformations before finally being represented on a map. A feedback cycle characterizes communication efficiency, whereby data are transformed into a map which is later interpreted as information and retrieved by the map user. The user is able to feedback to the cartographer so that any future transformations can be then improved. This enables the creation of a more useful approximation of reality through the inclusion of the feedback loop.

Following the concrete conceptualization of cartographic communication models, a number of academics with differing theoretical perspectives set out to challenge the information theory approach to cartographic scholarship. These critiques argue that information theory alone cannot account for the cultural or cognitive approach to maps or the collective processes involved in making them. A defect within the construction of these communicative models is the assumption that the map user or reader is the passive recipient of information; this is both an underestimate and oversimplifies the audiences role and the power of the receiver (map user). Therefore, the reductive nature of the cartographic communication models has obviously led to much debate within the literature (Montello, 2002).

Robinson and Petchenik (1977) suggested that maps have greater significance than simply a medium or mechanism of information transfer. For them, cartographic communication should be understood not only as a sequence of flows, which are quantifiable, but also as perceptual and cognitive processes which can be used to encode the geographical meaning of maps. Petchenik (1977) highlighted the obvious flaws in such communication models, suggesting that map readers are cognitively able agents who can and should actively create knowledge and meaning from maps.

From this point of view, Robinson and Petchenik (1977) went on to develop influential cartographic research agenda adopting a psychological approach. Their research was dominated by the exploration of the visual information process, including the understanding of the psychological effects of colour, the perceptibility and readability of typology and re-examination of previous mapping conventions. Therefore, without rejecting the cartographic communication theory altogether, they tried to reform it through a richer understanding of the underlying science. In Robinson *et al.*'s (1995: 18–19) *Elements of Cartography*, one of the most influential textbooks in cartography, they liken the field of cartography to a drama played by two actors (map maker and user) with two stage properties (the data and the map). The drama is then characterized through a series transformations that, ultimately, should lead to an effective map. Each transformation influences the map's effectiveness and corresponds to data collection; data abstraction and manipulation and, finally, information interpretation.

Another critique to the functional and reductionist perspectives of cartographic communication models came from those who emphasized the importance of semiotics. In this view, information is synthesized through the sign systems used on a map and their inner meaning, and by understanding them it is then possible to improve mapping practices. Cartographic icons act as representational conventions embedding meaning and cultural significance. The symbols transfer meaning to the user and 'iconic and linguistic codes are used to access the semantic field of geographical knowledge' (Wood and Fels, 1986: 81). Semiotic research thus concerns itself primarily with the perception of graphical symbols on the map, which is important, but, as Leonard Guelke's research affirmed, they should not be the primary principle of map making.

Thus, Guelke's, 1976 *Cartographica* article is part of the critique of cartographic communication models. Guelke proposes that the map will have different meanings for the end user and the cartographer, suggesting that there is a difference between the conceptual understanding of the map user and its creator. This, of course, depends upon the experience both actors will have with the phenomena being mapped but must surely be indicative of a potential problem of building the knowledge. This concern, whilst originally aired more than thirty years ago, still seems relevant today, as it is common for experienced GIS experts to create maps outside of their domain specialism, and they thus do not have all the necessary knowledge to understand how potential map users will enrol their products to solves tasks.

As much map making first morphed into the domain of the GIS specialist and has latterly been adopted by a wider range of 'amateur' mappers through Web 2.0 applications (Haklay, Singleton and Parker, 2008), mapping is more and more driven by the desire for technical innovation and the visual display of information rather than for actual constructive building of geographical knowledge (Jones *et al.*, 2009). As a result, rigorously designed, useful and usable static and Web-based dynamic maps are becoming a rarity. The distance between the map designer and the map user is, in some senses, becoming wider. Notice that even if the user is operating the technology that is making the map, their control over the default values and other aspects of cartographic symbology are often limited and constrained by the technology. As Guelke clearly stated, in many instances the user's conception of reality is actually more relevant and current than the cartographer or Web-mapper. The user is likely to have more intimate knowledge of the discipline, the phenomena and the place being mapped as the resultant representation is merely seen as a product by the specialist. Frequently, the paid GIS developer, for example, works in isolation from the users and the map is often simply transmitted over the Internet as a *fait accompli*, a fixed and final product. This seems to be a replication of the 1969 cartographic communication model by Koláčcný where the map is produced and then subsequently consumed albeit with/without appropriate symbology or colours. From this perspective Guelke's early article has real contemporary resonance.

In *Cartographic Communication And Geographic Understanding*, and in many of his following papers, Guelke is taking a cultural standpoint by stating that maps should contain geographic meaning as part of the ideas of epistemological idealism, whereby he 'sought to understand the development of the Earths' landscape by uncovering the thought behind it' (Guelke, 1974: 193). Understanding this approach in the most simplistic of terms if these ideas were extended to cartography – the map could become an object of cultural significance – and so could be used to reveal insights into the human actions of the cartographer. The map would then be an object that could be used to define identity and culture, as indeed many historical maps are used to do just that. Is this possible with the homogenous map making that technology produces?

Somewhat surprisingly, the advances in technology have not changed the affairs of cartographic communication significantly, as maps are now more than ever the domain of the technocrat – consider, for example, that with the vast majority of online mapping services users have little say on the visualization of the map and, even if they are using pushpins to add or annotate information, the range of symbols and their cartographic variability is limited. In many ways, the information realm is inherently reductive within this digital world of binary logic. Contemporary GIS maps reflect what data can and has been recorded in this binary fashion – perhaps what is more interesting are the features that cannot be summarized in such a reductive fashion. As it stands, the move to automate map production has leant itself to the creation of cartographic representations which are homogenous artefacts, failing to incorporate any enriching contextual data relating to places. The ability to extract geographical meaning in these types of maps can be challenging.

11.3 The Role of Technology and the Influence of Cartographers in Interactive Mapping

This attention to technology leads us to another interesting aspect of Guelke's original article. The paper uses eight different maps as exemplars, of which one is a reference map, and the remainder are thematic maps. Throughout the discussion the implicit assumption is that these are delivered as finished end products to a person (map user) who uses them to extract meaning about the study area and phenomena. This is a core aspect of the cartographic communication theory that is easy to overlook today. At the core of the communication conceptualization, the cartographer was seen as the primary actor who codes the geographical 'message' using the practices of map design and the product is then consumed by others.

At the time that this concept was evolving, computer maps were starting to emerge, and more than a decade before Guelke's 1976 article, Pivar, Fredkin and Stommel (1963) were discussing the significant potential of computers to be used to produce interactive maps that can be changed at will. However, the ability to manipulate maps interactively with an acceptable response time (that is, within a second or two) remained beyond the reach of most users of geographical information until the 1990s. With the advent of personal computers came the ability to support interactive cartographic composition by professional users of GIS. Notice how Figure 10.3 from Guelke's 1976 article is imitating, on paper, the interactivity of GIS by adding layers of information to achieve the full composition.

In the early stages of GIS development the ability to compose meaningful maps remained within the realm of geography or cognate disciplines, and it could be expected that although GIS users were not all cartographers, they would have some knowledge and training in map making, and hence know about the cartographic communication models to some extent. However, by the early 2000s, as Unwin (2005) suggests, many of these users of GIS were 'accidental geographers', with no specific cartographic design training, mapping experience and often a naïve and unproblematic approach to geographic information representation and manipulation. Whilst, as we have noted, in most cases digital maps are unchangeable by end users, the current provision of Web-based tools (such as the innovative GeoCommons.com service) does put the power of map composition in the hand of more map producers and consumers than ever before. The producer and consumer can now often be the same person. In these cases, the problems of understanding and sense making are not any more addressed by carto-graphers acting as facilitators and arbitrators, although because of the lack of basic knowledge in map design, this producer-user might be constructing a representation that will lead them to draw wrong conclusions about the real situation. Technology and automation, thus, replaced the conscious cartographer.

As Guelke (1976: 135) notes 'the more appropriate the context of specific infor-mation to the needs or purpose of the user, the more meaningful it becomes' he then moves on to discuss the gulf between the cartographer's conceptualization of the problem and the user needs. However, despite technological advancements the problems that are highlighted in the 1976 article have not been resolved. First of all, in many technological representations of cartography, the reference map is unchanged by its user. For example in Web-mapping systems the base map is pre-rendered, so the key decisions of cartographic design are not made available to the map user. In systems where cartographic manipulation is necessary, such as desktop GIS products, the map creator is confronted with a host of 'default' values for projection, scale, orientation, layout and symbology that have been coded in the system. Some 'defaults' can be changed if you know where to look in the interface, whereas others are subtle and latent in the software. In addition, the user needs to understand the meaning of the geographical information that they are using. Often a vector-based system with point, line and area formats is manipulated to extract meaning. The system user is then required to understand the semantic meaning of the information and to consider the most appropriate visualization suitable for the map's end purpose. This can be a difficult choice and, for many map users, a cartographic representation created under such circumstances will be more difficult to understand than one prepared with thought and care by a trained cartographer. Such confusion by end users can be illustrated by the myriad of maps produced with incorrect projection systems by the mainstream media.

Thus, Guelke's (1976: 144) call for 'emphasis on meaning as the primary principle of map making' is relevant in the era of computer generated maps, and the tools that are provided to manipulate these maps are mostly geared towards the manipulation of symbols, and not to the process of helping the understanding of the information and knowledge construction.

11.4 Theoretical Engagement With Technical Aspects of Geography

Guelke's *Cartographica* article from 1976 is also relevant for its contribution to the disciplinary debate on the role of techniques and methods more broadly in geographical scholarship. The use of statistics or spatial analysis in geography can only be seen as unproblematic under a simplistic interpretation of positivism. The researchers that have developed and used such techniques have also reflected on their power and relevance to geographic enquiry (famously David Harvey, 1970, who remains influential to this day). Yet, the breadth of geography and its composition from people who, as result of their subdisciplinary interests, are exposed to different ideas and readings, means that from time to time researchers who are coming from different points of view will engage in a discussion about the meaning of a specific technique and its impact on the praxis of geography.

What makes such interactions important to the discipline as a whole is their role in revealing deeper meaning within the technical methodology that might have been ignored, or more often overlooked, by their users. Whilst, at first glance, such exchanges seem to be based on miscomprehension and an inability to communicate across philosophical divides, they have a long lasting value in progressing the understanding of techniques and methodologies in Anglo-centric academic geography. The famous exchange of Taylor (1990) and Openshaw (1991) might seem initially acrimonious, with a lack of empathy from each protagonist about the epistemology and ontology of the other.[1] Yet, through such hard-fought intellectual exchanges some core themes about the technical methodology emerge and these issues reverberate and can indeed lead to new research approaches and alliances. As Schuurman (2000) has charted, the debate that was started with Taylor's theoretically driven engagement with GIS has led to the development of the research strand of GIS and society which later on evolved into critical GIS.

11.5 The Role of Cartography in Geographic Understanding

A final point of conclusion for this essay considers the role of interaction within the communication process. Up until the 1970s, the map user was of central importance. All maps were static and captured one point in time and were presented to the map user as a finished article. The map was generally produced on paper and so there was very little if any technological mediation by the user. The only point at which

---

[1] See also follow-up commentaries by Taylor and Overton (1991) and Openshaw (1992).

technology was really introduced was during its printing stage. The user in the communication process is the end user – the map reader. Therefore, throughout the twentieth century, the map required no interaction from the end user, apart from folding, and the main cognitive process was centred on the reading and interpretation of the information.

Since the rise of GIS and subsequent Web-mapping technologies this has changed. Computers are now common mediators in the process of message transference. The end user frequently encounters the message mediator and is forced to interact and engage with the technology. Even with an application as relatively straightforward as Google Maps, the user (who has to become more than just the map user), has to interact with a computer, access the Web, navigate using the browser and interact with the application interface where finally they can engage with the map, zooming to the most appropriate map scale or pan around to find the location of interest. It is the same for in-car satellite navigational devices, where the driver must become both a fairly adept computer system user and the map reader, with several layers of choices requiring significant decisions to be made to achieve a usable product. In contemporary digital cartography the user has to interact with technology acting as a mediator before engaging with the geographical meaning of the cartography. This aspect is not accounted for in the traditional cartographic communication models. The computer as a cartographic mediator should certainly be a relevant and prominent feature of the communication process, if we wish to understand cartography in this way.

The final noteworthy aspect of the 1976 *Cartographica* article is that some of its themes are long-standing. While Guelke's piece was written at a time when cartography was growing as a subdiscipline within geography, and together with other papers of that era demonstrates the optimism of a fresh subdiscipline with the vivacity of youth, unfortunately this type of optimism frequently wanes with time – which is exactly what has happened. Even more noteworthy is that while the ICA and cartographic journals are still active, the number of academics who identify themselves as cartographic researchers seems to be continually shrinking. Cartography departments within universities are few and far between now when once they were commonplace. Furthermore, the role of cartography within the wider geographic discipline is challenged to the point that there is a struggle to maintain the role of maps within geography (Dodge and Perkins, 2008). Yet, the construction of geographic under-standing though possible with text, image and multimedia (and necessary, as Guelke noted) is not necessarily as rhetorically potent as it is with well designed map representations. Cartographic visualization and other spatial representations remain the most powerful tool for facilitating geographic knowledge exploration and under-standing. This theme continues to thrive in the area of geovisualization and the wider research into human–computer interaction and GIS.

References

Board, C. (1967) Maps as models, in *Models in Geography* (eds. R.J. Chorley and P. Haggett), Methuen and Co., London, pp. 671–725.

Bunting, T. and Guelke, L. (1979) Behavioral and perception geography: A critical appraisal. *Annals of the Association of American Geographers*, **69**, 448–462.

Dodge, M. and Perkins, C. (2008) Reclaiming the map: British geography and ambivalent cartographic practice. *Environment and Planning A*, **40**(6), 1271–1276.

Guelke, L. (1971) Problems of scientific explanation in geography. *Canadian Geographer*, **15**(1), 38–53.

Guelke, L. (1974) An idealist alternative in human geography. *Annals of the Association of American Geographers*, **64**(2), 193–202.

Guelke, L. (1976) Cartographic communication and geographic understanding. *The Canadian Cartographer*, **13**(2), 107–122. (Reproduced as Chapter 10 of this volume.)

Haklay, M., Singleton, A. and Parker, C. (2008) Web mapping 2.0: The neogeography of the geoweb. *Geography Compass*, **3**, 2011–2039.

Harvey, D. (1970) *Explanation in Geography*, Arnold, London.

Jones, C.E., Haklay, M.E., Griffiths, S. and Vaughan, L. (2009) A less-is-more approach to geovisualization – enhancing knowledge construction across multidisciplinary teams. *International Journal of Geographical Information Science*, **23**(8), 1077–1093.

Kitchin, R., Perkins, C. and Dodge, M. (2009) Thinking about maps, in *Rethinking Maps: New Frontiers in Cartographic Theory* (eds. M. Dodge, R. Kitchin and C. Perkins), Routlege, London, pp. 1–25.

Koláčný, A. (1969) Cartographic Information – a fundamental concept and term in modern cartography. *The Cartographic Journal*, **6**(1), 47–49.

Montello, D.R. (2002) Cognitive map-design research in the twentieth century: Theoretical and empirical approaches. *Cartography and Geographic Information Science*, **29**(3), 283–304.

Muehrcke, P.C. (1972) Maps in geography. *Cartographica*, **18**, 1–41.

Openshaw, S. (1991) A view on the GIS crisis in geography, or using GIS to put Humpty Dumpty back together again. *Environment and Planning A*, **23**, 621–628.

Openshaw, S. (1992) Further thoughts on geography and GIS: a reply. *Environment and Planning A*, **24**, 463–466.

Petchenik, B.B. (1977) Cognition in cartography. *Cartographica*, **14**(1), 117–128.

Pivar, M., Fredkin, E. and Stommel, H. (1963) Computer-compiled oceanographic atlas: an experiment in man–machine interaction. *Proceedings of the National Academy of Sciences of the United States of America*, **50**, 396–398.

Poore, B. and Chrisman, N.R. (2006) Order from noise: Toward a social theory of geographic information. *Annals of the Association of American Geographers*, **96**(3), 508–523.

Robinson, A.H. and Petchenik, B.B. (1977) The map as a communication system. *Cartographica*, **14**(1), 92–110.

Robinson, A.H., Morrison, J.L., Muehrcke, P.C. *et al.* (1995) *Elements of Cartography*, 6th edn, John Wiley and Sons, Inc., New York.

Schuurman, N. (1999) Critical GIS: theorizing an emerging science. *Cartographica*, **36**(4), 1–99.

Schuurman, N. (2000) Trouble in the heartland: GIS and its critics in the 1990s. *Progress in Human Geography*, **24**(4), 569–590.

Taylor, P. (1990) Editorial comment: GKS. *Political Geography Quarterly*, **9**, 211–212.

Taylor, P.J. and Overton, M. (1991) Further thoughts on geography and GIS. *Environment and Planning A*, **23**, 1087–1090.

Unwin, D.J. (2005) Fiddling on a different planet? *Geoforum*, **36**(6), 681–684.

Wood, D. and Fels, J. (1986) Designs on signs/Myth and meaning in maps. *Cartographica*, **23**(3), 54–103. (Reproduced as Chapter 14 of this volume.)

12

A Conceptual Framework and Comparison of Spatial Data Models

Donna J. Peuquet[1]

Abstract

This paper examines the major types of spatial data models currently known and places these models in a comprehensive framework. This framework is used to provide clarification of how varying data models, as well as their inherent advantages and disadvantages, are interrelated. It also provides an insight into how these conflicting demands may be balanced in a more systematic and predictable manner for practical applications, and reveals directions for needed future research.

12.1 Introduction

The rapidly expanding range of available spatial data in digital form, and the rapidly increasing need for their combined use, have revealed two very basic and severe problems associated with the application of automated spatial data handling technology: a rigidity and narrowness in the range of applications and data types which can be accommodated, as well as, unacceptable storage and speed efficiency for current and anticipated data volumes.

A general lack of versatility of spatial data processing systems exists, both for individual systems capabilities to accommodate a broader range of applications, as well as for the incorporation of differing types of spatial data from a variety of sources. The primary example of the need for very flexible spatial databases is the current

[1] Originally published: 1984, *Cartographica*, **21**(4), 66–113.

At the time of publication: Peuquet was Associate Professor in the Department of Geography, University of California at Santa Barbara.

attempts to incorporate LANDSAT and other remote sensed imagery and cartographic data within the same database. Spatial data have been accumulating at an increasingly rapid rate over the past two decades. This represents a very major investment and an extremely valuable resource, which is in demand for a wide variety of research and decision making applications. Attempts to integrate these data into existing systems have, to date, proven extremely difficult, at best.

The problem of a lack of versatility and the difficulty of integration are compounded by the fact that current spatial databases are encountering severe problems with physical storage volumes and time needed for processing. The geographic database systems in existence, however, pale in comparison to the scope of the databases being actively planned by a number of federal agencies and private corporations. The United States Geological Survey (USGC) is envisioning a cartographic database containing all information from 55 000 map sheets covering the entire United States. If these sheets were scanned once at, for example, 250 pixels per map inch (which is not high precision by cartographic standards), the total data would be approximately $1.5 \times 10^{**}15$ pixels. Some common procedures on one of these digitized map sheets currently can take hours of computer time to execute. The USGS situation is, in turn, dwarfed by NASA's current plans for the development of a database incorporating all spacecraft data for the earth, as well as the other planets and bodies in our solar system.

These efficiency, versatility and integration problems are attributable in large part to the profound differences in the commonly used storage formats and, more basically, to a lack of fundamental knowledge concerning properties of spatial data and a lack of a unified body of knowledge on the design and evaluation of spatial data models.

This paper presents an overall taxonomy of digital data models for the storage and manipulation of geographic data and a review of selected data models within this taxonomic structure. This is intended to serve two purposes. The first is to provide a unified framework and some directions for continuing research in the area of spatial data handling techniques. The second is to help remedy the current state of confusion, which seems to exist amongst practitioners as to the options and tradeoffs involved in this diverse subject.

This paper is organized in six sections. The first section provides a general introduction to the nature of current shortcomings of spatial data model technology in view of current and anticipated needs. This is followed by the presentation of a uniform theoretical framework, drawn primarily from the computer science literature. The third section reviews the various types of spatial data models as they are currently used in digital, geographic data storage and processing applications, with specific examples. The fourth section discusses recent developments in spatial data models. Here, changes in data model requirements are discussed within the context of recent research. The emphasis is placed on new approaches and on specific new models which hold promise but have not yet been used in any large-scale practical application. The fifth section briefly discusses the special problems involved with handling space–time data given the context of current theory and recent developments. The final section addresses future developments and their implications. Of necessity, this final section is broader in scope and deals with a number of developments which are affecting the demands on, and capabilities of, spatial databases in the future.

12.2 Theoretical Framework

12.2.1 Levels of Data Abstraction

A data model may be defined as a general description of specific sets of entities and the relationships between these sets of entities. An entity is a thing which exists and is distinguishable; that is we can tell one entity from another. Thus, a chair, a person and a lake are each an entity (Ullman, 1983). An entity set is a class of entities that possesses certain common characteristics. For example, lakes, mountains and desks are each entity sets. Relationships include such things as 'left of', 'less than' or 'parent of'. Both entities and relationships can have attributes, or properties. These associate a specific value from a domain of values for that attribute with each entity in an entity set. For example, a lake may have attributes of size, elevation and suspended particulates, amongst others.

A comparable definition of a data model was given by Codd (1981), who stated that a data model consists of three components: a collection of object types, a collection of operators and a collection of general integrity rules. As Date states, 'Codd was the first to formulate the concept of a data model in his original 1970 paper within the context of the relational database model' (Codd, 1970). Date also asserts that: 'The purpose of any data model, relational or otherwise, is of course to provide a formal means of representing information and a formal means of manipulating such a representation' (Date, 1983: 182–183).

Since, as defined above, this is a human conceptualization and tends to be tailored to a given application, different users and different applications are likely to have different data models to represent the same phenomenon (Figure 12.1). As the word 'model' implies, the most basic characteristic of a data model is that it is an abstraction of reality. Each data model represents reality with a varying level of completeness.

Many data model designers realize that in order to determine how a collection of data is to be ultimately represented in digital form, the data need to be viewed at a number of levels. These levels progress from reality, through the abstract, user-orientated information structure, to the concrete, machine-orientated storage structure. There is, however, a lack of universal agreement as to how many levels of abstraction one should distinguish (Klinger, Fu and Kunii, 1977; Martin, 1975; Senko *et al.*, 1973; Senko, 1976; Tompa, 1977; Nyerges, 1980). These differences can in large part be attributed to context. For the purposes of the present discussion, four levels will be used (Figure 12.2):

Reality: the phenomenon as it actually exists, including all aspects which may or may not be perceived by individuals.

Data Model: an abstraction of the real world which incorporates only those properties thought to be relevant to the application or applications at hand, usually a human conceptualization of reality.

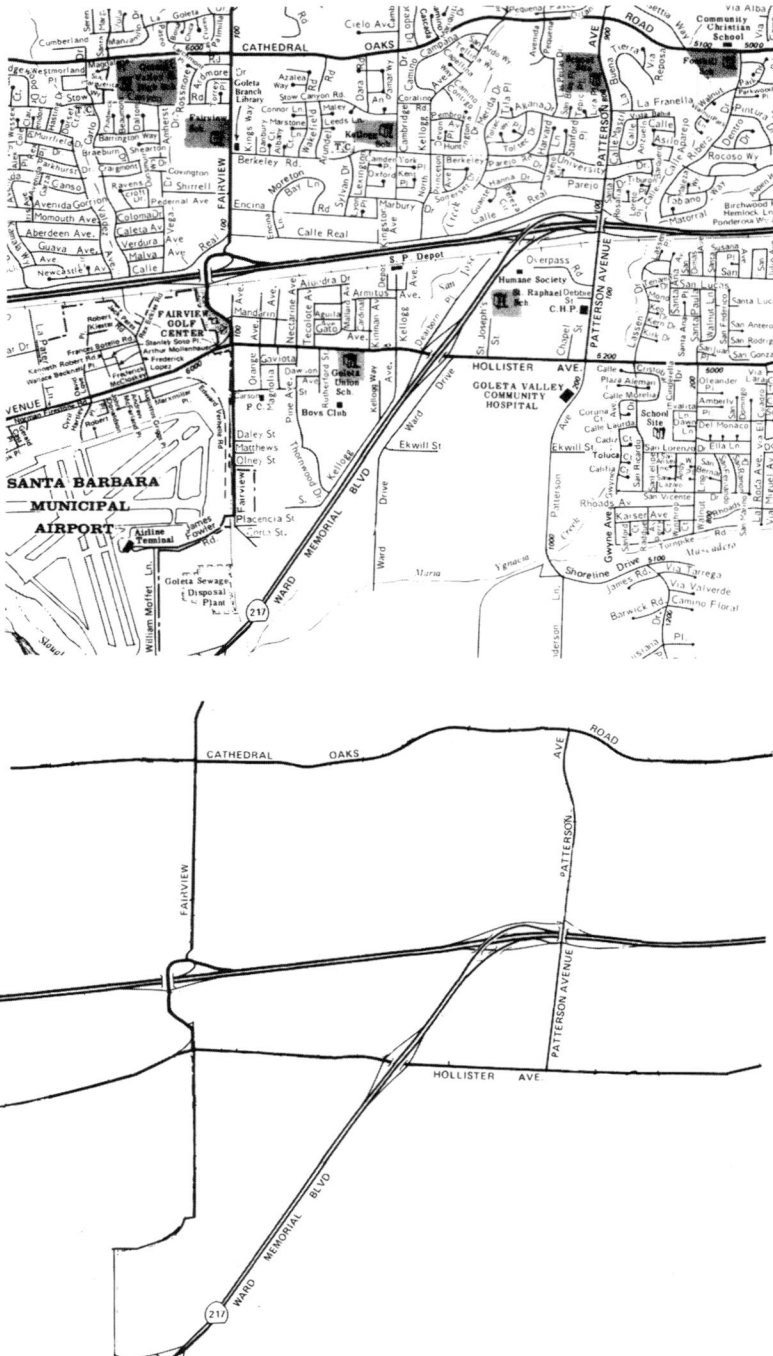

Figure 12.1   The overall database model (top) is likely to be confusing, overly complex for individual applications. Varying, simplified views of the data may be derived from the overall database model for specific applications (bottom).

record 323

record 324

record 322

record 321

record 325

End of file marker
End of chain marker

File structure

Polygons		Chain List	Chains						
Name	Pointer		Name	Points	Length	from	to	left	right
1	.	►11	►11	.	.	9	7	0	2
2	.	12	12	.	.	7	8	1	2
3	.	13	13	.	.	8	9	0	2
.	.								
.	.								

Data structure

Points
x y strings

►xy, xy, ...

Nodes

Name	x	y
7	x	y
8	x	y
►9	x	y

Data model

10 6
1 14
7 12
11 2 3
9 13 8 5
15 3 16 4
19 1
20
4 23 5
18
2
17

Real world

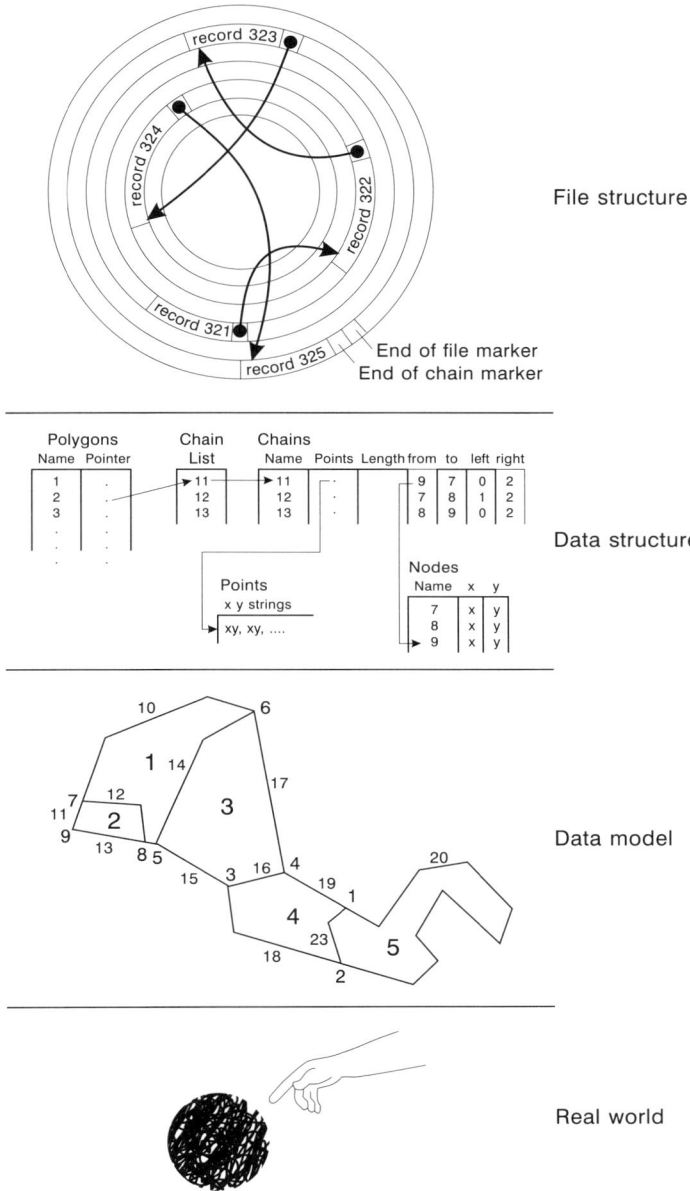

Figure 12.2 Levels of data abstraction.

Data Structure: a representation of the data model often expressed in terms of diagrams, lists and arrays designed to reflect the recording of the data in computer code.

File Structure: the representation of the data in storage hardware.

These last three views of data correspond to the major steps involved in database design and implementation. The overall process is one of progressively refining general statements into more specific statements. Within a level, the process of stepwise refinement would be used to provide a smooth transition from one level to the next.

The term 'data model' is used again here, but in the narrower context of a specific level of data abstraction. This is the result of a considerable amount of confusion which existed within the computer science, image processing and geographic literature. The problem is a historical one. The term 'data structure' was commonly used as the generic term or used synonymously with 'data model'. However, with the development of systematic software design techniques and the easing of restrictions of the computing environment due to software and hardware technological advancements, the 'nuts and bolts' of language and hardware implementation is no longer a dominating force in database design. Thus, the term 'data model', in this context of levels of data abstraction, has evolved to connote a human conceptualization of reality, without consideration of hardware and other implementation conventions or restrictions.

A data structure is built upon the data model, and details the arrangement of the data elements. This can therefore be described as a structural model, with individual elements within each group organized into lists and arrays, and the relationships explicitly defined. This is equivalent to the mathematician's broad definition of a graph (Mark, 1979). Relationships between objects, or data elements, may be expressed explicitly or implicitly. Explicit relationships are written into the data structure as data elements themselves. Implicit relationships can be indicated by the relative position of the individual data elements. Derivation of some implicit relationships may require computation through analysis of some or all of the data. An example would be nearest neighbour of a point amongst points distributed irregularly in space.

A file structure defines the physical implementation mechanism (i.e. the storage model). This is the translation of the data structure into a specific hardware/software environment.

12.2.2 General Concepts

Since no model or abstraction of reality can represent all aspects of reality, it is impossible to design a general-purpose data model that is equally useful in all situations. This is particularly true when dealing with complex phenomena. For example, some spatial data models, when implemented in a digital environment, are good for plotting, but very inefficient for analytic purposes. Other data structures may be excellent for specific analytical processes, but may be extremely inefficient for producing graphics.

Varying approaches have been used in the design of spatial data models. To provide an example of the range of approaches which have been used, Bouillé's approach attempted to derive a data model which included all identifiable entities and their relationships into what he terms 'phenomenon-based design' in deriving the 'phenomenon structure' (Bouillé, 1978). Data models and subsequent data structures derived from such an approach, in attempting nearly complete representation of reality, tend to become like reality usually is – extremely complex. The result would most often be a level of complexity far beyond that which is useful or efficient in a computer context, and would contain many entities and relationships which are not essential to the application at hand.

Mark, on the other hand, adopts a philosophy that the data structure or data model design should be driven by its intended use and exclude any entities and relationships not relevant to that use (Mark, 1979). This results in a data model which tends to be a far from complete representation of reality, but instead contains only the essential elements necessary for a particular task. Such a minimalist approach, compared to the phenomenon-based design process of Bouillé, tends to produce models of minimum complexity.

These two views toward data model design represent two opposite extremes in the basic trade-off involved in the data modelling process. The more perfectly a model represents reality (i.e. the more completely all entities and possible relations are incorporated), the more robust and flexible that model will be in application. However, the more precisely the model fits a single application, excluding entities and relations not required to deal with that application, the more efficient it will tend to be in storage space commutation and ease of use.

The selection or design of a data model must, therefore, be based both on the nature of the phenomenon that the data represents and the specific manipulation processes which will be required to be performed on the data. This fact has been apparent to some degree to designers and builders of geographic data handling systems and geographic databases; however, the precise mechanisms of the trade-offs involved between the various options available have never been discussed in depth.

The process of deriving an optimum balance between these two positions is best accomplished in practice by using both of these approaches simultaneously in a 'both ends toward the middle' process. This is a process which has, unfortunately, not yet been formalized.

12.2.3 The Nature of Geographic Data

The term 'spatial' data applies to any data concerning phenomenon areally distributed in two-, three- or N-dimensions. This includes such things as bubble chamber tracks in physics and engineering schematics. Geographic data, more specifically, are spatial data which normally refer to data pertaining to the earth. These may be two-dimensional, modelling the surface of the earth as a plane, or three-dimensional to describe subsurface or atmospheric phenomena. A fourth dimension could be added for time-series data, as well. In the context of the present discussion, the term 'geographic' may also apply to data pertaining to other planets and objects in space.

There are several types of spatial data, and the differences between them become obvious when they are displayed in graphic form, as shown in Figure 12.3. The first is point data, where each data element is associated with a single location in two- or three-dimensional space, such as the locations of cities of the United States. The second is line data. With this data type, the location is described by a string of spatial coordinates. These can represent either: (a) isolated lines where individual lines are not connected in

Point Data
Cities

Line Data
a) Tree structure:
River system

b) Disconnected lines:
Faults

c) Network structure:
Major roads

Polygon Data
a) Isolated polygones:
Selected metropolitan areas

b) Adjacent polygons:
State outlines

c) Nested polygons:
The average percentage of possible subshine

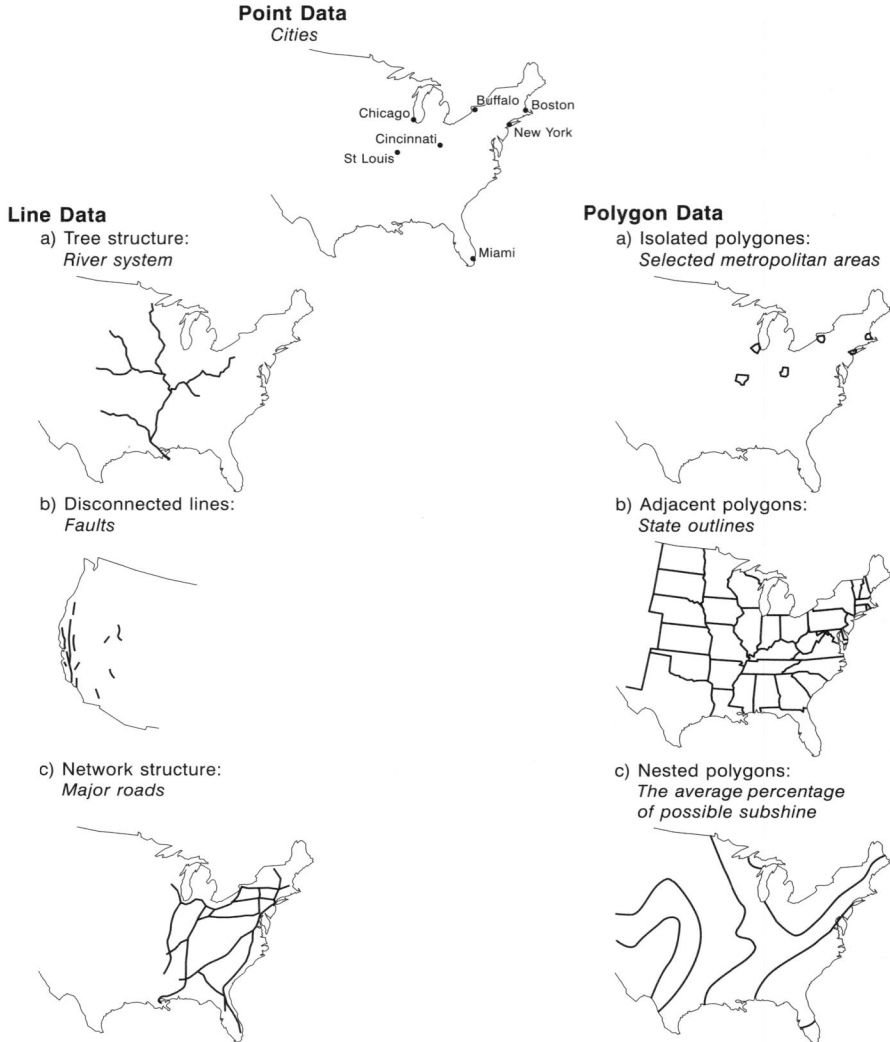

Figure 12.3 Examples of spatial data types.

any systematic manner, such as fault lines, (b) elements of tree structures, such as river systems, or (c) elements of network structures, as in the case of road systems.

The third type is polygon data, where the location of a data element is represented by a closed string of spatial coordinates. Polygon data are thus associated with areas over a defined space. These data can themselves be any one of three types: (a) isolated polygons, where the boundary of each polygon is not shared in any part by any other polygon, (b) adjacent polygons, where each polygon boundary segment is shared with at least one other polygon, and (c) nested polygons, where one or more polygons lie entirely within another polygon. An example of adjacent polygons is the state boundaries in a map of the United States. Contour lines on a topographic map are an example of nested polygon data. A fourth category of data is some mixture of the above types. This might include different line structures mixed together, line structures mixed with a polygon structure or with discrete points. For example, in a map of the United States a state may be bounded by a river which is both a boundary between adjacent polygons as well as part of the tree structure of a river network. These four categories of spatial data are known as image or coordinate data (IGU, 1975, 1976). This means that these data portray the spatial locations and configurations of individual entities. A spatial data entity may be a point, line, polygon or a combination of these. Each entity also has characteristics which describe it, called attribute or descriptor data. For example, the latitude and longitude coordinates of the city of Santa Barbara are part of the image data set, while its population would be part of the descriptor data set. Similarly, the coordinates which make up a spatial entity such as the outline of the State of California are image data, while statistics such as the total number of forested acres are descriptor data.

Spatial phenomena, and spatial data models, have a number of characteristics which significantly differentiate them from one-dimensional or list-type models. Firstly, spatial entities have individual, unique definitions which reflect the entities' location in space. For geographic data, these definitions are commonly very complex, given the tendency of natural phenomena to occur in irregular, complex patterns. Particularly for geographic data, these definitions are recorded in terms of a coordinate system. This coordinate system may be one of a number of types: latitude and longitude, UTM street address, and so on. These coordinate systems may not necessarily have precise, mathematical transformations, such as street address to latitude and longitude.

The relationships between spatial entities are generally very numerous and, in fact, given the nature of reality or our perceptions of it, and the limitations of the modelling process, it is normally impossible to store all of them. The definitions of these relationships, and the entities themselves in the case of geographic data, also tend to be inexact and context dependent. This is true of even very basic spatial relationships such as 'near' and 'far', or 'left' and 'right'.

The combination of these properties (multidimensionality, fuzzy entities and relationship definitions and complex spatial definitions) make the modelling of geographic data uniquely difficult. The models themselves tend to be complex and the resultant data files tend to be not very compact.

An additional problem arises in the transformation of a conceptual data model into data structure and file structure views for computer implementation. Graphic input devices, such as digitizers, transform area, line and point structures into numeric, computer-readable form by recording spatial coordinates of map entities. There is a basic problem underlying this transformation. Spatial data are by definition two- or three-dimensional. How then can these data be represented in computer memory which is usually linear, or one-dimensional in nature, while preserving these implicit spatial interrelationships? If they are simply listed in a continuous linear stream, coordinates of the entities contain neither the topology inherent in line networks or adjacent polygons, nor spatial relationships, such as 'above' or 'left of'. These relationships are data in themselves and are often of primary importance, particularly to geographers, when examining spatial data (Dacey and Marble, 1965). The coordinates must therefore be structured so as to preserve these two- or three-dimensional relationships and yet be capable of being stored in linear or list fashion within the computer.

12.2.4 Form versus Function

The performance versus representational fidelity trade-off mentioned in the 'general concepts' section above, impacts directly upon the storage, manipulative and retrieval characteristics of the data structure and physical file structure. It is necessary to examine these trade-offs using a specific set of usage-based criteria, so that the overall quality or suitability of a specific data model can be evaluated within a particular context. The general criteria are:

1. completeness
2. robustness
3. versatility
4. efficiency
5. ease of generation

Completeness may be thought of in terms of the proportion of all entities and relationships existing in reality which are represented in the model of a particular phenomenon. Robustness is the degree to which the data model can accommodate special circumstances or unusual instances, such as a polygon with a hole in it. Efficiency includes both compactness (storage efficiency) and speed of use (time efficiency). Ease of generation is the amount of effort needed to convert required data in some other form into the form required the data model.

In varying degree, each of these factors enters into consideration for any given application. The relative importance of each factor is a function of the particular type of data to be used and the overall operational requirements of the system. For example, if the database to be generated will be very large and must perform in an interactive context, compromises would likely be necessitated with the first three factors because overall efficiency and ease of generation would predominate.

It is possible to quantitatively measure the performance of several of these criteria, such as speed and space efficiency for a particular data model. It is not possible, however, to provide quantitative measures for the more abstract factors of data completeness, robustness or versatility. This, combined with the fact that we still have little knowledge of the performance characteristics of a wide range of spatial processing algorithms and how they interact with other algorithms and varying data models, indicates that the spatial data modelling process is much more an art than a science. Experience and intuition will remain primary factors in the interpretation of vague system requirements specifications and the construction of satisfactory data models, particularly for complete and integrated geographic information systems.

Additional comments on the process of balancing trade-offs in spatial data model design will be made at the beginning of the later 'recent developments in spatial data models' section.

12.3 Examples of Traditional Geographic Data Models

12.3.1 Basic Types

Geographic data have traditionally been presented for analysis by means of two-dimensional analogue models known as maps (Board, 1967). The map has also provided a convenient method of spatial data storage for later visual retrieval and subsequent manual updating, measuring or other processing. In order to update a map or display results of any manual procedure performed on the data, a new map must be hand drawn or the old one modified by hand. This process is laborious and time consuming, requiring both skill and precision on the part of the individual drafting the map.

Two other basic types of spatial data models have evolved for storing image data in digital form: vector and tessellation models (Figure 12.4). In the vector type of data model, the basic logical unit in a geographical context corresponds to a line on a map, such as a contour line, river, street, area boundary or a segment of one of these. A series of x–y point locations along the line is recorded as the components of a single data record. Points can be represented in a vector data organization as lines of zero length (i.e. one x–y location). With the polygonal mesh type of organization, on the other hand, the basic logical unit is a single cell or unit of space in the mesh. These two types are thus logical duals of each other.

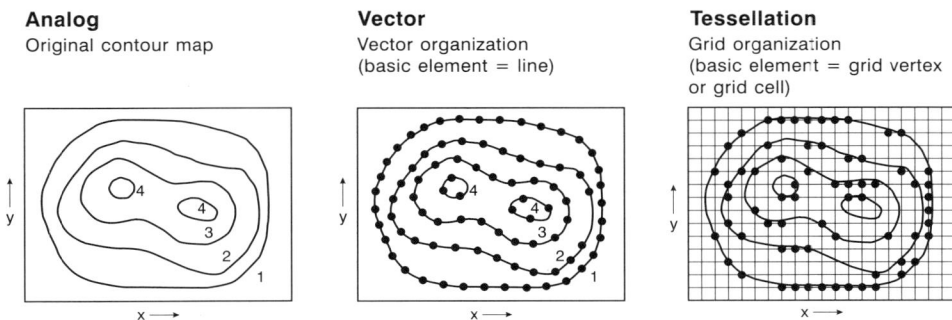

Figure 12.4 Basic types of spatial data models.

Common usage has usually considered the two basic spatial data model types to be raster, or grid, and vector. As this paper will show, however, the class of non-vector spatial data models encompasses much more than data models based on a rectangular or square mesh. This class includes any infinitely repeatable pattern of a regular polygon or polyhedron. The term used in geometry for this is a 'regular tessellation'. A tessellation in two dimensions is analogous to a mosaic, and in three dimensions to a honeycomb (Coxeter, 1973).

There also exists what can be viewed as a third type of spatial data model – the hybrid type. This class of data model is a recent development which possesses characteristics of both vector and tessellation data models.

Each of these approaches has also been used in fields other than geography to represent spatial data, such as scanner images in picture processing. The characteristics of each of these types of models and their trade-offs for representing geographic phenomena should become clearer through the discussion of some specific examples of some 'classic' geographic data models.

12.3.2 Vector Data Models

12.3.2.1 Spaghetti Model

The simplest vector data model for geographic data is a direct line-for-line translation of the paper map. As shown in Figure 12.5, each entity on the map becomes one logical record in the digital file and is defined as strings of x–y coordinates. This structure is very

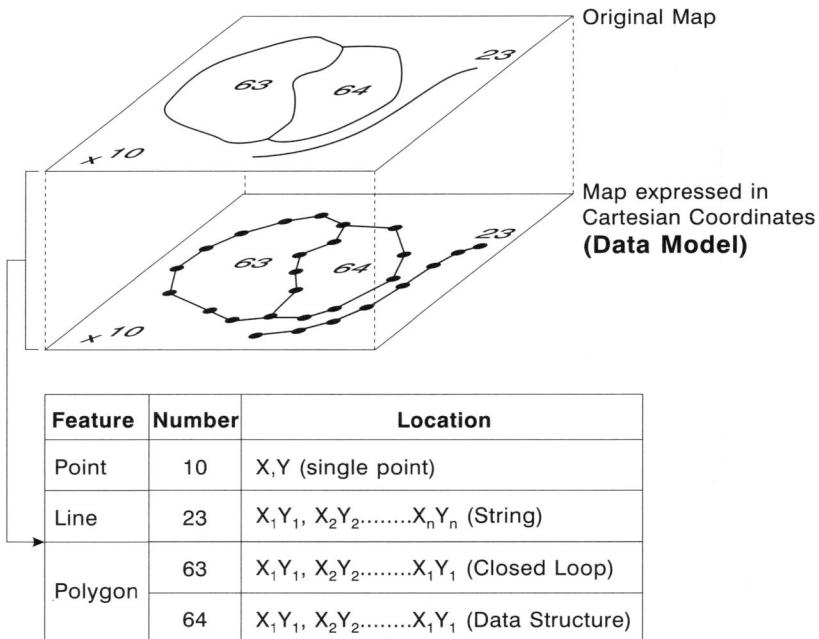

Feature	Number	Location
Point	10	X,Y (single point)
Line	23	$X_1Y_1, X_2Y_2........X_nY_n$ (String)
Polygon	63	$X_1Y_1, X_2Y_2........X_1Y_1$ (Closed Loop)
	64	$X_1Y_1, X_2Y_2........X_1Y_1$ (Data Structure)

Figure 12.5   The 'spaghetti' data model (adapted from Dangermond, 1982).

simple and easy to understand since, in essence, the map remains the conceptual model and the x–y coordinate file is more precisely a data structure. The two-dimensional map model is translated into a list, or one-dimensional model. Although all entities are spatially defined, no spatial relationships are retained. Thus, a digital cartographic data file constructed in this manner is commonly referred to as a 'spaghetti file'; that is a collection of coordinate strings heaped together with no inherent structure. A polygon recorded in this manner is represented by a closed string of x–y coordinates which define its boundary. For adjacent polygon data, this results in recording the x–y coordinates of shared boundary segments twice – once for each polygon.

The 'spaghetti' model is very inefficient for most types of spatial analyses, since any spatial relationships which are implicit in the original analogue document must be derived through computation. Nevertheless, the lack of stored spatial relationships, which are extraneous to the plotting process, makes the spaghetti model efficient for reproducing the original graphic image. The spaghetti model is thus often used for applications that are limited to the simpler forms of computer-assisted cartographic production. Corrections and updates of the line data must rely on visual checks of graphic output.

12.3.2.2 Topologic Model

The most popular method of retaining spatial relationships amongst entities is to explicitly record adjacency information in what is known as a topologic data model. A simplified example of this is shown in Figure 12.6. Here, the basic logical entity is a straight line segment. A line segment begins or ends at the intersection with another line or at a bend in the line. Each individual line segment is recorded with the coordinates of its two endpoints. In addition, the identifier, or name of the polygons on either side of the line is recorded. In this way, the more elementary spatial relationships are explicitly retained and can be used for analysis. In addition, this topological information allows the spatial definitions of points, lines and polygon-type entities to be stored in a non-redundant manner. This is particularly advantageous for adjacent polygons. As the example in Figure 12.6 shows, each line segment is recorded only once. The definitions and adjacency information for individual polygons are then defined by all individual line segments which comprise that polygon on the same side, either the right or the left.

12.3.2.3 GBF/DIME

The GBF/DIME (Geographic Base File/Dual Independent Map Encoding) model is the best known model built upon this topological concept. It was devised by the US Census Bureau for digitally storing street maps to aid in the gathering and tabulation of Census data by providing geographically referenced address information in computerized form (US Department of Commerce, 1970). Developed as an improvement of the Address Coding Guides, the initial GBF/DIME files were created in the early 1970s.

In GBF/DIME file, each street, river, railroad line, municipal boundary and so on, is represented as a series of straight line segments. A straight line segment ends where two

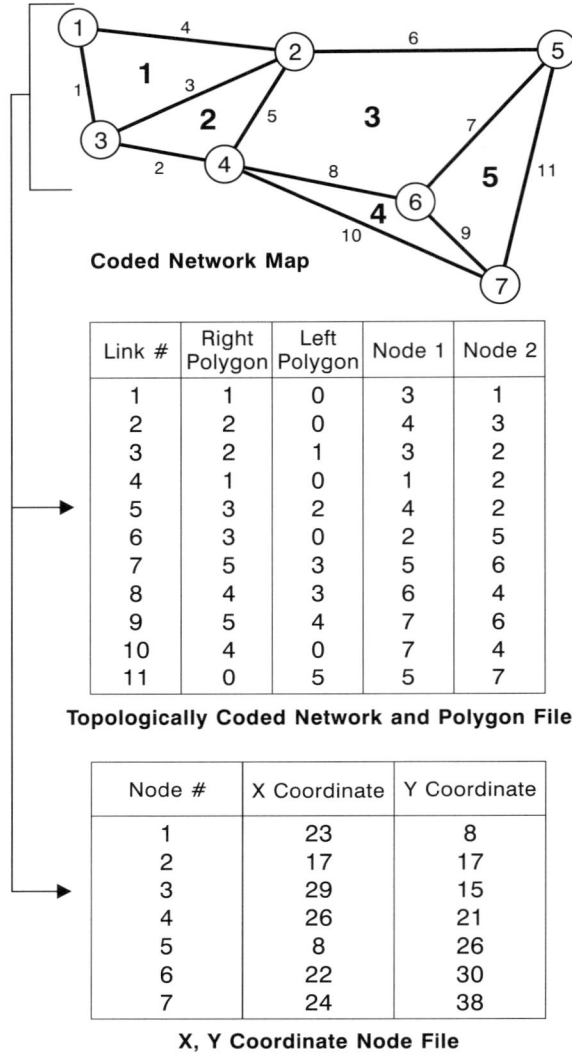

Coded Network Map

Topologically Coded Network and Polygon File

Link #	Right Polygon	Left Polygon	Node 1	Node 2
1	1	0	3	1
2	2	0	4	3
3	2	1	3	2
4	1	0	1	2
5	3	2	4	2
6	3	0	2	5
7	5	3	5	6
8	4	3	6	4
9	5	4	7	6
10	4	0	7	4
11	0	5	5	7

X, Y Coordinate Node File

Node #	X Coordinate	Y Coordinate
1	23	8
2	17	17
3	29	15
4	26	21
5	8	26
6	22	30
7	24	38

Figure 12.6   The topological data model (from Dangermond, 1982).

lines intersect or at the point a line changes direction. At these points and at line endpoints, nodes are identified (Figure 12.7).

As shown in Figure 12.8, each GBF/DIME line segment record contains Census tract and block identifiers for the polygons on each side. The DIME model offers a significant addition to the basic topological model in that it explicitly assigns a direction to each straight line segment by recording a From node (i.e. low node) and a To node (i.e. high node). The result is a directed graph which can be used to automatically check for missing segments and other errors in the file, by following the line segments which comprise the boundary of each census block (i.e. polygon) named in the file. This walk

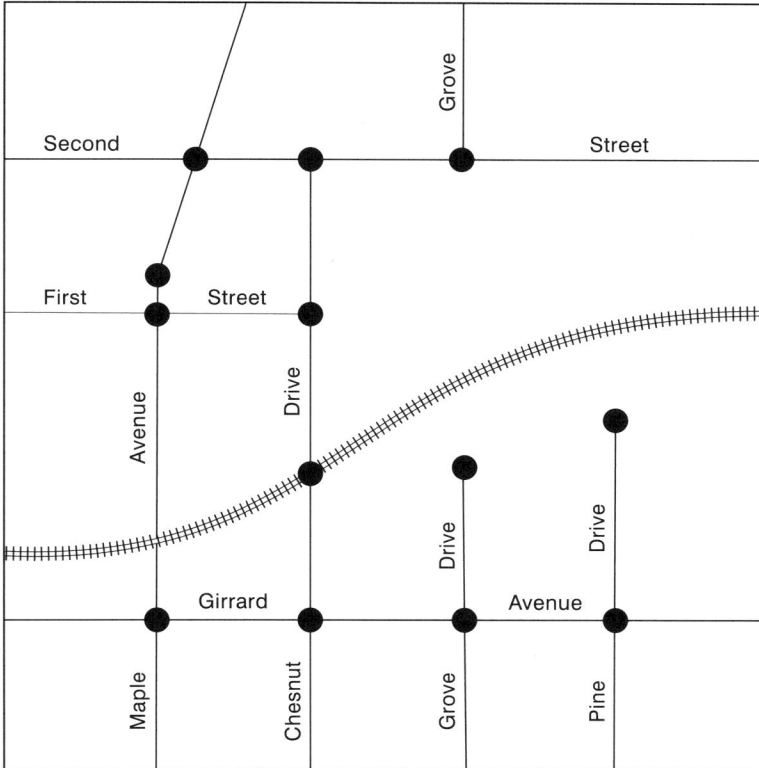

Figure 12.7 Graphic elements of a DIME file.

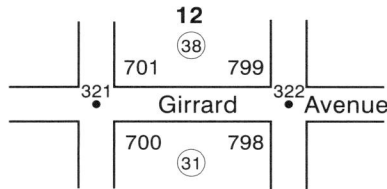

Street Name	Girrard
Street Type	Avenue
Left Addresses	701-799
Right Addresses	700-798
Left Block	38
Left Tract	12
Right Block	31
Right Tract	12
Low Node	321
X-Y Coordinate	155 000 - 232 000
High Node	322
X-Y Coordinate	156 000 - 234 000

Figure 12.8 Contents of a sample DIME file record.

around each polygon is done by matching the To node identifier of the current line segment with the From node identifier of another line segment via a search of the file. If line segment records cannot be found to completely chain around a polygon in this manner, a line segment is missing or a node identifier is incorrect.

Another feature worth noting is that each line segment is spatially defined, according to the definition of the model, using both street addresses and UTM coordinates. This is in recognition of the fact that some locational systems (e.g. street addresses), which may be needed for some types of applications, cannot be directly derived from conventional Cartesian or polar coordinate systems.

The main problem with the DIME model, like the previous two models described, is that individual line segments do not occur in any particular sequence order. To retrieve any particular line segment a sequential, an exhaustive search must be performed on the entire file. To retrieve all line segments which define the boundary of a polygon, an exhaustive search must be done as many times as there are line segments in the polygon boundary!

12.3.2.4 POLYVRT

POLYVRT (POLYgon conVERTer) was developed by Peucker and Chrisman (1975) and implemented at the Harvard Laboratory for Computer Graphics in the late 1970s. This model overcomes the very major retrieval inefficiencies seen in simpler topologic structures by explicitly and separately storing each type of data entity separately in a hierarchical data structure (Figure 12.9). To make the distinctions between types of entities both logically and topologically meaningful, a chain is denoted as the basic line entity. A chain is defined as a sequence of straight line segments which begins and ends at a node. A node is defined as the intersection point between two chains. The point coordinate information to define each chain is not stored as part of the chain record. Instead, a pointer to the beginning of this information within a separate Points file is recorded. Similarly, pointers are given within the Polygons file to the individual chains which comprise it. Note that the individual chain records contain the same explicit direction and topology information used within GBF/DIME: From and To nodes as well as the left and right adjacent polygons. If a chain defines an outer boundary of the entire area, such as for chain 13 in Figure 12.9, this outer area is denoted as polygon 'o'.

This structure provides a number of advantages over GBF/DIME for retrieval and manipulation. Firstly, the hierarchical structure allows selective retrieval of only specific classes of data at a time. A second advantage of the POLYVRT model is that queries concerning the adjacency of polygons need only deal with the polygon and chain portion of the data. Only the individual chains which bound the polygons of interest are retrieved. The actual coordinate definitions are not retrieved until explicitly needed for such operations as plotting or distance calculations.

The number of line or chain records in a POLYVRT database depends only upon the number of polygons present in the data and not on the detail of their boundaries. In computer implementation, this physical separation allows a much greater efficiency in

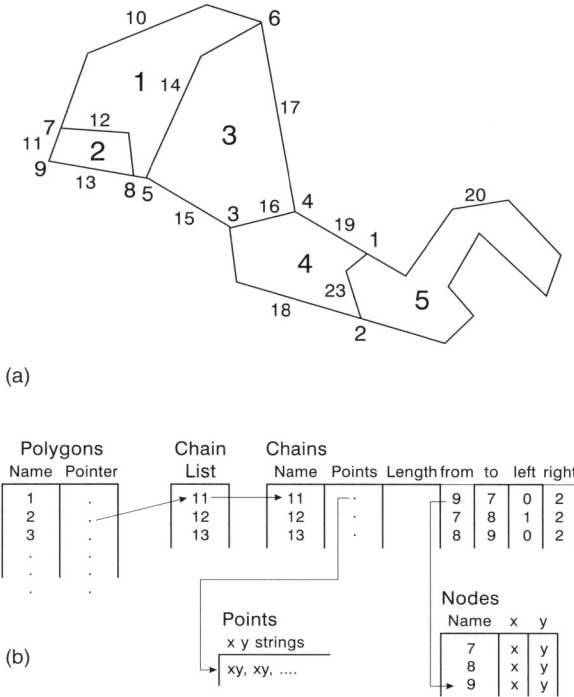

Figure 12.9   (a) The POLYVRT data model; (b) the POLYVRT data structure.

needed central memory space as well as speed for many operations. This gives POLYVRT a significant advantage for use with entities which have highly convoluted boundaries. However, this physical separation also causes the need for a link or pointer structure. These non-data elements add a significant amount of extra bulk to the model. The amount of overhead this generates usually cannot be tolerated for databases containing a large number of entities. The other major disadvantage is that incorrect pointers can be extremely difficult to detect or correct. The initial generation of this structure can also be cumbersome and time consuming.

On the other hand, the POLYVRT approach has considerable versatility. Peucker and Chrisman represent a POLYVRT data model and its corresponding data structure which are tailored to represent a set of adjacent polygons. The model can also be augmented for the representation of more complex data. It does not violate the basic concept of the model to add another level to the hierarchy, such as an additional level of polygons. Using this modified POLYVRT to represent a map of the United States, for example, the higher polygons could be states and the lower polygons could represent counties.

Other changes to the basic POLYVRT structure open other possibilities. Various types of polygons in the same level of the hierarchy can be defined by a prefix added to the polygon identifier. Additional information concerning polygons, nodes and chains can be encoded into their respective identifiers in a similar manner.

Peucker and Chrisman discuss how the POLYVRT structure could be used for topographic data by encoding all ridges and channel lines as chains, and encoding the peaks, passes and pits as nodes. An auxiliary structure is used with this, representing a triangulated grid with the sample elevation points at the vertices of the triangle. This dual structure seems cumbersome but reveals still further possibilities for uses of a POLYVRT approach. It would be a great aid in overlaying two-dimensional with three-dimensional data if the two types were in the same or similar format.

12.3.2.5 *Chain-codes*

Chain code approaches are actually a method of coordinate compaction rather than a data model. They are included in this discussion for two reasons. Firstly, this methodology provides significant enhancements in compaction and analytical capabilities and, therefore, has been frequently integrated into spatial data models, including some which will be discussed below. Secondly, chain codes have had a major impact on spatial data models and spatial data processing to such an extent that they are commonly viewed as a data model in their own right.

The classical chain coding approach is known as Freeman–Hoffman chain codes (Freeman, 1974). This consists of assigning a unique directional code between 0–7 for each of eight unit-length vectors as shown in Figure 12.10. The eight directions include the cardinal compass directions, plus the diagonals. Using this scheme to encode line data upon a grid of given unit resolution results in a very compact digital representation. As also seen in this example, x–y coordinate information need only be recorded for the beginning of each line. Direction is inherent in this scheme, providing an additional compaction advantage for portraying directed data, such as stream or road networks.

Through the use of special code sequences, special topological situations such as line intersections can be noted. One of the special coding sequences is also used for providing a mechanism for run-length encoding. This eliminates the need for repeated direction codes for long, straight lines. The flag used to signal that one of these special codes follows is '04'. This directional chain code sequence would mean that the line retraces itself, a meaningless sequence in most cases. It thus can be used as a convenient flag. The reader is referred to Freeman for his complete listing. These codes can, of course, be augmented or changed to suit a particular application.

There have been several variations of this coding scheme derived. The first, also described by Freeman (1979) is to use a 4, 16 or 32 vector notation on the same square lattice. The 4-directional encoding scheme allows representation of each code with two instead of three bits, and is sufficient in cases where the data tend to consist of long lines which are orthogonal to one another, such as in some engineering applications. Sixteen or 32-direction coding allows for more accurate encoding of arbitrary-shaped curves. This smooths out the stair-casing effect introduced by the directional approximations necessary when fewer directions are used for encoding (Figure 12.11). Similarly, there is a direct relationship between the number of directional vectors and the unit vector length for any given desired encoding accuracy for arbitrarily-shaped lines. In terms of compaction, this obviously presents a trade-off between the number of direction-vector

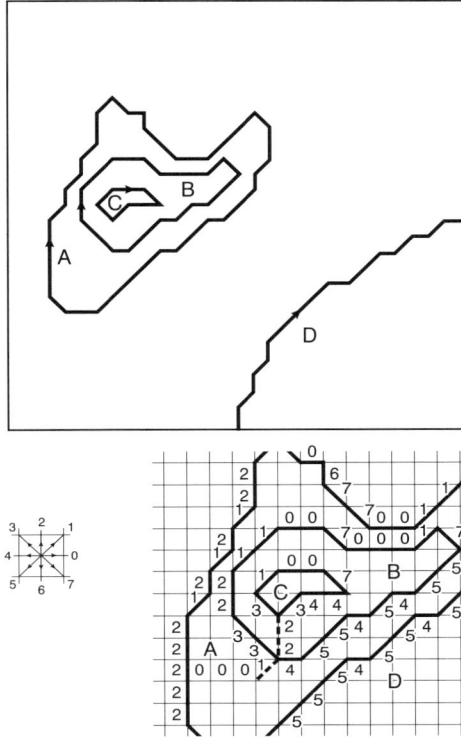

Figure 12.10 Contour map and the resulting chain coded lines (adapted from Freeman, 1974).

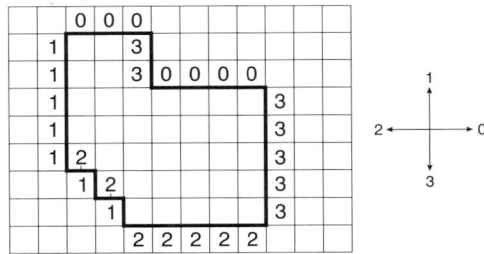

4 - directional coding

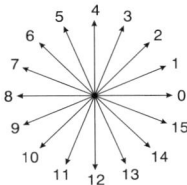

16 - directional coding

Figure 12.11 4- and 6-directional chain coding.

codes required to represent a given line and the number of bits required to represent each code.

The second well known variation on the Freeman–Hoffman chain coding scheme is Raster Chain codes, or RC codes, introduced by Cederberg (1979). This scheme uses only half of the standard eight direction vectors as shown in Figure 12.12. This was designed to process scan-line-formatted data in raster order (each scan line in sequence, top to bottom and left to right) to produce chain coded vector-formatted data. Since processing in this order never encounters 'backwards' vectors relative to the processing direction, only half of the eight standard direction codes are needed. This restricted directionality does, however, have the effect of segmenting the directional continuity of arbitrary shaped lines. If directional continuity of vector data is needed, the conversion of raster chain codes to Freeman–Hoffman chain codes is a straightforward process of 'flipping' or reversing the directionality of selected vector segments. For closed polygons, the selection of vector segments to be reversed is based on the Jacobsen Plumbline algorithm. This conversion process was described in detail by Chakravarty (1981).

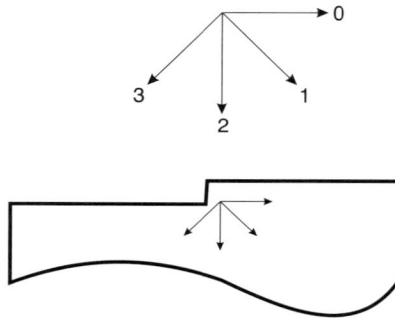

Figure 12.12   The raster chain coding scheme (adapted from Cederberg, 1979).

A third variation on the chain coding concept is its use on a hexagonal rather than a square lattice (Scholten and Wilson, 1983).

The primary disadvantage of chain codes is that no spatial relationships are retained. It is, in fact, a compact spaghetti-format notation. Another disadvantage is that coordinate transformations, particularly rotation, are more difficult with chain coded data.

As previously mentioned, the primary advantage of the chain coding approach is its compactness. Chain coding schemes are frequently incorporated into other schemes for the purpose of combining the compaction advantage of chain codes with the advantages of another data model. The use of incremental directional codes instead of Cartesian coordinates results in better performance characteristics than the simple spaghetti data model. The standard method of operation for vector plotters is to draw via sequences of short line segments using (usually) eight possible direction vectors. Vector plotter hardware thus seems to be tailor-made for chain coded data. Graphic output on these devices requires no coordinate translation, making the process very efficient.

The use of unit vector direction codes is also advantageous for a number of measurement and analytical procedures, such as distance calculations and shape analyses. Algorithms for many of these procedures for chain coded data were developed and documented by Freeman (1974, 1979).

12.3.3 Tessellation Models

As stated in the beginning of this section, tessellation, or polygonal mesh, models represent the logical dual of the vector approach. Individual entities become the basic data units for which spatial information is explicitly recorded in vector models. With tessellation models, on the other hand, the basic data unit is a unit of space for which entity information is explicitly recorded.

12.3.3.1 Grid and Other Regular Tessellations

All three possible types of regular tessellations have been used as the basis of spatial data models. Each has differing functional characteristics which are based on the differing geometries of the elemental polygon (Ahuja, 1983). These three are square, triangular and hexagonal meshes (Figure 12.13).

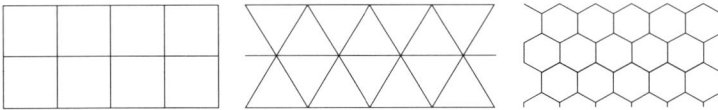

Figure 12.13   The three regular tessellations.

Of these, the regular square mesh has historically been the most widely used, primarily for two very practical reasons: (1) it is compatible with the array data structure built into the FORTRAN programming language, and (2) it is compatible with a number of different types of hardware devices used for spatial data capture and output. Fortunately, a number of higher-level computing languages are currently available which provide a great deal of flexibility in representing data through both additional intrinsic structures and user-defined structures. The ability to easily mix languages within the same program has also facilitated the programming task in general.

In the earliest days of computer cartography, the only graphic output device commonly available was the line printer (Tobler, 1959). Each character position on the line of print was viewed as a cell in a rectangular grid. Later devices for graphic input and output, particularly those designed for high speed, high volume operation, process data in rectangular mesh form. These include raster scanners, also known as mass digitizing devices, and colour refresh CRTS. Remote sensing devices, such as the LANDSAT MSS, capture data in gridded form as well (Peuquet and Boyle, 1984).

The tremendous data volumes being accumulated through the use of these grid-orientated, data input devices is in itself generating significant inertia toward using data in that form, rather than converting it to vector form.

The primary advantage of the regular hexagonal mesh is that all neighbouring cells of a given cell are equidistant from that cell's centre point. Radial symmetry makes this model advantageous for radial search and retrieval functions. This is unlike the square mesh where diagonal neighbours are not the same distance away as neighbours in the four cardinal directions from a central point.

A characteristic unique to all triangular tessellations, regular or irregular, is that the triangles do not all have the same orientation. This makes many procedures involving single-cell comparison operations, which are simple to perform on the other two tessellations, much more complex. Nevertheless, this same characteristic gives triangular tessellations a unique advantage in representing terrain and other types of surface data. This is done by assigning a z-value to each vertex point in the regular triangular mesh (Figure 12.15). The triangular faces themselves can represent the same data via the assignment of slope and direction values.

Regular triangular meshes, however, are rarely used for representation of this type of data. Irregular triangular meshes are used instead, although Bengtsson and Nordbeck (1964) have shown that the interpolation of isarithms or contours is much easier and more consistent given a regular mesh. Perhaps a contributing factor in the almost total lack of use of the regular triangular mesh for surface data is simply that such data are normally not captured in a regular spatial sampling pattern. An irregular triangular mesh has a number of other advantages, which will be discussed later in this paper.

In terms of processing efficiency on general procedures to compute spatial properties, such as area and centroid calculations, or to perform spatial manipulations, such as overlay and windowing, the algorithms initially devised for operation on square grids can easily be modified to work in the case of a triangular or hexagonal mesh. These, in fact, have the same order of computational complexity (Ahuja, 1983).

12.3.3.2 Nested Tessellation Models

Regular square and triangular meshes, as described above, can each be subdivided into smaller cells of the same shape, as shown in Figure 12.14. The critical difference between square, triangular and hexagonal tessellations on the plane is that only the square grid can be recursively subdivided with the areas of both the same shape and orientation. Triangles can be subdivided into other triangles, but the orientation problem remains. Hexagons cannot be subdivided into other hexagons, although the basic shape is approximated. These hexagonal 'rosettes' have ragged edges (Figure 12.13). Ahuja (1983) describes these geometrical differences in detail.

There are several very important advantages of a regular, recursive tessellation of the plane as a spatial data model. As a result, this particular type of data model is currently receiving a great deal of attention within the Computer Science community for a growing range of spatial data applications (Samet, 1984). The most studied and used of these models is the quadtree, based on the recursive decomposition of a grid (Figure 12.16).

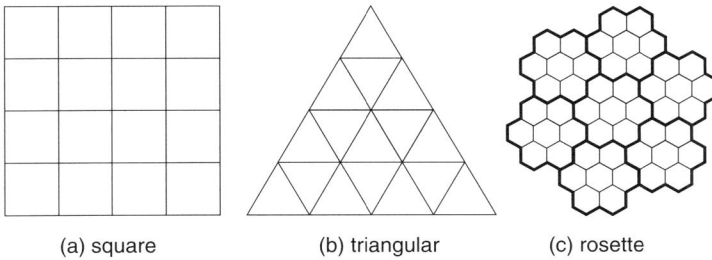

(a) square   (b) triangular   (c) rosette

Figure 12.14 The three regular tessellations in recursively subdivided form.

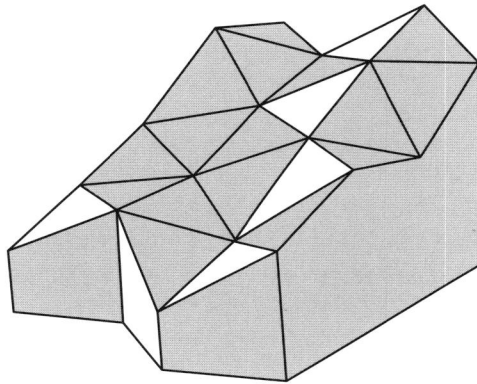

Figure 12.15 A regular triangulated network representing surface data (adapted from Bengtsson and Nordbeck, 1964).

The advantages of a quadtree model for geographical phenomena, in addition to the advantages of a basic standard model, include:

1. Recursive subdivision of space in this manner functionally results in a regular, balanced tree structure of degree four. This is a hierarchical, or tree, data model where each node has four sons. Tree storage and search techniques are two of the more thoroughly researched and better understood topics in computer science. Techniques are well documented for implementation of trees as a file structure, including compaction techniques and efficient addressing schemes.

2. In cartographic terms, this is a variable scale scheme based on powers of two and is compatible with conventional Cartesian coordinate systems. This means that scale changes between these built-in scales merely require retrieving stored data at a lower or higher level in the tree. Stored data at multiple scales also can be used to get around problems of automated map generalization. The obvious cost of these features, however, is increased storage volume.

3. The recursive subdivision facilitates physically distributed storage, and greatly facilitates browsing operations. Windowing, if designed to coincide with areas represented by quadtree cells, is also very efficient. These are features which are very advantageous for handling a very large database.

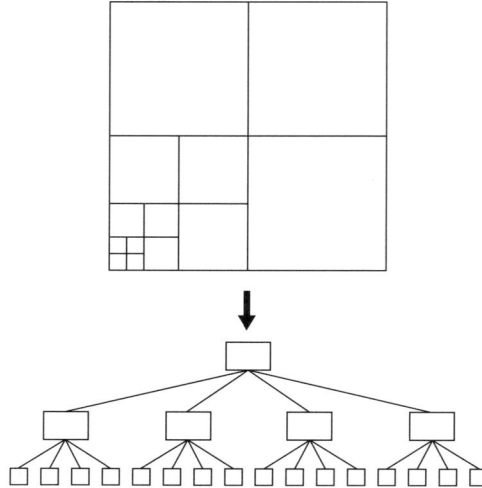

Figure 12.16   The quadtree data model.

Advantages 1 and 3 also hold for the other two types of tessellations, taking into consideration that a recursive hexagonal tessellation has a branching factor of seven instead of four. Although all recursive tessellations can be viewed as having the variable scale property, the triangular and hexagonal versions do not have direct compatibility with Cartesian coordinate systems.

The following is a brief discussion of the major types of quadtrees. A comprehensive discussion of quadtrees and all of its variant forms, as well as an extensive bibliography, has been given by Samet (1984).

Besides the general data model described above, the term quadtree has also acquired a generic meaning, to signify the entire class of hierarchical data structures based on the principle of recursive decomposition, many of which were developed in parallel. The 'true', or region quadtree, was first described within the context of a spatial data model by Klinger (1971; Klinger and Dyer, 1976), who used the term Q-tree. Hunter (1978) was the first to use the term quadtree in this context. Finkel and Bentley (1974) used a similar partition of space into rectangular quadrants. This model divides space based on the location of ordered points, rather than regular spatial decomposition (Figure 12.17). Although this was also originally termed a quadtree, it has become known as a point quadtree in order to avoid confusion. It is an adaptation of the binary search tree for two-dimensions (Knuth, 1973).

A data model related to the quadtree is the pyramid, which was developed within the field of image understanding (Tanimoto and Pavildis, 1975). A pyramid is an exponentially tapering stack of discrete arrays, each one $\frac{1}{4}$ the size of the previous

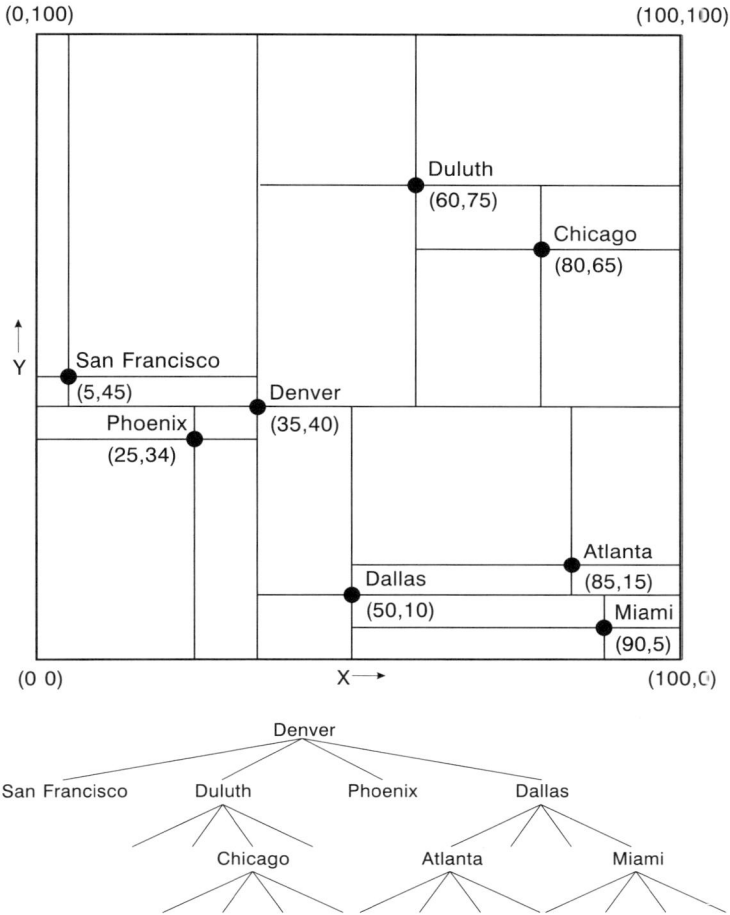

Figure 12.17   The point quadtree data model (from Samet, 1984).

without the explicit inter-level links of a tree structure. Because the pyramid does not have a strictly recursive structure, scales based on other than powers of two can be defined.

12.3.3.2.1 Area quadtrees   The quadtree concept and all derivative algorithms may be extended into multiple dimensions (Reddy and Rubin, 1978; Jackins and Tanimoto, 1980; Jackins and Tanimoto, 1983). The oct-tree (branching factor = 8) or three-dimensional quadtree is probably the best known of these. Individual quadtrees representing different classes of data can also be spatially registered to form multiple layers, as can be done in a gridded database. This is known as a 'forest' of quadtrees.

The recursive decomposition based on the hexagonal tessellation, or septrees (branching factor = 7), retains the deficiency that a hexagon cannot be subdivided into smaller hexagons. This means that the smallest hexagonal resolution unit in a given implementation must be predetermined. Conversely, higher-level resolution units formed by an aggregation of hexagons can only approximate a hexagon (Figure 12.14). Algorithms for septrees have been developed by Gibson and Lucas (1982). This work has capitalized on the radial symmetry of hexagonal tessellation by basing these procedures on a base 7 addressing scheme, which they named Generalized Balanced Ternary, or GBT (Figure 12.18). Vectors, distance measurements and several other procedures can be performed directly on the GBF addresses without conversion.

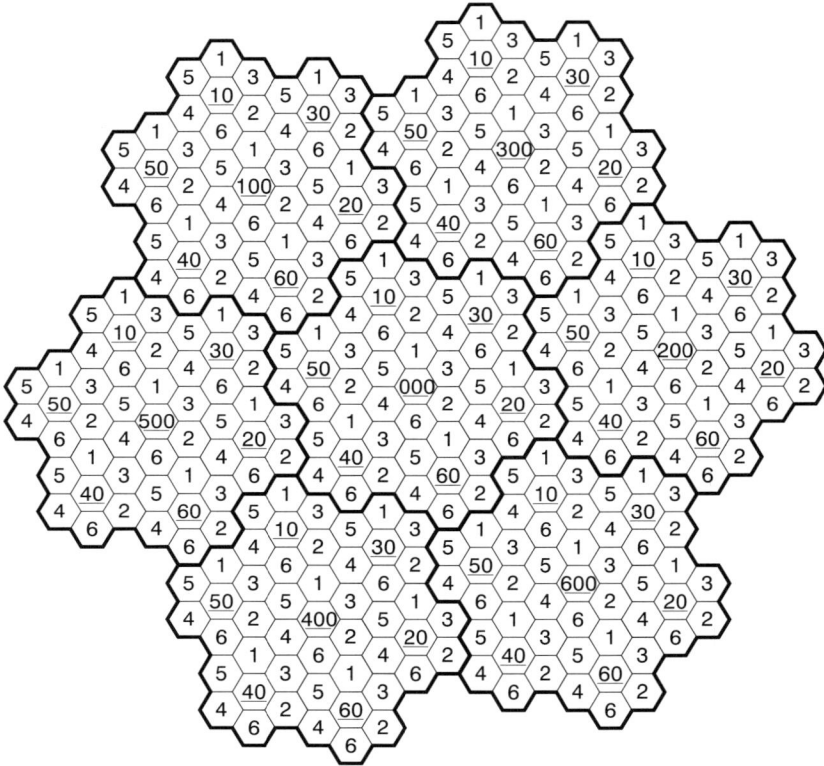

Figure 12.18   A nested hexagonal tessellation with a hierarchical, base 7 indexing scheme.

Recursive decomposition based on the triangular tessellation is the other alternative. This model is called a triangular quadtree, since each triangle is subdivided into four smaller triangles, yielding a tree with a branching factor of four. Again, this model retains all of the inherent advantages and disadvantages of the regular triangular tessellation with the added advantages associated with a hierarchical structure. Although a direct addressing scheme analogous to those for square and hexagonal tessellations is easily derived, such a scheme would not have any advantage in addition to allowing direct retrieval of individual stored data elements.

Generally, most developmental work on quadtree-type data models and associated algorithms has been based on classical tree storage and traversal techniques, which are based on pointers. The alternative of using direct addressing techniques has been explored by a number of researchers in addition to Lucas (Abel and Smith, 1983; Gargantini, 1982). To distinguish these from the pointer-based approach, they have been termed linear quadtrees. This term is derived from the fact that, by using direct addressing structures, the data can be physically organized in linear fashion; that is as a list.

12.3.3.2.2 Point quadtrees   As stated above, point quadtrees base the subdivision on the location of ordered data points rather than regular spatial decomposition. A point quadtree takes one data point as the root and divides the area into quadrants based on this point (Figure 12.17). This is done recursively for each ordered data point, resulting in a tree of degree 4. Since the arrangement of data points in the tree is determined by relative location amongst the points, yielding a regular data decomposition rather than a regular areal decomposition, they are useful in applications which involve search and nearest neighbour operations. One disadvantage to point quadtrees is that the shape of the tree is highly dependent on the order in which the points are added. Additions and deletions are therefore impossible except at the leaves of the tree.

A problem with multiple dimensions with any type of quadtree structure is that the branching factor becomes very large (i.e. 2k for k dimensions), which in turn would require much storage space. The k–d tree of Bentley is an improvement on the point quadtree by avoiding a large branching factor (Bentley, 1975). The k–d tree divides the area into two parts instead of four at each point, yielding a tree of degree 2 (Figure 12.19). The direction of this division is rotated amongst the coordinates for successive levels of the tree. Thus, in the two-dimensional case, the data space could be divided in the x direction at even levels and the y direction at odd levels.

12.3.3.3 Irregular Tessellations

There are a number of cases in which an irregular tessellation holds some advantages. The four most commonly used types for geographical data applications are square, triangular and variable (i.e. Thiessen) polygon meshes. The basic advantage of an irregular mesh is that the need for redundant data is eliminated and the structure of the mesh itself can be tailored to the areal distribution of the data. This scheme is a variable resolution model in the sense that the size and density of the elemental polygons vary over space.

An irregular mesh can be adjusted to reflect the density of data occurrences within each area of space. Thus, each cell can be defined as containing the same number of occurrences. The result is that cells become larger where data are sparse, and small where data are dense.

The fact that the size, shape and orientation of the cells is a reflection of the size, shape and orientation of the data elements themselves is also very useful for visual inspection of various types of analyses.

Perhaps the irregular tessellation most frequently used as a spatial data model is the triangulated irregular network (TIN) (Figure 12.20). TINs are a standard method of

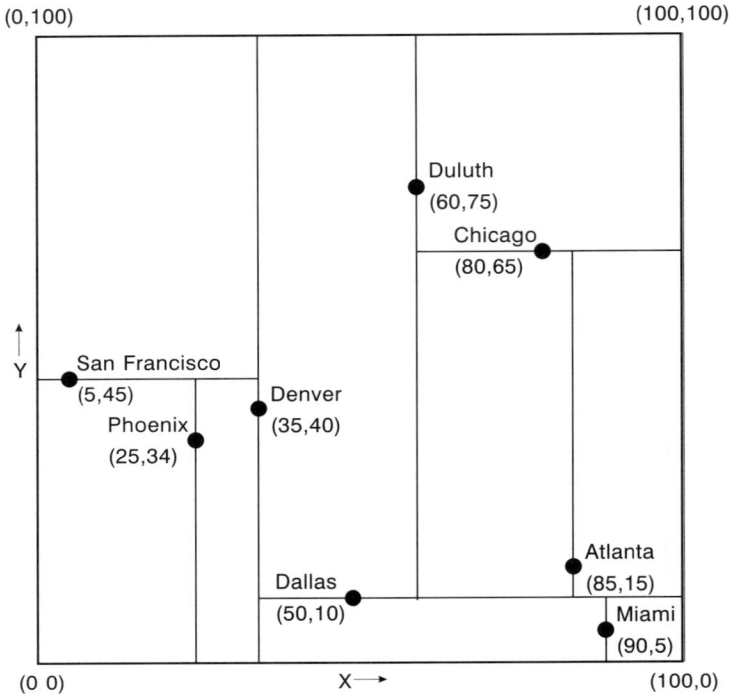

Figure 12.19 The k–d tree data model (from Samet, 1984).

representing terrain data for landform analysis, hill shading and hydrological applications. There are three primary reasons for this. Firstly, it avoids the 'saddle point
problem', which sometimes arises when drawing isopleths based on a square grid
(Mark, 1975). Secondly, it facilitates the calculation of slope and other terrain-specific
parameters. Thirdly, the data are normally recorded at points distributed irregularly in
space.

A major problem associated with irregular triangulated networks is that there are
many possible different triangulations which can be generated from the same point set.
There are thus also many different triangulation algorithms. Any triangulation

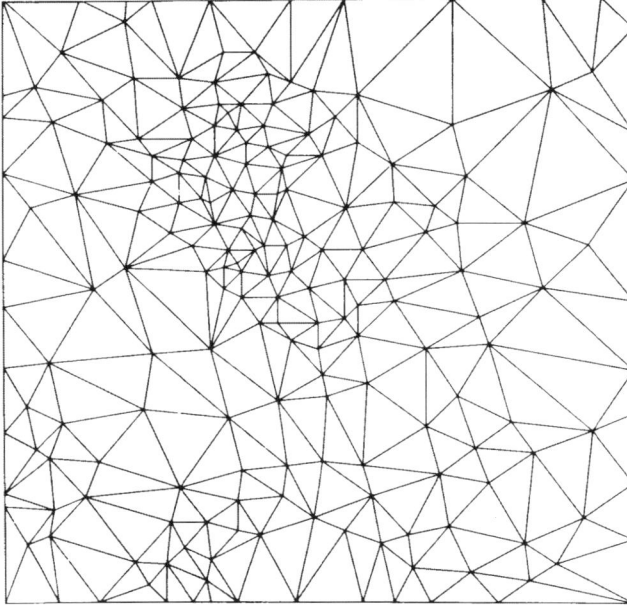

Figure 12.20 A Triangulated Irregular Network (TIN).

algorithm will also require significantly more time than subdivision of a regularly spaced point set.

Thiessen polygons, also called Voronoi diagrams or dirichlet tessellations, are the logical dual of the irregular triangulated mesh. Thiessen polygons are constructed by bisecting the side of each triangle at a 90° angle, the result, as shown in Figure 12.21, is an irregular polygonal mesh where the polygons are convex and have a variable number of sides.

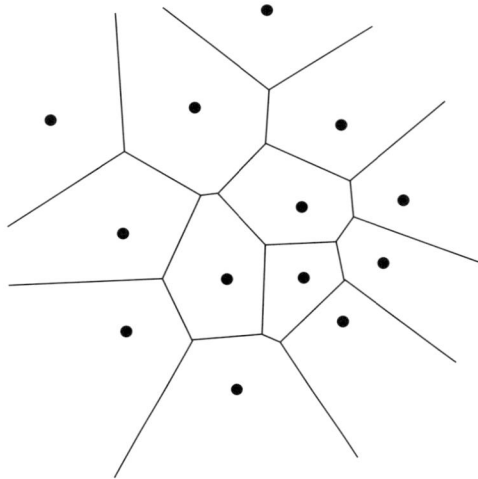

Figure 12.21   An example of Thiessen polygons.

Rhynsburger (1973) has described the following alternate logical derivation. Given a finite number of distinct points that are at least three in number and distributed in some manner on a bounded plane, each point begins to propagate a circle at a constant rate. This growth continues until the boundary of a circle encounters another circle or the boundary of the plane. The analytical derivation of Thiessen polygons has been studied by a number of people (Rhynsburger, 1973; Kopec, 1963; Shamoas, 1977).

Thiessen polygons are useful for efficient calculation in a range of adjacency, proximity and reachability analyses. These include closest point problems, smallest enclosing circle (Shamoas, 1977), the 'post office' problem (Knuth, 1973) and others.

The first documented practical application of Thiessen polygons was in the determination of recipitation averages over drainage basins by Thiessen (1911), for whom Thiessen polygons were later named (Rhynsburger, 1973). Two extensions of the basic concept have also been developed. The first of these is to assign a positive weight to each of the points which represents the point's power to influence its surrounding area, to produce a weighted Voronoi diagram.

This was described by Boots (1979) and has particular advantages for marketing and facility location siting problems. Drysdale and Lee (1978) have also generalized the Voronoi diagram to handle disjoint line segments, intersecting line segments, circles, polygons and other geometric figures.

Although it is seen that various irregular polygonal tessellations are each uniquely suited to a particular type of data and set of analytical procedures, they are very ill-suited for most other spatial manipulation and analytical tasks. For example, overlaying two irregular meshes is extremely difficult, at best. Generating irregular tessellations is also a complex and time consuming task. These two factors make irregular tessellations unsuitable as database data models except in a few specialized applications.

12.3.3.4 Scan-Line Models

The parallel scan-line model, or raster, is a special case of the square mesh. The critical difference with the parallel scan-line model is that the cells are organized into single, contiguous rows across the data surface, usually in the x direction, but do not necessarily have coherence in the other direction. This is often the result of some form of compaction, such as raster run-length encoding. This is a format commonly used by 'mass digitizing' devices, such as the Scitex drum scanner.

Although this model is more compact than the square grid, it has many limitations for processing. Algorithms which are linear or parallel in nature (i.e. input to a process to be performed on individual cells does not include results of the same process for neighbouring cells) can be performed on data in scan-line form with no extra computational burden in contrast to gridded data. This is because null cells (i.e. cells containing no data) must also be processed in the uncompacted, gridded form. Many procedures used in image processing fall into this category. Other processes which do depend upon neighbourhood effects, require that scan-line data be converted into grid form.

12.3.3.5 Peano Scans

A family of curves which generate a track through space in such a way that n-dimensional space is transformed into a line, and vice versa, was discovered in 1890 by the mathematician Giuseppe Peano (Peano, 1973). These curves, also known as space-filling curves, preserve some of the spatial associativity of the scanned data space on the single dimension formed by the scan. Figure 12.22 (left) shows an example of a simple two-dimensional Peano curve. With this particular version, all changes of direction are right angles. Figure 12.22 (right) shows a similar Peano scan in three-dimensions.

Peano scans possess several properties which can be useful in some spatial data handling applications. These were summarized by Stevens, Lehar and Preston (1983):

1. The unbroken curve passes once through every locational element in the dataspace.
2. Points close to each other in the curve are close to each other in space, and vice versa.
3. The curve acts as a transform to and from itself and n-dimensional space.

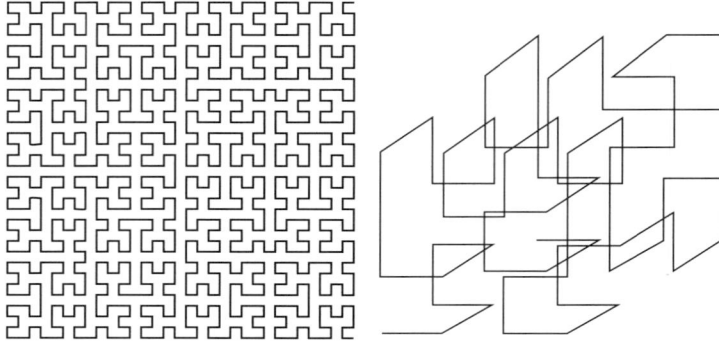

Figure 12.22 A right-angle Peano curve in two and three dimensions (from Stevens, Lehar and Preston, 1983).

The first known practical application of Peano curves as a digital geographic data model was as the areal indexing scheme within the Canada Geographic Information System (CGIS) (Tomlinson, 1973). This database divided gridded areal data into 'unit frames'. The frame size was determined for convenience of retrieval and processing. Each unit frame in the system was assigned a unique number starting at the origin of their coordinate system. From that point, frames are sequenced so that they fan out from the origin as frame number increases. This arrangement is shown in Figure 12.23 for the first 64 unit frames. This numbering scheme, named the Morton matrix after its designer (Morton, 1966), is in reality the trace of a Z-shaped (or N-shaped) Peano scan (Figure 12.24). This spatial indexing scheme uses the property that areas close together on the earth will likely have a minimum separation in a sequential digital file. This has the effect of reducing search time, especially for small areas. An additional benefit of this

	0	1	2	3	x →			
0	0 0000	2 0010	8 1000	10 1010	32	34	40	42
1	1 0001	3 0011	9 1001	11 1011	33	35	41	43
2	4 0100	6 0110	12 1100	14 1110	36	38	44	46
3	5 0101	7 0111	13 1101	15 1111	37	39	45	47
y	16	18	24	26	48	50	56	58
↓	17	19	25	27	49	51	57	59
	20	22	28	30	52	54	60	62
	21	23	29	31	53	55	61	63

Figure 12.23 The Morton matrix indexing scheme.

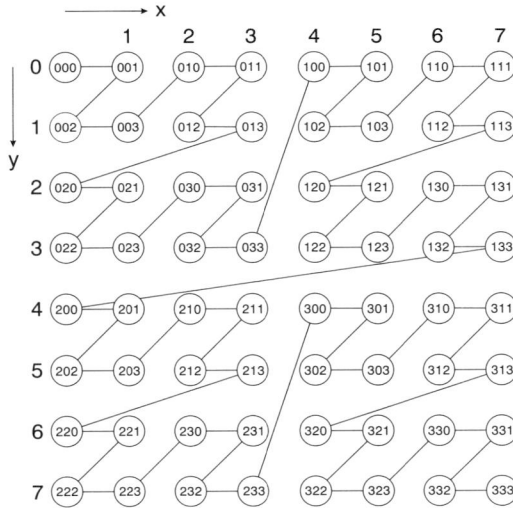

Figure 12.24 The relationship between the Morton matrix indexing scheme and the Z-shaped Peano curve.

addressing structure is that Morton matrix addresses can be directly computed by interleaving the binary representation of the geographic x and y coordinates (Figure 12.25). This addressing scheme was examined more fully by White (1981) and Tropf and Herzog (1981). This work was built upon the recursive properties of Peano curves and the direct correspondence between the Z-shaped Peano curve and quadtree structures. It should also be noted that the quadtree addressing scheme used by Abel and Smith (1983) and Mark and Lauzon has one important difference from the Morton matrix scheme. The Morton matrix scheme is level specific. In other words, the quadtree level cannot be determined from the address code itself. This level must

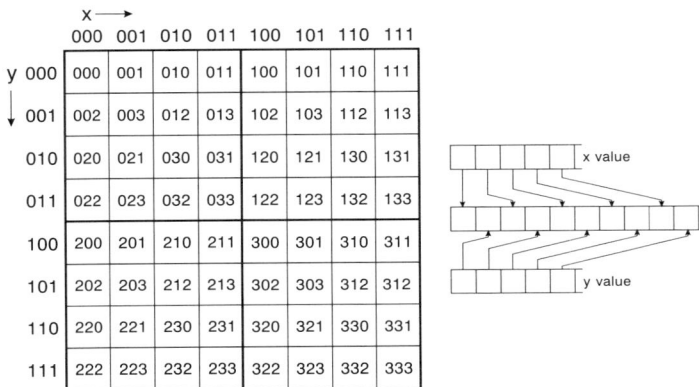

Figure 12.25 Bit-wise interlaced indexing scheme.

be explicitly indicated by an additional code. This is not the case with the scheme used by Abel and Smith, in which the digits of the addresses correspond to the levels of the quadtree model (Figure 12.26).

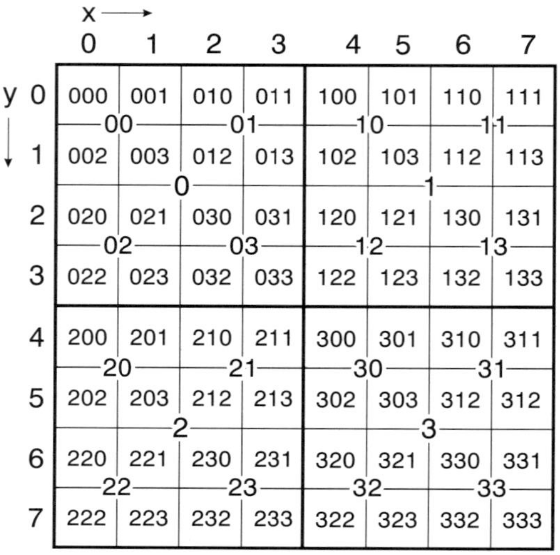

x →

y	0	1	2	3	4	5	6	7
0	000	001	010	011	100	101	110	111
	—00—		—01—		—10—		—11—	
1	002	003	012	013	102	103	112	113
	—0—				—1—			
2	020	021	030	031	120	121	130	131
	—02—		—03—		—12—		—13—	
3	022	023	032	033	122	123	132	133
4	200	201	210	211	300	301	310	311
	—20—		—21—		—30—		—31—	
5	202	203	212	213	302	303	312	312
	—2—				—3—			
6	220	221	230	231	320	321	330	331
	—22—		—23—		—32—		—33—	
7	222	223	232	233	322	323	332	333

Figure 12.26   Hierarchical, base 5 quadtree indexing scheme.

The properties of Peano curves have also been shown to have significant utility for image processing applications (Stevens, Lehar and Preston, 1983). These applications include data compression in the spatial and spectral domain, histogram equalization, adaptive thresholding and multispectral image display. The reason for this is that, as stated above, these and many other techniques for manipulating and analysing imagery data are sequential and single channel operations, that is they are linear in nature. Peano curves can thus be used to collapse these multidimensional data into a single dimension. The property of preserving some of the spatial relationships in the one-dimensional Peano scan data allows improved interpretation, and thus improved results, from these procedures.

12.3.4 Relative Merits

In summary, each of the three basic types of data model (paper map, vector, tessellation) has advantages and disadvantages which are inherent in the model itself. Individual models, such as the ones discussed above, can overcome these only to a limited degree, and always only by some sort of trade-off. Vector data models are direct digital translations of the lines on a paper map. This means that the algorithms also tend to be direct translations of traditional manual methods. The repertoire of vector-mode algorithms is thus both well developed and familiar. The primary drawback of vector-type data models is that spatial relationships must be either explicitly recorded or computed. Since there is an infinite number of potential spatial relationships, this means that the essential relationships needed for a particular application or range of applications must be anticipated.

Conversely, spatial interrelationships are 'built-in' for regular, tessellation-type data models. Grid and raster data models are also compatible with modern high-speed graphic input and output devices. The primary drawback is that they tend to be not very compact. Regular tessellations tend to force the storage of redundant data values. Redundant data values can be avoided by the use of a wide variety of compaction techniques. Another drawback is that the algorithm repertoire is less fully developed. It is assumed that this latter drawback will diminish or disappear as the current increase in the use of raster and other tessellation-type models continues (Peuquet, 1979).

From a modelling perspective, vector and tessellation data models are logical duals of each other. The basic logical component of a vector model is a spatial entity, which may be identifiable on the ground or created with the context of a particular application. These may thus include lakes, rivers, roads and entities, such as 'the 20-foot contour level'. The spatial organization of these objects is explicitly stored as attributes of these objects. Conversely, the basic logical component of a tessellation model is a location in space. The existence of a given object at that location is explicitly stored as a locational attribute.

From this perspective, one can clearly see that neither type of data model is intrinsically a better representation of space. The representational and algorithmic advantages of each are data and application dependent, even though both theoretically have the capability to accommodate any type of data or procedure.

12.4 Recent Developments in Spatial Data Models

12.4.1 The Problems of Very Large Databases and
Data Interchange

As stated in the beginning of this paper, current and anticipated spatial data volumes have generated a two-faceted problem:

1. existing data structures are too inefficient and inflexible to meet current requirements; and

2. format conversions between different data structures to satisfy the current range of required applications produces significant processing overhead.

The rate of increase in data volumes and demands for fast performance has meant that storage and speed advances in computing hardware technology can no longer be relied upon to provide a cost-effective solution. Even if this brute-force approach were economically feasible, the amount of inefficiency often present with 'traditional' data structures represents unnecessary overhead which may, to a large degree, be alleviated by further developing our knowledge of spatial data models. This would at least result in reduced overall costs for spatial data handling on a practical level, and a cleaner solution from a theoretical standpoint.

As the size of any database becomes very large, several important problems arise which must be dealt with:

1. efficiency,
2. heterogeneity,
3. accuracy and security.

Dealing with these problems in order to maintain a functional database is critical, since a large database always represents a large investment of time and resources, and often is an integral part of the day-to-day operation of the owner organization.

Inefficiencies, even major and obvious ones, can often be tolerated if the database is small or infrequently used. In these cases, it is often more cost-effective to absorb the extra time and computing costs than to bear the expense of careful initial construction or retroactive fixing of a system. For any type of large database, however, overall space and time efficiency becomes a critical factor. Even in a governmental context where internal and external use of the database often is not expected to be self-supporting, inefficiencies in a large and frequently used system tend to multiply into a major drain of resources.

The problem in obtaining efficiency is that, as stated in the beginning of this paper, the current state-of-the-art does not allow optimally efficient spatial databases to be built in a predictable manner. Very little is known about the performance characteristics of many individual spatial data models and algorithms. Even less is known about how to combine groups of algorithms and data models in a complex system for optimal performance.

The problem of data heterogeneity becomes a frequent and major consideration in dealing with large databases and is the cause of most data interchange problems. The usual need in the earth sciences is to combine different types of data, from various organizational sources, captured through varying equipment and techniques, for varying purposes and to varying quality standards and resolutions.

Combining different types of data with different spatial resolutions can be achieved through the use of some of the new spatial data modelling approaches, such as quadtrees. Dealing with variability with the other factors can only be done via coordination and standardization amongst the various data capture organizations and user groups. The overall accuracy and error characteristics of a database resulting from the combination of a number of data layers of differing error characteristics overlaid on top of each other is little understood, and most often impossible to calculate. What is understood, however, is that the overall error rate is multiplicative rather than additive. This problem could be significantly eased through improved documentation and coordination.

12.4.2 Hybrid Vector/Tessellation Models

One approach to the storage and processing trade-offs between tessellation and vector data structures is to store the spatial data in (usually) raster or grid form, perhaps with only minor modification from its raw scanner raster output form. The data are then converted to vector format when advantageous for performing a given analytic or manipulative process. Frequently, the result is then converted back again for graphic output. This conversion approach is the most commonly used because it is conceptually so straightforward. What is soon discovered, however, is that these data structure conversions can quickly become a bottleneck within a system as the volume of data and frequency of use increases (Peuquet, 1981a, 1981b). Tessellation-to-vector conversion requires some type of intricate line-following procedure, because cartographic lines are characteristically both convoluted and topologically complex. These conversion procedures represent significant system overhead, which must be avoided, or at least minimized.

Another approach is to develop new tessellation or specifically raster-orientated algorithms for processes which currently have only vector-orientated solutions. Theoretically, this could eventually eliminate the need for vector data structures and result in the development of exclusively tessellation-orientated geographic information systems. This would be particularly desirable in applications where raster-formatted areal data, such as LANDSAT imagery, is used in conjunction with map line data.

A wide variety of efficient raster-orientated data analysis and manipulation procedures have already been developed, primarily within the field of image processing (Peuquet, 1979). However, some spatial analytical processes seem to be intrinsically sequential or vector-orientated and cannot be restated in a parallel manner (Lee, 1980). These include some commonly performed analytical procedures, such as network shortest-path and optimal-routing problems. Other processes may prove to be so much more efficient when performed in vector mode that the additional time overhead for data structure conversion would be cancelled out. For these large-volume raster-based systems, the high overhead of 'forced' vectorization of data could thus be greatly reduced but not totally eliminated in a raster-orientated geographic information system if any intrinsically vector-orientated procedures are required by the user.

A possible solution to this dilemma is the development and use of hybrid types of spatial data models for geographic data which incorporate characteristics of both structures. The Vaster data model outlined below is the first such hybrid model and was developed by Peuquet (1981a, 1981b). This was developed as a raster–vector hybrid because of the prevalence of this particular data model dilemma, although it could be easily modified for hexagonal or triangular tessellations.

12.4.2.1 The Vaster Data Model

The basic logical unit of the raster-type model is the scan line, whereas a map line segment is the basic logical unit of the vector-type model. A map line segment is customarily defined to be a single uninterrupted map line. A line interruption is defined as the occurrence of the end of a map line, or the intersection with another line or the map boundary.

In contrast, the basic logical unit of the Vaster structure is the swath. Each swath spans a constant, known range over y and would correspond to a group of contiguous scan lines if the data were organised in raster format. Each swath contains a raster component and a vector component, as depicted in Figure 12.27. Both components are recorded at the same grid resolution. The leading edge of each swath (minimum y value) is recorded in raster format as a single scan line and functions as the index record for the swath, containing an identifier and x-coordinate for each map line intercept. The encoding used is similar to the structure developed by Merrill (1973). The raster encoding scheme used in the Vaster structure, however, contains all map line intersections within the same record. This is done to allow for efficient linkages between the raster and vector portions of the swath and to allow types of line structures other than nested polygons. The data contained in the remainder of the swath are recorded in vector format. All vectors contained in each swath are arranged in scan line intercept sequence in order of ascending x; polygons internal to a swath are listed separately. Each line intercept noted in the index record functions as the endpoint of each vector line segment within the swath, in sequence. Note that there is no scan line record at the end of the swath, since the next scan line record is functionally the leading edge of the next swath. In digital

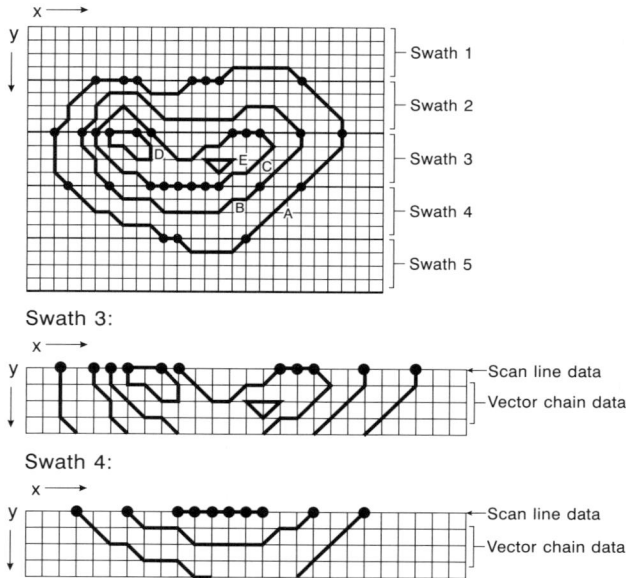

Figure 12.27   (top) Vaster Organization (logical record-swath); (bottom) Individual swaths.

form, the raster record in each swath contains an ordered sequence of line identifier–x-coordinate pairs.

Each map line in the remainder of the swath, as shown in Figure 12.28, is recorded in a chain code notation tailored to fit the special requirements of the hybrid Vaster structure.

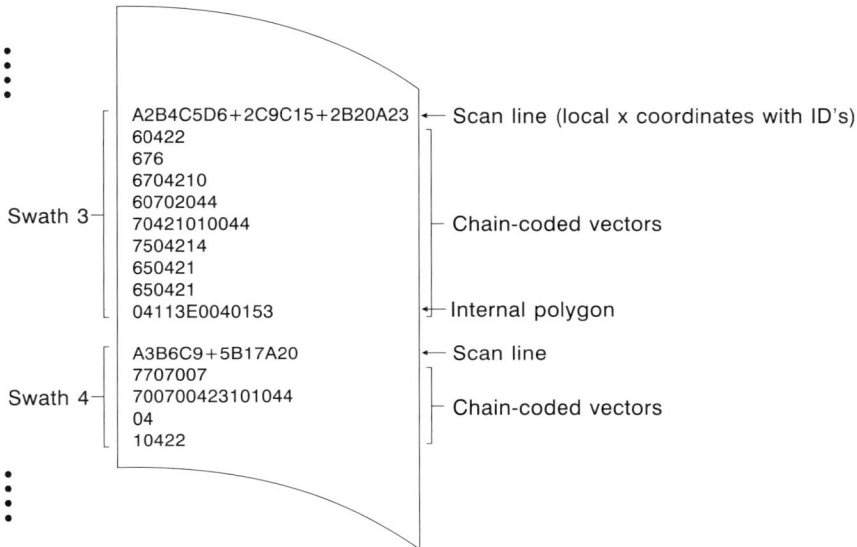

Figure 12.28 Vaster-formatted data in digital form.

Given the presence of raster-formatted data within the Vaster data structure and with the remainder of the data locationally keyed to these raster-formatted positions, the spatial ordering and preservation of spatial relationships inherent in the raster structure is retained to a significant degree. In addition, the presence of both raster- and vector-formatted data in the Vaster structure offers retrieval and processing advantages for cartographic database applications. Functionally, the Vaster format contains a two-level hierarchy. The raster portion of swaths may be used alone, without need to reference the more detailed vector file for many common processes or portions of processes, thereby speeding up performance considerably. A significant advantage of the use of this hybrid data structure for a very large database is the ability to make use of only the raster file as a set of generalized data for quick browsing or sampling of the database. This separation of use requires that the raster portion of the database be stored as a separate file in order to maximize speed in accessing the raster data as well as for more efficient data retrieval based on spatial criteria.

Many queries could use only the coarser, raster structured portion of the database where approximate solutions are desired, as may be the case in many centroid, area, perimeter or arc-length calculations. Most spatial relation queries, such as point inclusion and relative position, could also be answered from raster-formatted data, using raster-orientated algorithms, which are generally more efficient for this type of task because they ignore the bulk of the data. These algorithms would operate in the conventional manner, as described by Merrill (1973). The margin of error between approximate measures derived from the raster data alone is a function of both the width of the swath (i.e. the sampling frequency) and the sinuosity of the map lines themselves. If this margin of error is unacceptable in the case of spatial relation queries, only a small portion of the vector-formatted data need be referenced. For example, to determine containment of a point within an enclosed polygon where the y-coordinate of the given point falls between two scan-line records (that is, within a swath), then only the chain code boundaries of the particular polygon in question within the pertinent swath is converted to raster format so that the Jacobsen Plumbline algorithm can be performed using the same y value as the given point (Merrill, 1973).

There are other processes for which raster data organization will increase the efficiency of vector algorithms given the preservation of spatial ordering in the Vaster format. The two foremost examples are mosaicking and overlay of Vaster-coded data when these procedures are desired at full resolution. The approach to both of these processes is to first perform mosaicking or overlay on the raster-formatted portion of the data in each swath. This serves to 'align' the remaining chain coded short line segments for easier vector-mode processing.

The Vaster structure may also be used as a full detail, vector, chain coded file. Any algorithms applicable to vector chain codes can be used in the normal manner after reconnecting the short chain segments of individual swaths by reference to the index (scan-line) records. Speed is thus maximized when a Vaster-formatted spatial data base is used for orientation, and approximate information over large areas and more detailed analyses, based on chain coded vectors, are reserved for relatively small areas.

The Vaster data format would seem to provide a 'best of both worlds' solution to the efficient storage and handling of spatial data. Large volumes of map data can be digitized via scanners, converted only part-way to vector format, and then be used by a wide variety of both raster-orientated and vector-orientated algorithms in the conventional manner. Conversion to either full raster or vector format would require much less time than conventional raster-to-vector and vector-to-raster procedures. Overlay with data stored in raster format, such as LANDSAT imagery, is thus greatly facilitated.

The price paid for this, however, is the problem of determining the optimum data storage resolution. The problem has two related aspects. The first is a data sampling problem analogous to the grid cell size problem. How narrow must the swaths be to provide a satisfactory interval between raster-formatted records for the first-level data resolution? Each significant map line should be intersected at least once by a scan line, thus avoiding occurrences of map entities which are completely contained within the vector portion of a single swath. Closely spaced scan lines (i.e. narrow swaths relative to map line density and sinuosity), while avoiding such occurrences, cause multiple scan-line intersections of the same continuous map line, which can inflate data volumes.

A second aspect of the problem arises from the hybrid nature of the Vaster structure. What is the raster–vector data volume ratio which would provide optimum overall performance for a given group of raster- and vector-orientated algorithms for a given task? It is the joint use of rasters and vectors which takes this question a level beyond that of a standard system optimization problem. The two aspects of this complex problem are also highly interrelated.

12.4.2.2 Strip Tree Model

Strip trees, as developed by Ballard (1981), is a method for representing map vectors by means of a hierarchy of bounding rectangles (Figure 12.29). This is also a hybrid data model, since the basic logical entity of the model is the cartographic line (i.e. vectors), but the lines themselves are not explicitly recorded. This representation is designed to allow such operations as union, intersection and length of curves to be performed efficiently.

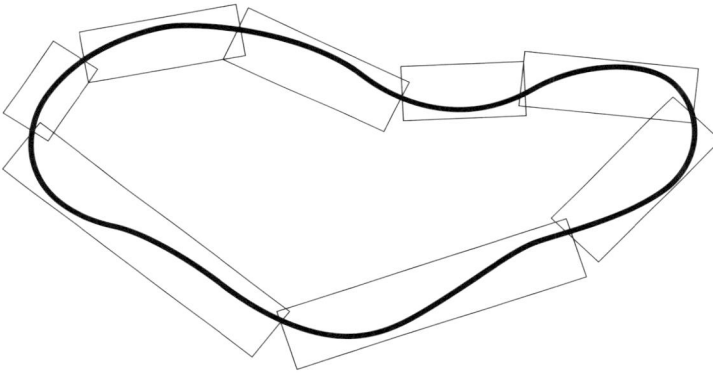

Figure 12.29   The strip tree model.

The representation of curves consists of a binary tree structure where lower levels in the tree correspond to finer resolutions (Figure 12.30). The tree structure was derived from using Duda and Hart's (1973) method for digitizing lines and retaining all intermediate steps in the digitizing process.

The idea of representing a cartographic line by a hierarchy of strips was first presented by Peucker (1976). He was able to use this data model to find line intersection and point-in-polygon. Burton used a similar but more general form where curves are divided into a hierarchy of bounding rectangles of a single orientation. These are then represented in a binary tree hierarchy (Burton, 1977).

As defined by Ballard, an individual strip is defined to consist of four elements, E1, E2, W1, and W2, where E1 (location XE 1 YE2) and E2 (location XE2 YE2) denote the beginning and end of the strip, respectively, and W1 and W2 denote the right and left distances of the strip borders from the directed line segment. These definitions are depicted in Figure 12.30.

A sequence of nested strips between the endpoints of a line is derived by finding the smallest bounding rectangle between those two points. Next, a point is picked which

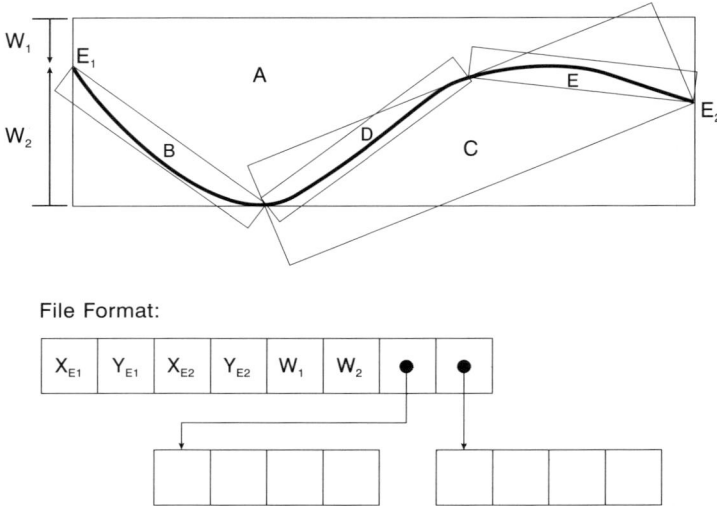

Figure 12.30   Hierarchical structure of strip tree model.

touches one of the two sides of the rectangle and the process is repeated for each of the two sublists (Figure 12.30). This results in two subtrees which are sons of the root node. The process terminates when the lowest level strips have a width equal to, or less than, a given resolution. The data record for each strip thus contains the x and y coordinates for each of the endpoints, W1, W2, and the pointers to the two descendant strips.

The binary tree resulting from this process is called a strip tree. Other operations which can be performed on strip trees include testing the proximity of a point, displaying a curve at different resolutions, intersecting two trees, the union of two trees, point-in-polygon, and area union and intersection.

This model, thus, could be viewed as the vector counterpart to quadtrees, since each exploits a hierarchical structure based on squares or rectangles with similar operational advantages. The primary difference in this respect between quadtrees and strip trees is that in order to do many of these operations, such as overlay, the quadtrees must be spatially registered. This is not the case with strip trees. Strip trees can be arbitrarily translated and scaled since they can be defined in terms of points which are grid independent.

12.5 Future Developments in Spatial Data Handling

12.5.1 Global Databases

A large-area database which covers, for example, all of the United States, or all of North America, can adjust for the curvature of the earth by conversion of the pertinent portion of the database to the appropriate map projection. Thus, a data file stored in latitude–longitude coordinates can be converted to, say, an Albers Conic Equal-Area if the data for the United States plotted on a small-scale map are desired. Equal-area projection coordinates are also necessary if large-area areal calculations are to be performed.

Vector-type data models, employing a spherical coordinate system such as latitude–longitude, pose no special problems on a global scale. The overwhelming proportion of geographic data being generated, however, is grid or raster based, such as LANDSAT imagery.

It is, unfortunately, topologically impossible to cover any spherical surface, such as the earth, with a square or rectangular grid. Although it is possible to convert stored coordinates for a portion of the database to an appropriate projection for analysis or map display, the real problem is to design a single, integrated global data model for representation of the database. The model should not have any spatial gaps or overlaps on the globe. For a tessellation-type model, it is assumed that a regular tessellation of an equilateral polyhedron is desired. Given these general requirements, we are limited to five shapes. This fact was known to the ancient Greeks and its proof may be found in any textbook on solid geometry. These five shapes are called the platonic solids, having been discussed by Plato. They consist of: (1) the regular tetrahedron, the four faces of which are equilateral triangles; (2) the regular hexahedron, or cubic; (3) the regular octahedron, the eight faces of which are equilateral triangles; (4) the regular dodecahedron, the twelve faces of which are regular pentagons; and (5) the regular icosahedron, the twenty faces of which are equilateral triangles (Fisher and Miller, 1944; Coxeter, 1973) (Figure 12.31).

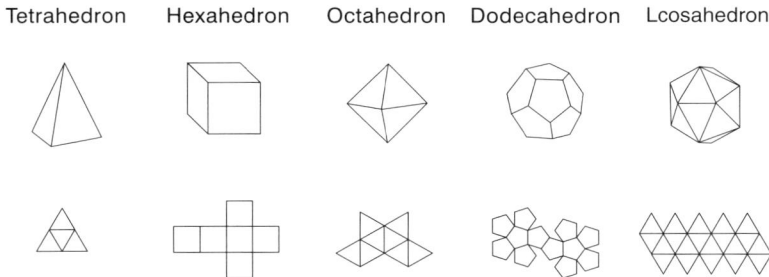

Figure 12.31 The five Platonic solids.

For purposes of a global data model, the dodecahedron is not a good choice because the pentagons cannot be recursively subdivided into other pentagons. The tetrahedron and hexahedron have faces which are too large in relation to the globe and would introduce significant areal and shape distortions. With a choice between the octahedron and icosahedron, the latter is the obvious choice; they both have triangular faces, but the individual flat faces of the icosahedron are smaller and would cause less distortion in representing portions of a spherical surface.

Using the 20 faces of the icosahedron, continued recursive subdivision into smaller triangles provides less and less distortion. Fuller and Sadao (1982) have published a projection based on this polyhedron, termed the 'dymaxion' projection. This projection is not widely used and is regarded by cartography professionals as a curiosity. The primary reason for this is that when laid out on a flat piece of paper, it presents a very distorted view of the earth because of the pattern of interruptions. Nevertheless, when viewed as a 'virtual map' in computer storage where visual perceptions are no longer of consequence, the icosahedron has a number of useful properties (Dutton, 1983; Fuller and Sadao, 1982). These present intriguing possibilities for use as a general purpose global data model. A regular spherical tessellation retains all of the desirable properties of a planar tessellation including implicit spatial relationships; geographic location is implied by location in the database, variable scale and rapid search.

The major drawback of this model for a high-resolution global database is that the number of triangles at each successive subdivision would increase geometrically, producing a data volume explosion of unmanageable size. Upon closer inspection, however, it is noticed that the size of each triangle must also decrease geometrically with successive subdivisions. The result is that small triangles are achieved after a surprisingly few subdivisions. For example, after eleven subdivisions (i.e. level twelve of a triangular quadtree) each triangle has an edge length of only 17.2 miles. This yields a cumulative total of 1 417 176 faces (Dutton, 1983).

The global triangular tessellation has other interesting properties as a global cartographic projection. When projected on a sphere, the original twenty triangles remain equal-area and equilateral. The edges of these triangles also form a set of great circle arcs. However, there is shape distortion from the original planar triangles created. A measure of the variations in linear scale can be calculated by comparing the projected lengths of the equal great circle arc segments with the lengths of those segments before projection. Also, the vertices of the twenty triangles provide a measure of angular distortion or departure from conformality. Since, as a regular network, these triangles are evenly distributed over the surface of the globe, it allows a fair comparison to be made between different parts of the same projection (Fisher and Miller, 1944).

A further problem is that subdivision into smaller triangles on the sphere does not produce equal-area triangles. Nevertheless, the methods for calculating the exact size of any individual triangle at any position and level in a recursive hierarchy, as well as the length of its sides and the angles of its vertices, are well known from the field of spherical trigonometry. By being selective in further subdividing the sphere based on the 30 edges of the icosahedron, there are some unique possibilities. If six great circle arcs are drawn from the centres of each of the original twenty triangles, three passing through the mid point of each edge and three to each corner, the mesh is equally subdivided into 120 right-angled triangles (Figure 12.32). This complicated pattern would include, amongst others, a pattern of thirty equal quadrilaterals and a series of overlapping hexagons, in addition to the original twenty triangles. The possibilities of this as a compound tessellation are tantalizing as a best-of-all-worlds approach, but the operational geometry is complex.

12.5.2 Space–Time Data

By far, the major emphasis in the development of geographic data models has been on two-dimensional models. This was a result of many factors including: (1) demand;

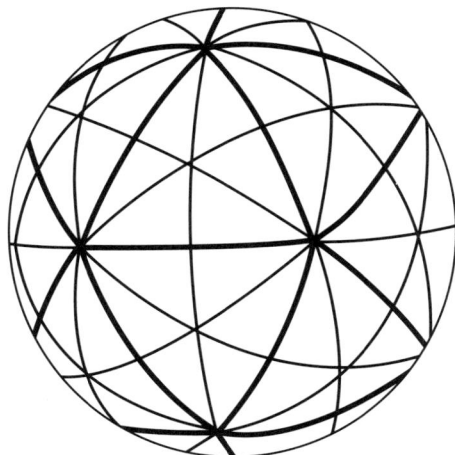

Figure 12.32 Fifteen symmetrically arranged great circles on the sphere.

(2) the state-of-the-art in analytical techniques in a number of fields; (3) limitations in computing capacity; and (4) the state-of-the-art in data modelling techniques. A significant amount of work on multidimensional models for specialized applications, such as 3-D TIN models for terrain analysis, has also been done. Only within the past few years has there been broad-based attention on the development and techniques for handling of time-series data. On the surface, it would seem that extending two-dimensional and three-dimensional models into the fourth dimension should be a straightforward process. This turns out not to be true in practice, however.

Using the example of spacecraft data, an individual instrument collects data at frequent time intervals, and often over the span of years. This type of time-series data collection generates enormous data volumes. The problem encountered in this area derives from two factors. (1) Time is different in nature in comparison with the space dimensions. It is entropic and runs in only one direction. (2) Very little is known about the entities and processes portrayed, particularly for non-earth data. Inferences, therefore, cannot be made from 'similar' occurrences in other locations. These two factors make compression via sampling of the time dimension without reducing or obscuring the information contained in the data virtually impossible. Since valid time-series data have been virtually non-existent in many fields until recently, detailed analyses of how various phenomena and their interrelationships change over time could not be made. Compression can, therefore, not be accomplished by exploiting derived or previously known statistical dependencies that exist between separate time samples. Compression by discarding data which would not be of interest to the user is also difficult because of lack of experience with this type of data. A primary means of analysis, such as was the case for the *Voyager I* flyby, is, therefore, to graphically portray complete sequences of time-series data as 'movies' to be subjected to expert visual interpretation. Methods for digitally storing and analysing space–time data are thus an area currently in need of further investigation.

12.6 Summary and Conclusions

As stated in the introduction of this paper, geographic databases currently in existence are experiencing severe problems of inefficiency, in terms of both compaction and speed of use, as well as rigidity and narrowness in the range of applications and data types which can be supported by a single database. These efficiency, versatility and integration problems can be traced in large part to the profound differences in the storage formats commonly used for spatial data handling. The basic problem, however, is a lack of understanding of the nature of spatial data, and a lack of a unified body of knowledge of the design and evaluation of spatial data structures.

This problem is particularly critical for many national-level governmental agencies worldwide, since they collect and use very large volumes of data of a wide range of types, from diverse sources, with varying accuracy, resolution and format characteristics as a basic and essential part of their function. These data sets must also frequently satisfy a wide range of scientific applications, the data needs for which are simultaneously vague and changing over time. Satisfying the unexpected request in this situation becomes the rule rather than the exception.

This paper has shown, through an examination of current knowledge and spatial data models in a comprehensive framework, that major advances in the performance of spatial data models and geographic information systems are both possible and probable within the next few years, which will allow us to meet these efficiency, versatility and integration needs.

The taxonomy presented here focused primarily on two-dimensional spatial data models. Extension of these models to three dimensions is straightforward. However, as discussed above, this does not carry through to the fourth dimension, time. This is the one area currently identified in spatial data models and computer spatial data handling where we have barely scratched the surface.

The taxonomy given in this paper has provided clarification of how varying data models are conceptually interrelated. Many of the gaps in the options available have been filled by very recent research, particularly in the areas of nested tessellations and hybrid data models, and by recent recognition of long known principles that were developed in other disciplines. This taxonomy can thus provide a framework for future work in the systematic analysis and comparison of performance characteristics of different types of spatial data models.

This knowledge on performance characteristics can serve as guidelines for building geographic information systems in the future in a more predictable and systematic manner. Performance characteristics can also serve as a guide for direction in future refinements and developments in spatial data structures and algorithms by revealing deficiencies.

References

Abel, D.J. and Smith, J.L. (1983) A data structure and algorithm based on a linear key for a rectangle retrieval problem. *Computer Vision, Graphics Image Processing*, **24**, 4–14.

Ahuja, N. (1983) On approaches to polygonal decomposition for hierarchical image decomposition. *Computer Vision, Graphics Image Processing*, **24**, 200–214.

Ballard, D. (1981) Strip trees: a hierarchical representation for curves. *Communications of the ACM*, **24**, 310–321.

Bengtsson, B. and Nordbeck, S. (1964) Construction of isarithms and isarithmic maps by computers. *Nordisk TidsckriftforInformations-Behandling*, **4**, 87–105.

Bentley, J.L. (1975) Multidimensional search trees used for associative searching. *Communications of the ACM*, **18**, 509–517.

Board, C. (1967) Maps as models, in *Models in Geography* (eds P. Haggett) Methuen, London, pp. 671–725.

Boots, B.N. (1979) Weighting Thiessen polygons. *Economic Geography*, **56**(3), 248–259.

Bouillé, F. (1978) Structuring cartographic data and spatial processing with the hypergraph-based data structure. *Proceedings, First International Symposium on Topological Data Structures for Geographic Information Systems*, Harvard University, Cambridge, MA.

Burton, W. (1977) Representation of many-sided polygons and polygonal lines for rapid processing. *Communications of the ACM*, **20**, 166–171.

Cederberg, R. (1979) Chain-link coding and segmentation for raster scan devices. *Computer Graphics Image Processing*, **10**, 224–234.

Chakravarty, I. (1981) A single-pass, chain generating algorithm for region boundaries. *Computer Graphics Image Processing*, **15**, 182–193.

Codd, E.F. (1970) A relational model of data for large shared data banks. *Communications of the ACM*, **13**, 378–387.

Codd, E.F. (1981) Data models in database management. *Proceedings, Workshop on Data Abstraction, Databases and Conceptual Modelling*.

Coxeter, H.S.M. (1973) *Regular Polytopes*, Dover Publications, New York.

Dacey, M. and Marble, D. (1965) Some comments on certain technical aspects of geographic information systems. *Technical Report No. 2 of ONR Task No. 389-142* (Office of Naval Research, Geography Branch).

Dangermond, J. (1982) A classification of software components commonly used in geographic information systems. *Proceedings, US–Australia Workshop on the Design and Implementation of Computer-Based Geographic Information Systems*, 70–91.

Date, C.J. (1983) *An Introduction to Database Systems*, 11th edn, Addison-Wesley, Reading, MA.

Drysdale, R. and Lee, D. (1978) Generalized Voronoi diagram in the plane. *Proceedings, 16th Annual Allerton Conference on Communications Control and Computers*, 833–839.

Duda, R.C. and Hart, P.E. (1973) *Pattern Classification and Scene Analysis*, Wiley-Interscience, New York.

Dutton, J. (1983) Geodesic modelling of planetary relief. *Proceedings of Auto-Carto VI*, 186–201.

Finkel, R.A. and Bentley, J.L. (1974) Quad trees: a data structure for retrieval on composite keys. *Acta Informatica*, **4**, 1–9.

Fisher, I. and Miller, O.M. (1944) *World Maps and Globes*, Essential Books, New York.

Freeman, H. (1974) Computer processing of line-drawing images. *Computing Surveys*, **6**, 57–97.

Freeman, H. (1979) Analysis and manipulation of line-drawing data. *Proceedings, NATO Advanced Study Institute on Map Data Processing*, Maratea, Italy.

Fuller, R.B. and Sadao, S. (1982) *Spaceship Earth Edition of the Dymaxionsky-Ocean Map*, Buckminster Fuller Institute, Philadelphia.

Gargantini, I. (1982) An effective way to represent quadtrees. *Communications of the ACM*, **25**, 905–910.

Gibson, L. and Lucas, D. (1982) Vectorization of raster images using hierarchical methods. *Computer Graphics and Image Processing*, **20**, 82–89.

Hunter, G.M. (1978) *Efficient Computation and Data Structures for Graphics*. Unpublished PhD Dissertation, Department of Electrical Engineering and Computer Science, Princeton University, Princeton, NJ.

IGU Commission on Geographic Data Sensing and Processing (1975) Information Systems for Land Use Planning. Report prepared for Argonne National Laboratory.

IGU Commission on Geographic Data Sensing and Processing (1976) Technical Supporting Report D, US Department of the Interior, Office of Land Use and Water Planning.

Jackins, C.L. and Tanimoto, S.L. (1980) Oct-trees and their use in representing three-dimensional objects. *Computer Graphics and Image Processing*, **14**, 249–270.

Jackins, C.L. and Tanimoto, S.L. (1983) Recursive decomposition of Euclidean space. *IEEE Transactions on Pattern Analysis and Machine Intelligence*, **5**, 533–539.

Klinger, A. (1971) Patterns and search statistics, in *Optimizing Methods in Statistics* (ed. J.S. Rustagi), Academic Press, New York, pp. 303–337.

Klinger, A. and Dyer, C. (1976) Experiments on picture representation using regular decomposition. *Computer Graphics and Image Processing*, **5**, 68–105.

Klinger, A., Fu, K.S. and Kunii, T.L. (1977) *Data Structures, Computer Graphics, and Pattern Recognition*, Academic Press, New York.

Knuth, D. (1973) *The Art of Computer Programming, Volume III: Sorting and Searching*, Addison-Wesley, Reading, MA.

Kopec, R.R. (1963) An alternative method for the construction of Thiessen polygons. *Professional Geographer*, **15**, 24–26.

Lee, D.T. (1980) Two-dimensional Voronoi diagrams in the Lp metric. *Journal of the ACM*, **27**, 604–618.

Mark, D.M. (1975) Computer analysis of topography: a comparison of terrain storage methods. *Geografiska Annaler*, **57a**, 179–188.

Mark, D.M. (1979) Phenomenon-based data-structuring and digital terrain modelling. *Geo-Processing*, **1**, 27–36.

Martin, J. (1975) *Computer Data-Base Organization*, Prentice Hall, Englewood Cliffs, NJ.

Merrill, R.D. (1973) Representation of contours and regions for efficient computer search. *Communications of the ACM*, **16**, 69–82.

Morton, G.M. (1966) A Computer Oriented Geodetic Data Base; and a New Technique in File Sequencing. Unpublished Report.

Nyerges, T.L. (1980) *Modeling the Structure of Cartographic Information for Query Processing*. Unpublished PhD Dissertation, Ohio State University.

Peano, G. (1973) *Selected Works* (ed. H.C. Kennedy), University of Toronto Press, Toronto.

Peucker, T. and Chrisman, N. (1975) Cartographic data structures. *American Cartographer*, **2**, 55–69.

Peucker, T. (1976) A theory of the cartographic line. *International Yearbook for Cartography*, **16**, 134–143.

Peuquet, D. (1979) Raster processing: an alternative approach to automated cartographic data handling. *American Cartographer*, **6**, 129–139.

Peuquet, D. (1981a) Cartographic data, part I: the raster-to-vector process. *Cartographica*, **18**, 34–48.

Peuquet, D. (1981b) An examination of techniques for reformatting digital cartographic data, part II: the vector-to-raster process. *Cartographica*, **18**, 21–33.

Peuquet, D. and Boyle, A.R. (1984) Raster scanning, processing and plotting of cartographic documents, forthcoming.

Reddy, D.R. and Rubin, S. (1978) Representation of three-dimensional objects. *Report #CMU-CS-78-113*, Computer Science Department, Carnegie-Mellon University.

Rhynsburger, D. (1973) Delineation of Thiessen polygons. *Geographical Analysis*, **5**, 133–144.

Samet, H. (1984) The quadtree and related hierarchical data structures. *ACM Computing Surveys*, **6**(2), 187–260.

Scholten, D.K. and Wilson, S.G. (1983) Chain coding with a hexagonal lattice. *IEEE Transactions on Pattern Analysis and Machine Intelligence*, **5**, 526–533.

Senko, M.E. *et al.* (1973) Data structures and accessing in data base systems. *IBM Systems Journal*, **12**, 30–93.

Senko, M.E. (1976) DIAM 11 and levels of abstraction. *Proceedings, Conference on Data: Abstraction, Definition and Structure*, **8**, 121–140.

Shamoas, M. (1977) *Computational Geometry*. Unpublished PhD Dissertation, Yale University.

Stevens, R.J., Lehar, A.F. and Preston, F.H. (1983) Manipulation and presentation of multidimensional image data using the Peano scan. *IEEE Transactions on Pattern Analysis and Machine Intelligence*, **5**, 520–526.

Tanimoto, S. and Pavildis, T. (1975) A hierarchical data structure for picture processing. *Computer Graphics and Image Processing*, **4**, 104–119.

Thiessen, A.H. (1911) Precipitation averages for large areas. *Monthly Weather Review*, **39**, 1082–1084.

Tobler, W. (1959) Automation and cartography. *Geographical Review*, **49**, 526–534.

Tomlinson, R.F. (1973) A Technical Description of the Canada Geographic Information System. Unpublished Report.

Tompa, F.W. (1977) Data structure design, in *Data Structures, Computer Graphics and Pattern Recognition* (ed. T.L. Kuni), Academic Press, New York, pp. 3–30.

Tropf, H. and Herzog, H. (1981) Multidimensional range search in dynamically balanced trees. *Angewandte Informatik*, 71–77.

US Department of Commerce (1970) The DIME geocoding system, in *Report No. 4, Census Use Study*, Bureau of the Census.

Ullman, J. (1983) *Principles of Database Systems*, Computer Science Press, Kockville, MD.

White, M. (1981) N-trees: large ordered indexes for multi-dimensional space. *Mimeo.*

13

Reflection Essay: *A Conceptual Framework and Comparison of Spatial Data Models*[1]

Jeremy Mennis

Temple University, Philadelphia, USA

13.1 Raster Versus Vector

In the introductory course in geographic information science that I teach, before I delve into the details of spatial data models, I brief the students on the 'raster versus vector debate' in the history of the field of geographic information science. I remind my students that whichever side of the debate they are on, they are to remain civil and not engage in verbally or physically abusive behaviour to students who may take up the opposite perspective. This is a joke that usually elicits some laughs; the idea that one could get emotionally worked up over data models seems utterly preposterous. What could there be to debate about! And anyway, why would one type of encoding necessarily exclude the other?

And yet, there was indeed an active academic exchange of ideas regarding how spatial information should be encoded in a computer in the 1970s and into the 1980s. This exchange, or debate, as it were, serves as the intellectual context for this article. The debate centred around which form of data encoding for GIS was 'best': raster or vector? Most GIS software at the time (and there were really only a handful) were singular in

---

[1] I was 15 years old when Peuquet's (1984) *Cartographica* article was published, and it would be another 11 years before I ever heard the words 'geographic information systems'. Thus, this essay is not a reflection of my personal experience of cartography and geographic information science at the time this article was published, but rather a reflection of the discovery of a body of literature that immediately preceded my own scholarship, its context in the history of the field, and its role in the development of the field since its publication.

Classics in Cartography: Reflections on Influential Articles from Cartographica Edited by Martin Dodge
© 2011 John Wiley & Sons, Ltd

their modelling approach, and so the decision made at the beginning of the software development process to use one or the other model was of great consequence. Of note is that many of the arguments and developments in spatial data modelling at that time were driven by technical issues: data volume and compression ratios, the form of data capture (e.g. scanning, digitizing), efficiency of data retrieval, and the ease of editing and updating information (though see Sinton, 1978; Chrisman, 1977 for early theoretical perspectives). The argument over data structures concerned which of these two dominant approaches, raster or vector, dealt best with these technical challenges, with the caveat given that having both would be nice but the overhead for conversion, or simultaneous storage in both formats, was simply too great.

The vector model had the advantage of closely mimicking the fundamental graphical elements of a paper map – points, lines and polygons. The raster model, on the other hand, was much more closely aligned with the way computer graphics were internally stored and plotted. Eventually, particular encoding strategies were adopted for different analytical domains. The vector model, particularly its topologic format, became the dominant encoding strategy for information about people and the features they create: census data, transportation infrastructure, jurisdictional units, cadastres and so on. All of these phenomena seem to have a natural affinity for representation as points, lines or polygons. They are easily identifiable entities with discrete boundaries to which attribute data can be attached (e.g. a land parcel and its owner). Raster, on the other hand, seemed the natural choice for encoding environmental characteristics that are continuous in nature – elevation, temperature and the like. It also has the advantage of being consistent in format with imagery derived from remote sensing.

Of course, the technical encoding issues that drove the raster versus vector debate have not disappeared – they are perhaps just as central now as they were then to developing useful spatial data handling computer applications. The debate has, however, been transformed from an issue that was primarily technical to more theoretical issues. Certainly, methods of data compression have advanced, the available algorithms for both data models have expanded dramatically, and other substantial technical advances have been achieved since the 1970s. At the same time, however, these technical achievements have only increased the scope of the domains to which GIS has been applied. For instance, as the ability to store ever more data has increased, so too has the volume of data that is becoming available due to geospatial technologies such as remote sensing. And as spatial data handling algorithms have advanced in sophistication, so too has the complexity of the analytical domains to which they are applied. It is safe to say that the state of the GIS technology continues to be pushed by its applications.

And yet, something else important and tangible has changed in academic GIS research since the 1970s. Much current research in cartography and geographic information science does not necessarily seek to overcome these technical challenges, such as data storage, so much as transcend them. This new kind of research recognizes that that the core issues of spatial data modelling are not necessarily defined by the current state of technological achievement, but rather concern the deeper theoretical issues associated with geographic representation, that is how we model geographic reality across a variety of media – cognitively, visually and computationally

(Egenhofer *et al.*, 1999). This emphasis on representation can serve to develop richer and more robust computational representations of geographic phenomena, as well as to provide collaborative and interactive computational environments that are suited to how people actually learn and interact in the 'real' world. In other words, data modelling in GIS concerns not only how to encode points, lines and polygons (or grid cells) in a computer, but how to capture, computationally encode, understand, and learn from, the semantically rich and complex world in which we live.

13.2 The Conceptual Turn and its Directions

I see the importance of Peuquet's, 1984 *Cartographica* article as the metaphoric exclamation point (or perhaps more the understated period) on the raster versus vector debate. It is primarily a review paper, summing up raster and vector data models neatly and thoroughly – the article weighs in at a hefty 47 pages – in a series of pros and cons for each approach. The article also offers a 'hybrid' data model – the 'vaster model,' that seeks to exploit the advantages of both raster and vector models. In hindsight, however, I see Peuquet's *Cartographica* article as the demarcation of an important point in the development of the field of geographic information science that I refer to here as 'the conceptual turn'.

The conceptual turn encompasses a paradigm shift (though probably 'paradigm' with a little 'p') from a preoccupation with the technological challenges of GIS research to a focus on more theoretical topics. It is associated with the coining of the term 'geographic information science,' as opposed to 'geographic information systems' (Goodchild, 1992), and with the critique of GIS from many scholars in Geography as a positivist and reductionist enterprise that serves to reinforce conventional power structures (Harley, 1989, reproduced as Chapter 16 of this volume; Pickles, 1995). For spatial data modelling research, the conceptual turn as embodied by Peuquet (1984) is at once a summation of a debate about representation that was primarily technical in nature (i.e. raster versus vector), as well as a foreshadowing towards issues of representation that are more theoretically orientated.

Perhaps the most immediately identifiable shift in the conceptual turn following the publication of Peuquet's (1984) article relates to the transformation of the raster–vector debate to a discussion of object versus field representation, or, as Couclelis (1993: 65) put it, 'Is the world a jig-saw puzzle of polygons, or a club sandwich of data layers?' This cognitive perspective posits that different spatial models are a reflection of different human conceptualizations of geographic phenomena. There was a rein-terpretation, of sorts, of spatial data modelling generally that sought to distinguish the cognitive conceptualization of a phenomenon from its computational encoding, and elucidate the process of abstraction by which a conceptualization is ultimately represented within a digital device. The process of data modelling, as depicted in Peuquet (1984, Figure 12.2) is expanded substantially by Burrough (1992: 396) in Figure 13.1 to show that there are actually many possible pathways and decisions a GIS data modeller must make, consciously or unconsciously, on the way to representing something in GIS.

Reflection Essay

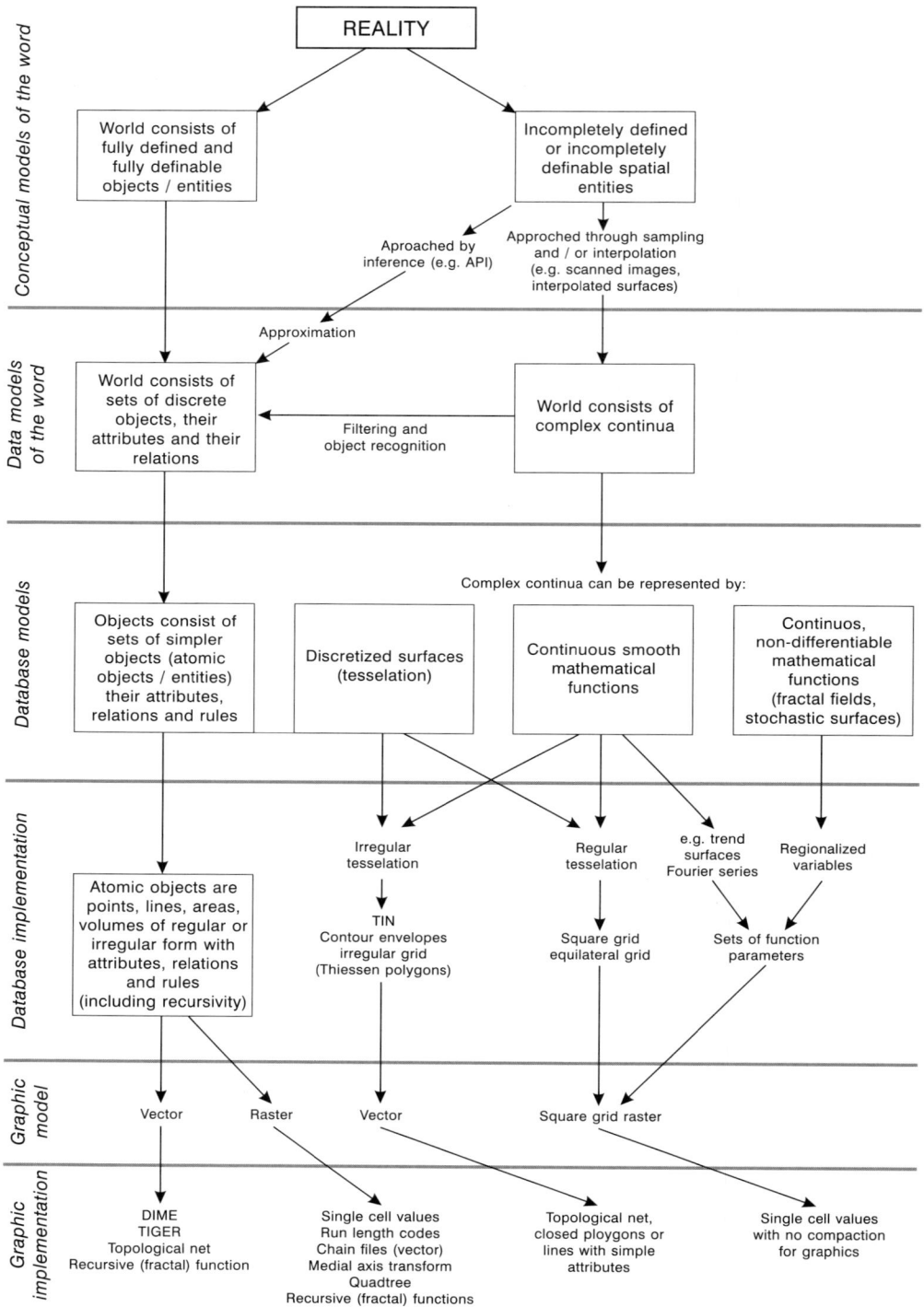

Figure 13.1 Schematic overview of stages of proceeding from conceptual model to graphic implementation (redrawn from Burrough, 1992: 396).

As geographic information science researchers turned their attention to more theoretical matters, they began to look outside their own fields and disciplines for perspectives and inspiration. Cognitive science provided a broad foundation for linking the representation of geographic phenomena in a computer to representation in the mind (MacEachren, 1995; Peuquet, 2002). The integration of research in cognitive science also allowed for scholars in geographic information science to interact with traditions in behavioural geography that had flowered in the 1970s. Researchers also looked to philosophy, and particularly ontology, to develop formalizations for the representation of geographic phenomena and their interrelationships (Fonseca *et al.*, 2002). More recently, researchers in geographic information science have incorporated research in social theory to interpret the role of cartographic and computational representations of geographic reality in reproducing and influencing societal norms (Crampton and Krygier, 2005).

Of course, technological innovation in GIS did not stop during this time. Rather, researchers sought to appropriate, incorporate and develop technological solutions to the theoretical challenges they were highlighting. Perhaps the most important development was the incorporation of object-orientated computing in the late 1980s and 1990s, which takes the form of a conceptual modelling approach, a type of programming language and elements of database management system (DBMS) software. The object-orientated approach was used to capture many of the principles of cognitive science, such as categorization and hierarchical information storage, within GIS databases. Another important technological trend was (and is) the continued integration of GIS with mainstream DBMS technologies, which allows GIS researchers to leverage the increasingly sophisticated representational capabilities developed in the field of information science and technology.

I am admittedly biased, but throughout these developments I have always considered the issue of spatial data modelling to be the key foundational concept underlying academic research and debate in geographic information science. Data modelling is the basis for encoding, and thus computational representation, in GIS, and it is difficult to identify a theoretical debate that does not concern the issue of representation at some level. Nowhere is this more evident than in the 'GIS tool or science' debate of the 1990s and critiques of GIS that have continued into the present (Wright, Goodchild and Proctor, 1997). At first blush, one might think such critiques have little to do with spatial data modelling. However, one of the primary arguments made by those critical of GIS is that the representational capability (i.e. the data models) of GIS is too crude to adequately capture the important, and often subtle, characteristics of geographic situations, such as (to name just a few examples) principles of social justice, the feelings of connection and self-identity associated with place, and lived experiences of individuals. By failing to capture such important characteristics, and by virtue of its 'scientific' and objective depiction of reality, GIS representations can, in fact, reinforce the absence of those characteristics in decision making and policy.

And yet, representational capability has been, arguably, the central research project in GIS scholarship since its inception. There have been substantial advances. Extensions to representation in time and three-dimensional space have been developed. There are data modelling approaches for representing complex geographic entities that may be related to each other through composition, categorization or other relationships, often using object-orientated modelling techniques. Ontological approaches for formalizing the representation of geographic domains have been developed. Methods have also been developed for encoding and processing very large volumes of spatial data, as well as new types of data (e.g. multidimensional data, data gathered from sensor networks, moving object data, georeferenced qualitative data).

There have also been some attempts at what might be called a 'unified theory' of representation in GIS, a conceptual framework or general principles to guide all kinds of representation in GIS. For example, in later work Peuquet (1988, 1994) and Mennis, Peuquet and Qian (2000) offer a vision for spatial data modelling that integrates object and field representations with temporal representation, as informed by principles of cognition. A distinction is also made here between observational data (encoded as field-like data) and a person's conceptualization of geographic entities (encoded as object-like data). Goodchild, Yuan and Cova (2007) describe a set of data types, such as the geo-atom and geo-dipole, from which aggregations can be formed to represent various types of geographic phenomena that may be conceptualized as objects or fields. These authors aim to elucidate an exhaustive categorization of abstract types of dynamic geographic phenomena based on a typology of geometry, movement and internal structure. All of these efforts can be seen to be a direct descendent of the raster–vector debate, recast in a more abstract and theoretical light.

Getting back to my introductory GIS course, it is perhaps interesting to note that I hardly mention any of these more advanced research topics to my students. Rather, when it comes to introducing spatial data models in GIS, I describe the raster and vector data models in a manner largely similar to that described in Peuquet (1984). A review of a variety of introductory GIS courses, as well as introductory GIS textbooks, suggests a similar pattern – raster and vector continue to serve as the primary approaches to teaching GIS data modelling. It is ironic – are we (the teachers of GIS) reproducing the raster versus vector debate in the next generation of GIS professionals, and its orientation towards technical as opposed to theoretical issues, at the same time we seek so strenuously with our research to depict GIS as a theoretical field? About this, Donna Peuquet had this comment: 'In a way, I personally feel that I am doing the field of GIScience a disservice by perpetuating this view [i.e. data modelling as raster versus vector] in the young and (previously) unprejudiced minds of future GIScientists in my intro GIS classes. Wouldn't it be better for the field as a whole to let them in, right away, on the latest and best that we know in terms of representation – even on a practical, implementational level?' With this sentiment in mind, I offer the following three 'big' challenges to spatial data modelling research below.

13.3 Three Big Challenges for the Future of Spatial Data Modelling

13.3.1 Representing Process

Certainly, the representation of time in GIS has continued to evolve as a data modelling challenge. Whereas in early research on time in GIS, the focus was on encoding information that had a temporal element, such as the date of a parcel transaction in a cadastral database, research in temporal GIS has turned to the representation of much more sophisticated representational challenges. For example, how do we encode data about features that move over space, such as cars, phones and people? How do we handle properties of features that may change over time? How can issues of identity change be treated, where features may be created, cease to exist, merge with other features, or change identity over time?

Even more challenging is the representation of processes. It is one thing to develop a data encoding strategy to efficiently store the movement of a car – this may be challenging from a technical viewpoint due to, say, large data volume, but is easily conceptualized. A car is a discrete and easily identifiable object that can be conceptualized as a point feature moving through a coordinate (perhaps network) space. Its movement can, therefore, be represented as a string of time-stamped coordinates. More complex spatio-temporal phenomena may be considered processes, and these would seem to be much more difficult to represent (Yuan and Stewart, 2008). Consider, for example, processes like a landslide, gentrification or a disease epidemic. Like a moving car, each of these processes is conceptualized as an entity that may be defined and described linguistically. But their representation in a GIS environment is far more challenging than that of a moving car. Processes such as these are composed of many parts, each of which is changing over time. The spatial and temporal boundaries of these processes are ambiguously defined. In fact, the process itself may only come into being through the interpretation of the properties of its component parts.

Herein lies a central challenge of representing processes, for it is not simply enough to develop a data model that encodes the geometry and/or identity of a process, but the data model should reflect a process's behaviours and how it may be recognized from the properties of its component parts. It is critical to separate the representation of the observational data, where a set of spatio-temporal observations may be seen as capturing the physical manifestation of a particular process, from the characteristics and behaviours of the process itself. When, for example, should the characteristics of the transmission of a disease amongst a group of individuals be recognized as an epidemic? Or, to give another example, at what point do we recognize that a neighbourhood is gentrifying? Is there simply a threshold relating to changes in the demographics of property owners? Are there markers of gentrification relating to certain types of commercial enterprises (e.g. fancy coffee shops and bookstores)?

13.3.2 Representing Meaning

As spatial data models have evolved, so has the sophistication of the type of information people wish to represent. Research in semantic data models and ontology have sought to formalize many 'higher level' geographic concepts, and their interrelationships, in the context of spatial database representation. However, it appears spatial data modelling has hit a wall with representing social and psychological phenomena that are more difficult to define, intangible or contextual. I consider such phenomena to have characteristics of social or psychological 'meaning', where meaning concerns how people feel about a place or geographic entity. Meaning can encompass, for instance, place characteristics such as 'sacredness', emotional attachments people may have, such as love and fear, and perceptions people have about places, such as safety and danger. Meaning can also encompass social interpretations of geographic situations, such as fairness, equity, discrimination and power relationships.

Some researchers may feel this is beyond the scope, or even possibility, of computational representation. Or perhaps one thinks that emotional responses and social justice considerations can be inferred from geographic data, but not encoded or reasoned with computationally. The integration of qualitative methods and GIS offers some promise here (Cope and Elwood, 2009; Kwan and Knigge, 2006). The traditional means of capturing the kinds of meaning I refer to here is through qualitative methods such as ethnography and interviewing. These types of data can provide rich indicators relating the meaning of places and geographic situations to individuals. One of the key elements of such a representation is the idea that meaning is not an intrinsic property of a geographic object or a location; rather, meaning is a relationship between an individual and a geographic entity or location. Because meaning is a relation between a place and a person, multiple representations for the same phenomenon are necessary. Developing data models to relate qualitative data that capture an individual's experiences and personal meanings associated with places and geographic phenomena would provide a substantial way forward for representing meaning in GIS.

13.3.3 Representing Learning

Another way of viewing representation of geographic phenomena in GIS is not as a static structure but as a process of learning and knowledge refinement, where the computational implementation of that dynamic process should be guided by what we know about cognitive processes of geographic learning and knowledge construction. Such an approach accounts for the fact that the semantics of geographic phenomena are indeed fluid and changing, that the semantics associated with geographic representations is derived from cognition of geographic phenomena, and that concepts can be mapped onto measurements referenced in space and time. Developing GIS software environments that support learning in this manner would have substantial implications for both teaching and the use of GIS for problem solving and decision making.

However, the means to generate mappings between complex geographic phenomena, and their ontological and semantic character, with fields of observational measurements, is not well established. It is here that a set of standard spatio-temporal processing functions would be particularly useful, because the mapping between a 'higher level', conceptually-defined object, event, process or other phenomenon with the underlying field from which such ontological constructs are composed or detected need not be encoded as a simple set of pointers (or relations in a relational data model); rather, the semantics may be defined as the functional methodology for how an object may be recognized and categorized from patterns embedded in field-based data. Thus, advancing representation of dynamic geographic phenomena in GIS depends not only on developing data models to encode spatio-temporal data, nor solely on creating high level conceptual models or database schema for representing spatio-temporal phenomena, but also on developing computational environments that support multiple representations and the ability to interactively manipulate the functional mappings between those lower level data structures and high level semantic or conceptual representations.

Maybe, in the future, GIS will look less like an analytical toolbox and more like a game[2] that facilitates learning – the mutant child of current commercial GIS packages, agent-based simulation, the Wii (home video game) and The Matrix (the movie). I can envision a GIS where a gamer receives a set of spatio-temporal observational data, and can populate their world with semantic objects of their own creation, complete with rules, relationships and behaviours. Then one can play the game like a simulation, interactively 'driving' the agents/objects, and learning how they interact to produce certain outcomes under a range of different scenarios. Or maybe in reverse, one can populate a world with objects, and generate 'observational data' under various game conditions. Many of the technologies necessary for such a GIS game already exist – gestural interfaces, three-dimensional visualization, cellular automata and multi-agent-based simulation software, ontology and knowledge databases, and futuristic video game chairs with built in speakers. I think my students will like it, even more than points, lines, polygons and grid cells.

Acknowledgements

Thanks to Donna Peuquet for her helpful comments on this essay, and, of course, for writing the article under discussion in the first place. All thoughts expressed here are, of course, my own.

Further Reading

Burrough, P.A. (1992) Are GIS data structures too simple minded? *Computers and Geosciences*, **18**, 395–400. (This article represents a substantial advance in the way researchers think about the process of abstraction through which particular GIS representational structures are chosen.)

Cope, M. and Elwood, S. (2009) *Qualitative GIS: A Mixed Methods Approach*, Sage, London. (Though recently published, this book illustrates the cutting edge of approaches to incorporating the representation of 'meaning', in the form of emotional content and the lived experience of individuals, in the context of GIS.)

Couclelis, H. (1993) People manipulate objects (but cultivate fields): beyond the raster–vector debate in GIS, in *From Space to Territory: Theories and Methods of Spatio-Temporal Reasoning* (eds A.U. Frank, F.I. Campari and U. Formentini), Springer, Berlin, pp. 65–77. (This paper represents a transitive moment in the transformation of the raster–vector debate to one concerning object and field representations of geographic phenomena.)

Goodchild, M.F., Yuan, M. and Cova, T. (2007) Towards a general theory of geographic representation in GIS. *International Journal of Geographical Information Science*, **21**(3),

---

[2] Thanks to C. Dana Tomlin for first alerting me to the idea of 'GIS-as-video-game'.

239–260. (This article focuses on the development of a comprehensive representational framework for object- and field-like phenomena in GIS.)

Peuquet, D.J. (2002) *Representations of Space and Time*, Guilford Press, New York. (This book substantially extends issues in spatial data modelling from a technical perspective to a more theoretical perspective by incorporating principles from philosophy and cognitive science.)

References

Burrough, P.A. (1992) Are GIS data structures too simple minded? *Computers and Geosciences*, **18**, 395–400.

Chrisman, N. (1977) *Impact of Data Structure on Geographic Information Processing*. Internal Report 7404, Laboratory for Computer Graphics and Spatial Analysis, Harvard University.

Cope, M. and Elwood, S. (2009) *Qualitative GIS: A Mixed Methods Approach*, Sage, London.

Couclelis, H. (1993) People manipulate objects (but cultivate fields): beyond the raster-vector debate in GIS, in *From Space to Territory: Theories and Methods of Spatio-Temporal Reasoning* (eds. A.U. Frank, F.I. Campari and U. Formentini), Springer, Berlin, pp. 65–77.

Crampton, J. and Krygier, J. (2005) An introduction to critical cartography. *ACME: An International E-Journal of Critical Geographies*, **4**(1), 11–33.

Fonseca, F., Egenhofer, M.J., Agouris, P. and Camara, G. (2002) Using ontologies for integrated geographic information systems. *Transactions in GIS*, **6**(3), 231–257.

Goodchild, M.F. (1992) Geographic data modelling. *Computers and Geosciences*, **18**(4), 401–408.

Goodchild, M.F., Yuan, M. and Cova, T. (2007) Towards a general theory of geographic representation in GIS. *International Journal of Geographical Information Science*, **21**(3), 239–260.

Egenhofer, M.J., Glasgow, J., Gunther, O. *et al.* (1999) Progress in computational methods for representing geographic concepts. *International Journal of Geographical Information Science*, **13**(8), 775–796.

Harley, J.B. (1989) Deconstructing the map. *Cartographica*, **26**(2), 1–20. (Reproduced as Chapter 16 of this volume.)

Kwan, M.-P. and Knigge, L. (2006) Qualitative research and GIS theme issue. *Environment and Planning A*, **38**(11).

MacEachren, A.M. (1995) *How Maps Work*, Guilford Press, New York.

Mennis, J.L., Peuquet, D.J. and Qian, L. (2000) A conceptual framework for incorporating cognitive principles into geographical database representation. *International Journal of Geographical Information Science*, **14**(6), 501–520.

Peuquet, D.J. (1984) A conceptual framework and comparison of spatial data models. *Cartographica*, **21**(4), 66–113. (Reproduced as Chapter 12 of this volume.)

Peuquet, D.J. (1988) Representations of geographic space: Toward a conceptual synthesis. *Annals of the Association of American Geographers*, **78**, 375–394.

Peuquet, D.J. (1994) It's about time: A conceptual framework for the representation of temporal dynamics in geographic information systems. *Annals of the Association of American Geographers*, **84**(3), 441–461.

Peuquet, D.J. (2002) *Representations of Space and Time*, Guilford Press, New York.

Pickles, J. (1995) *Ground Truth: The Social Implications of Geographic Information Systems*, Guilford Press, New York.

Sinton, D. (1978) The inherent structure of information as a constraint to analysis: mapped thematic data as a case study, in *Harvard Papers on GIS*, vol. **7** (ed. G. Dutton), Addison-Wesley, Reading, MA.

Yuan, M. and Stewart, K. (2008) *Computation and Visualization for the Understanding of Dynamics in Geographic Domains: A Research Agenda*, CRC/Taylor and Francis, New York.

Wright, D.W., Goodchild, M.F. and Proctor, J.D. (1997) Demystifying the persistent ambiguity of GIS as 'tool' versus 'science'. *Annals of the Association of American Geographers*, **87**(2), 346–362.

14

Designs on Signs/Myth and Meaning in Maps

Denis Wood and John Fels[1]

Abstract

Every map is at once a synthesis of signs and a sign in itself: an instrument of depiction – of objects, events, places – and an instrument of persuasion – about these, its makers and itself. Like any other sign, it is the product of codes: conventions that prescribe relations of content and expression in a given semiotic circumstance. The codes that underwrite the map are as numerous as its motives, and as thoroughly naturalized within the culture that generates and exploits them. *Intrasignificant* codes govern the formation of the cartographic icon, the deployment of visible language, and the scheme of their joint presentation. These operate across several levels of integration, activating a repertoire of representational conventions and syntactical procedures extending from the symbolic principles of individual marks to elaborate frameworks of cartographic discourse. *Extrasignificant* codes govern the appropriation of entire maps as sign vehicles for social and political expression – of values, goals, aesthetics and status – as the means of modern myth. Map signs, and maps as signs, depend fundamentally *on* conventions, signify only in relation to other signs, and are never free of their cultural context or the motives of their makers.

Spread out on the table before us is the *Official State Highway Map of North Carolina*. It happens to be the 1978–79 edition. Not for any special reason: it just came to hand when we were casting about for an example. If you don't know this map, you can well enough imagine it, a sheet of paper – nearly two by four feet – capable of being folded into a handy pocket or glove compartment sized four by seven inches. One side is taken up by an inventory of North Carolina points of interest – illustrated with photos of,

[1] Originally published: 1986, *Cartographica*, **23**(3), 54–103.

At the time of publication: Wood was Associate Professor of Design in the School of Design, North Carolina State University, USA. Fels taught cartography in the School of Natural Resources, Sir Sandford Fleming College, Lindsay, Ontario, Canada.

---

amongst other things, a scimitar horned oryx (resident in the state zoo), a Cherokee woman making beaded jewellery, a ski lift, a sand dune (but no cities), a ferry schedule, a message of welcome from the then governor and a motorist's prayer ('Our heavenly Father, we ask this day a particular blessing as we take the wheel of our car . . .'). On the other side, North Carolina, hemmed in by the margins of pale yellow South Carolina and Virginia, Georgia and Tennessee, and washed by a pale blue Atlantic, is represented as a meshwork of red, black, blue, green and yellow lines on a white background, thickened at the intersections by roundels of black or blotches of pink. There is about it something of veins and arteries seen through translucent skin, and if you stare at it long enough you can even convince yourself that blood is actually pulsing through them. Constellated about this image are, *inter alia*, larger scale representations of ten urban places and the Blue Ridge Parkway, an index of cities and towns, a highly selective mileage chart, a few safety tips and . . . yes, a legend (Figure 14.1).

14.1 Legends

It doesn't say so, of course, but it is all the same. What it says is, 'North Carolina Official Highway Map/1978–79'. To the left of this title is a sketch of the fluttering state flag. To the right is a sketch of a cardinal (state bird) on a branch of flowering dogwood (state flower) surmounting a buzzing honey bee arrested in mid-flight (state insect). Below these, four headings in red – 'Road Classifications', 'Map Symbols', 'Populations of Cities and Towns' and 'Mileages' – organize collections of marks and their verbal equivalents (thus, a red dot is followed by the words 'Welcome Center'). We will return to these in a moment, but for the sake of completeness it should be noted that below these one finds graphic and verbal scales (in miles *and* kilometres), as well as the pendent sentence, 'North Carolina's highway system is the Nation's largest State-maintained Network. Hard surfaced roads lead to virtually every scenic and vacation spot'.[2]

Clearly this legend – to say nothing of the rest of the map – carries a heavy burden, one that reflects aggressively the uses to which this map was put (Figure 14.2). We stress the plural because it is a fact, not so much *overlooked* (cartographers are not *that* naive), but nonetheless ordinarily ignored, denied, suppressed. For certainly in this case the first and primary 'user' was the State of North Carolina, which *used* the map as a promotional device (in *this* context 'used' comes naturally), as an advertisement more likely than most to be closely looked at, even carefully preserved (because of its other uses), and so one given away at Welcome Centers just inside the state's borders, at Visitor Centers elsewhere, from booths at the State Fair, and in response to requests from potential tourists, immigrants and industrial location specialists. This is all perfectly obvious in 'The Guide to Points of Interest' and the selection of photographs that decorate it (unless we have the emphasis backwards, and the 'Guide' is first of all a

---

[2] As will become more apparent below, it is not irrelevant that were our legend a photograph in the *National Geographic Magazine,* it is this pendent sentence which would be called the 'legend'. At the *Geographic*, caption writing is an art practiced by those in the Legends Division.

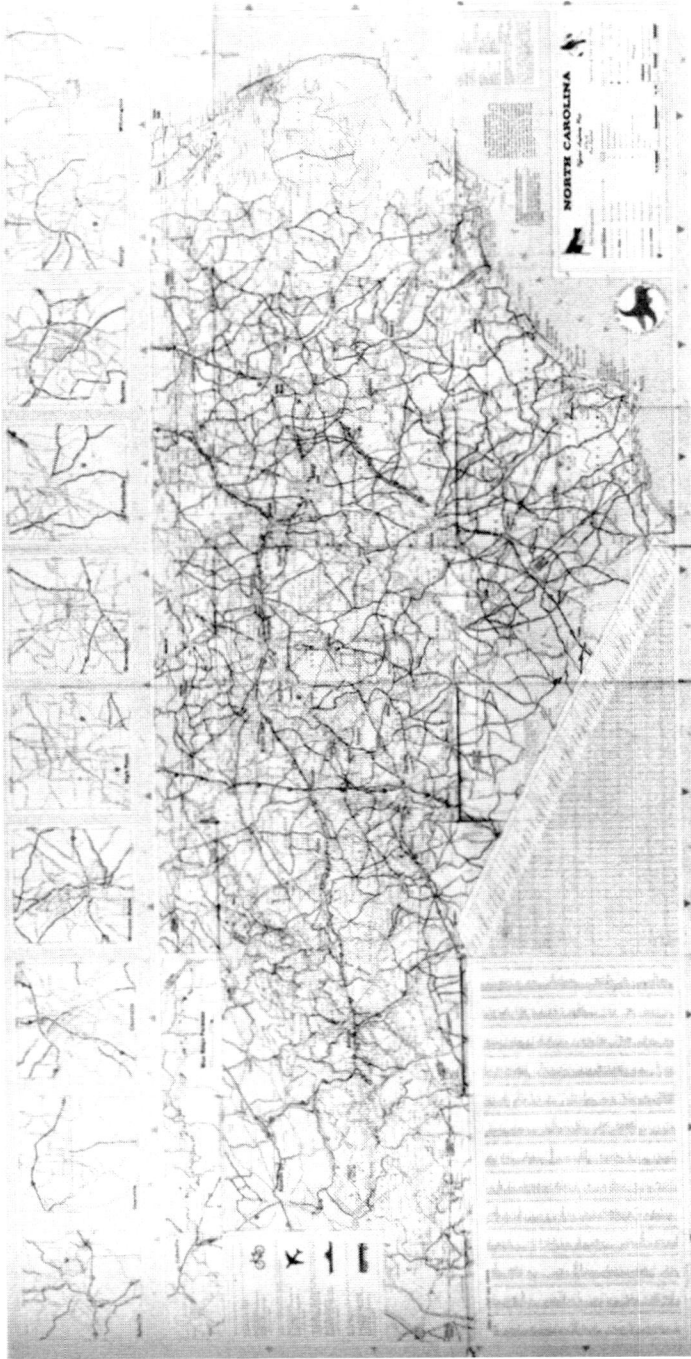

Figure 14.1 The '1978–79 North Carolina Transportation Map & Guide To Points of Interest'. That's what it says on the cover fold. The map image proper is titled on the legend block as the 'North Carolina Official Highway Map 1978–79', whereas the headline on the other side of the sheet reads, 'North Carolina Points of Interest'. Unfortunately the distinctions amongst the pale blue, yellow, pink and white are all but lost in this reproduction. (North Carolina Department of Transportation.)

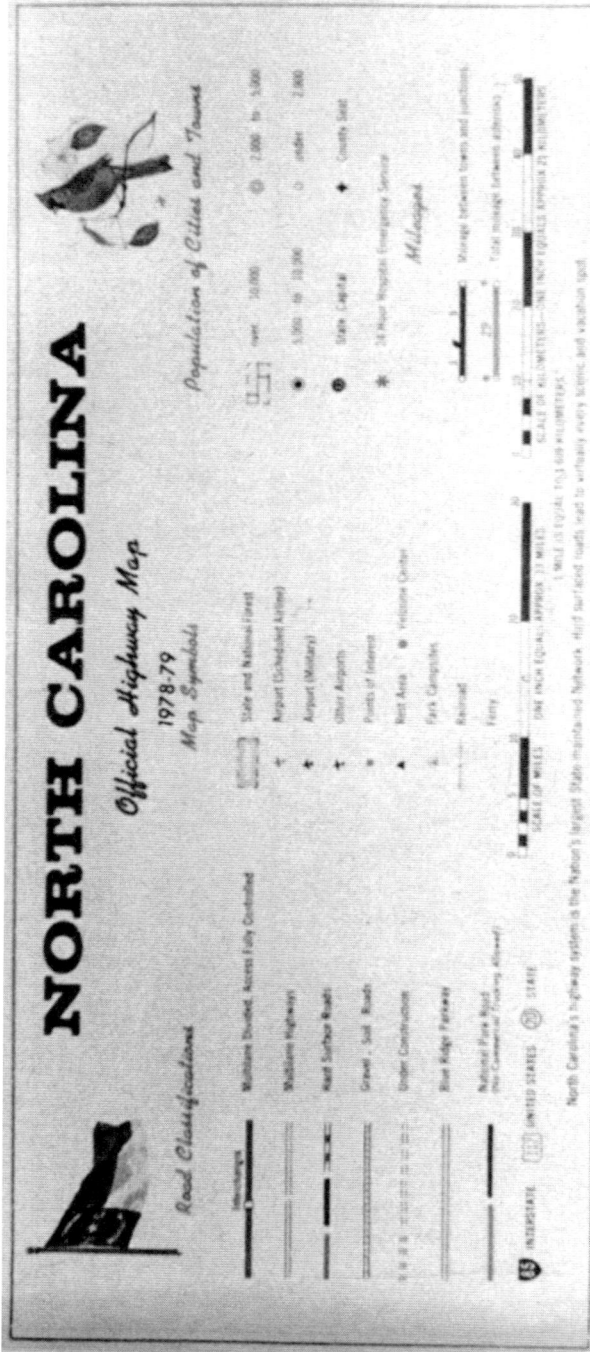

Figure 14.2   The legend block from the '1978–79 North Carolina Transportation Map & Guide To Points of Interest'. Again, it's too bad you can't appreciate the colour. (North Carolina Department of Transportation.)

way of justifying the photographs, like text in the *National Geographic Magazine*), but it is no less evident in the legend itself.

Nor is it here just a matter of the unavoidable presence of the state flag, flower, bird and insect – though there they are in children's encyclopaedia colours – but primarily of what *else* the map's makers have chosen for the legend and the ways they have chosen to organize it (for more than one principle of order operates under even seemingly straight forward subheadings such as 'Populations of Cities and Towns'). It is conventional to pretend, as Robinson and Sale have put it, that 'legends or keys are naturally indispensable to most maps, since they provide the explanations of the various symbols used' (Robinson and Sale, 1969: 270),[3] but that this is largely untrue hardly needs belabouring. Legends flare into cartographic consciousness not much earlier than thematic maps, are nonetheless still dispensed with more often than not, and never provide explanations of more than a portion of the 'symbols' found on the maps to which they refer. Their essential absence from, say, United States Geological Survey topographic survey sheets, or the plates of a *Rand McNally International Atlas,* makes this all too clear. That legends do exist for these maps – someplace in the book, or by special order – only serves to underscore, through their entirely separate, off-somewhere-else character, exactly how dispensable they really are.

Nor is this dispensability a result of the 'self-explanatory' quality of the map symbols, for, though Robinson and Sale might insist that, 'no symbol that is not self-explanatory should be used on a map unless it is explained in a legend' (Robinson and Sale, 1969: 270), the fact is that NO symbol *explains* itself, stands up and says, 'Hi, I'm a lock', or 'We're marsh', anymore than the *words* of an essay bother to explain *themselves* to the reader. Most readers make it through most essays (and maps) because as they grew up through their common culture (and *into* their common culture) they learned the significance of most of the words (and map symbols). Those they don't recognize they puzzle out through context, or simply skip, or ask somebody to explain. A few texts come with glossaries, though like map legends these are rarely consulted and readily dispensed with. But this familiarity with signs *on the part of the reader* never becomes a property of the mark, and even the most obvious, transparent sign remains opaque to those unfamiliar with the code.

It is not, then, that maps don't need to be decoded; but that they are by and large encoded in signs as readily interpreted by most map readers as the simple prose into which the marks are translated on the legends themselves. For, at best, legends less 'explain' the marks than 'put them into words', so that should the *words* mean nothing the legend is rendered less helpful than the map image itself, where the signs at least have a context and the chance to spread themselves a little (as anyone who has read a map in a foreign language can attest). One way to appreciate this while approaching an under-standing of the role legends actually play is to take a look at those signs on maps that don't make it onto the legend, of, for instance, this *North Carolina Official Highway Map.* Concentrating for the moment on the map image of the state proper, ignoring, that is, the

---

[3] It is instructive that despite their indispensability, legends are granted but a paragraph in the chapter on *design,* where they play the role of illustrations of the principles of figure-ground relationships. In light of the discussion, below, of the 'naturalization' function of myth, it is not surprising that Robinson and Sale (1969: 270) should have said, 'naturally indispensable'.

little maps of the state's larger cities, the inset of the Blue Ridge Parkway, the mileage chart (the instructions for which do happen to be pasted over the map image proper, though over South Carolina, just below Kershaw), the guide to other transportation information sources, the borders and rules, and the letters, numbers and other marks that facilitate the operations of the index of cities and towns – though to pretend that any of this is half as self-evident as the signs on the map image is to miss how laboriously we have learned to interpret the architecture of this picture plane, how much we have come to take for granted – still, ignoring all this, and all the words, and somehow managing to overlook that logo of the North Carolina Department of Transportation floating on the Atlantic some twenty miles due east of Cape Fear, it is nevertheless the case that eighteen signs deployed on the map image *do not appear on the legend.* That's half as many as do.

Why don't they? It's not, certainly, because they're self-explanatory. No matter how many readers are convinced that blue naturally and unambiguously asserts the presence of water, or that little pictograms of lighthouses and mountains explain themselves, signs are *not* signs for, dissolve into marks for, those who don't know the code. *Look* at these: where, in the eyes and eyebrows of Mt Sterling, can anyone see the mountain: or, in the pair of upended nail pullers, the lighthouse at Cape Fear? (Figures 14.3 and 14.4). Nor is there anything more 'self-evident' about the use of blue for water. Not only historically has water been rendered in red, black, white, brown, pink and green, but it disports in other colours on the obverse of this very map: in silver and white on the 'cover' photo of Atlantic surf; in tawny-pewter in the photograph of fishing boats at anchor; in warm silver-grey in a shot of the moonlit ocean off Wrightsville Beach; and in yellow-green in the photograph of the stream below Looking Glass Falls. Only in the

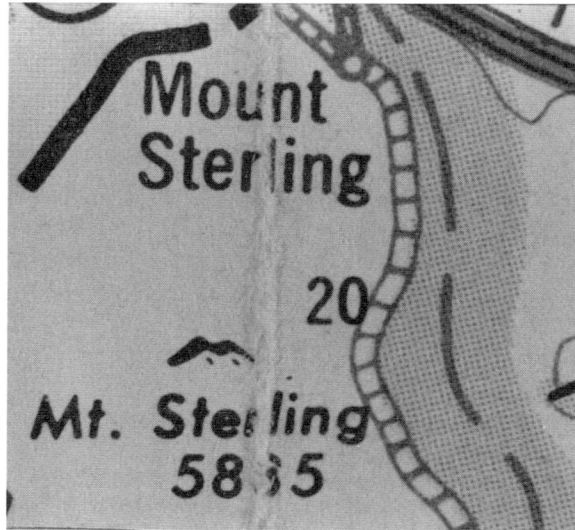

Figure 14.3   The eyes and eyebrows of Mt Sterling. Note the wear along the fold. The map has been many times folded and unfolded. (From the '1978–79 North Carolina Transporation Map & Guide To Points of Interest', North Carolina Department of Transportation.)

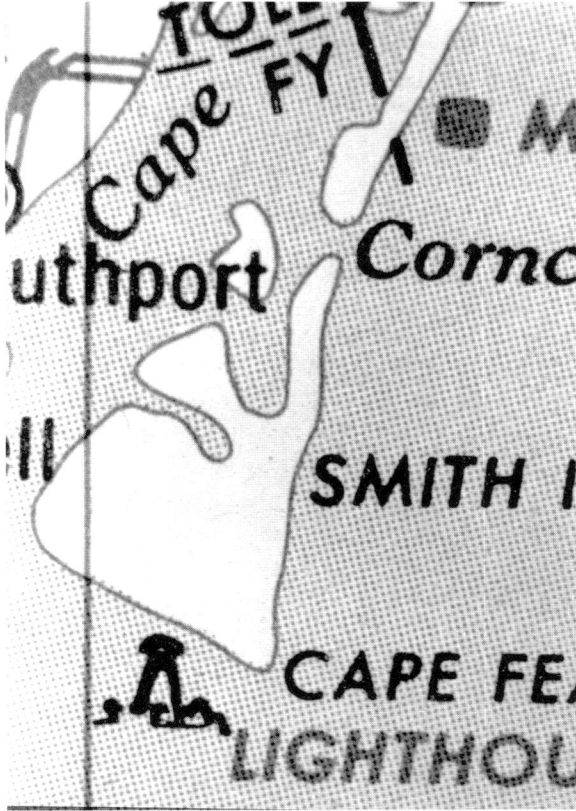

Figure 14.4   A pair of upended nail pullers trying to pass themselves off as the Cape Fear Lighthouse. (From the '1978–79 North Carolina Transportation Map & Guide To Points of Interest', North Carolina Department of Transportation.)

falls, where it indicates shadows, is there blue in any of these waters. This lack of any sort of 'necessary' or 'natural' coupling between blue and water proves fortuitous, for the colour used to represent water on the map *image* does double-duty as background for the sheet *as a whole,* and surely we were never intended to read the circumjacent margin for a circumfluent ocean. There's no way around it: each of these signs is a perfectly conventional way of saying what is said ('lighthouse', 'mountain', 'water') – which is why the *map seems* so transparent, so easy to read. But *were* the function of the legend to explain such conventions (or at least translate them into words), then these would belong on it as surely as those that are there.

And if these belong there, so do the yellow tint used for 'other states', the white used for 'North Carolina', the thick continuous green-with-dashed-red line that asserts 'National Park' and the thick continuous yellow-with-long-shortdashed-black line that stutters 'county' (so long as the border isn't along or over water). These all may be equally conventional, but they are less vernacular than the blue for water and so are open to greater misconstrual, especially on a map on which a long-short-short dashed black line mutters 'state', a continuous blue line murmurs 'coast' or 'bank', a fine dashed red line coughs at 'military reservation', a slightly thicker dashed red line says 'Indian reservation', and a still thicker one proclaims 'Appalachian Trail'. A fine dashed line in black whispers 'national wildlife refuge'. A continuous line in red hints, in degrees, at the graticule.

Yet, while all these signs are absent, *on* the legend we find interpretative distinctions made amongst the shapes and colours of the road signs of the Interstate, federal and state highway systems. Does the person really exist for whom the graticule is self-evident, yet the highway signs obscure? No, there is no such human being, though doubtless there are many immured in subtleties of the highway signage system to whom the graticule and its associated cabalism of degrees and minutes is a deep mystery. What becomes gradually clear is that if the purpose of the legend ever were 'explanation', everything is backwards: the things least likely to be most widely known are the very things about which the legend is reticent, while with respect to precisely those aspects both natives and travellers are most sure to be familiar, the legend is positively garrulous. Garrulous, not necessarily informative: the signs under 'Road Classifications' comprise less a system than a yardsale of marks, many of which remain, despite their inclusion on the legend, 'unexplained'. What is one to make, for instance, of the three marks given for 'Hard Surface Roads?' Are we to distinguish amongst solid red, solid black and enclosed, dashed blue? Or are these just three arbitrary ways of designating the same reality? Suggestions of system inevitably evaporate under the heat of attention: about the time you've concluded that red is the colour of federal highways, you run down us 74B in black; and by the time you've decided that unnumbered state roads are in enclosed, dashed blue, you realize you don't have the foggiest idea what that means. There are another three equally vague signs for highways under construction, and another two for multilane. There would seem to be an interest in portraying access (controlled or not), jurisdiction (federal or state), condition (constructed, under construction), composition (hard surface, gravel, soil) and carrying capacity (multilane or not) but not *enough* interest to force anybody to confront the graphic complexity implied by a five-dimensional code. Nor is this mess limited to the 'Road Classifications' portion of the legend. Of the seven signs under 'Populations of Cities and Towns', only four relate to population, and these do so without consistency. The state capital, county seats and '24 Hour Hospital Emergency Service' have individual designations confusingly related to the signs of population. Thus, the sign for 'State Capital' is circular like the signs for towns with less than 10 000 people; but the 'County Seat' sign is some kind of lozenge. The sign for 'Emergency Service' is a bright blue asterisk.

We can see your lips moving as you read this. They're saying, 'What a sad sack of a map! My undergraduates could do better'. But that's not true. Undergraduates would collapse if confronted with a task of this complexity. The design problems alone would give them fits (not to mention compilation etc. etc.), but the political realities would destroy them, the demands of interagency collaboration, for example (for while one side of our map was handled by the Department of Transportation, the other was produced by the Department of Commerce), the rigors of pleasing state senators and representatives, the imperative to manifest those miniscule but vital tokens of partisanship that distinguish the map of a Republican administration from that of the Democrats. Nor is it such a sad sack of a map. It's a fair example of the genre. It's indistinguishable, for instance, from the *Illinois Official Highway Map, 1985–86;* from the *Michigan Great Lake State Official Transportation Map* for 1974 (which makes up for the omission of its state insect by illustrating *inter alia* the state gem [greenstone], state fish [trout] and state stone [petoskey]); and it's a lot less weird than the *Texas – 1976 Official Highway Travel Map*, which in an attempt at shaded relief manages only to look badly singed. *All* the maps of the genre, and most other genres as well, are characterised by legends (like ours) which in a more or less muddled fashion put into words map signs that are so customary as to be widely understood without the words, while leaving the map images themselves *littered* with conventions it taxes professional cartographers to put into English.

14.2 Myths

Invariably the knee-jerk reaction is either to pooh-pooh the examples, no matter how many times multiplied, as bad (as in, 'Those are just *bad* maps!') or to call for a revolution in the design of their legends ('Rethinking Legends for the State Highway Map'). Both completely miss the point. *There is nothing wrong with the design of these legends: they are supposed to be the way they are.* This will be difficult for many to accept, but once it is understood that the role of the legend is less to elucidate the 'meaning' of this or that map element than to function as a sign in its own right, this conclusion is even more difficult *to evade.* Just as the bright blue asterisk signifies '24 Hour Hospital Emergency Service' so the legend as a whole is itself a signifier. As such, the legend refers not to the map (or at least not directly to the map), but back, through a judicious selection of map elements, to that to which the map image itself refers, to the state. *It is North Carolina that is signified in the legend, not the elements of the map image,* though it *is* the selection of map elements and their disposition within the legend box that encourages the transformation of the legend into a sign. It is a sign only a cartographer (or graphic designer) could fail to understand. Others receive in a glance, naively or otherwise, this sign of North Carolina's subtly mingled automotive sophistication, *urbanity* and *leisure* opportunity. Apprehended this way, the legend makes sense. The headings in red – heretofore so bizarre – appear now as headlines to a jingoist text. Under the fluttering flag, the words, 'Road Classifications'. *Plural.* North Carolina's road system is so rich, that one classification can't handle it. And across the legend, under the bucolic branch *cum* bird (read 'rural', read 'traditional values') and the bee if you can see it (read 'hard working' [read 'no unions']), the words, 'Populations of Cities and Towns'. Cities and towns *and* birds and bees. It is almost too much, though as it says on the 1986–1987 edition of this map, 'North Carolina has it all'.

It certainly has a lot of whatever it is. Look at those road signs! Their proliferation can no longer be seen as a manifestation of graphic and taxonomic chaos, though, but as a sign insisting that roads really *are* what North Carolina's all about. The sign's abundant density supports the presumption of the headline and justifies the proximity of the flag. That there are more signifiers than signifieds is no longer a mystery to be explained, but part of the answer to the question, 'Does North Carolina *really* have a lot of roads?' It's the graphic analogue to the assertion in black at the bottom of the legend box that reads

'North Carolina's highway system is the Nation's largest State-maintained Network'. What the roads connect, of course, are all those cities. It's wonderful the way it takes seven signs and four lines to unfold the complexities of what the cartographer can't help observing is but a four tier urban hierarchy. Again, it's the graphic equivalent of a remark from the governor's letter on the other side of the map about 'booming' cities. Hey: this is a *hip* state (though bucolic), urban, urbane, sophisticated (but built on traditional values). The whiff of sophistication is heightened by the kilometre scale, so *European,* almost risqué, though it's carefully isolated in the lower right hand corner of the legend under the heading, 'Mileages'. Roads and cities: *roads to and from cities,* that is, the very desideratum for anyone looking to locate, say, a plant somewhere in the South. Modern, in other words, up-to-date. But as the bird and branch and honey bee remind us, not off the wall.

And yet it's not all work either. In between, in between moments, in between the roads and the cities and towns, in between the *signs* for the roads and the cities and towns, under the innocuous heading 'Map Symbols' (which from its central position also casts its net over all the map signs on the legend), may be found the signs for fun, *clean* fun, *good* clean fun, but still fun: 'Park Campsites', 'State and National Forest', 'Welcome Center', 'Rest Area' and 'Points of Interest', to say nothing of the signs for still other ways of getting around, ferries, railroads and three kinds of airports. Led by that bright green forest sign that visually lies at the centre of the legend (read 'parks'), this heterogeneity speaks of caring for people ('*Welcome* Center', '*Rest* Area') and is the graphic version of the remainder of that black sentence that sums up the legend (and is counterposed at the bottom against 'North Carolina' at the top): 'Hard surfaced roads [for which there are three signs] lead to virtually every scenic and vacation spot'.

Wow! It's almost overdone. Had it been done up slick by some heavy duty design firm, it would have been overdone. But here, it's just hokey enough to seem sincere. *It is sincere.* We don't believe for a minute anyone sat down and cynically worked this thing out, carefully offsetting the presumptuousness of the overheated highway symbolism with the self-effacing quality of the children's encyclopaedia colours. But this is not to say that with this legend we are not in the presence of what Roland Barthes (1972: 109) has called 'myth' –a kind of 'speech' better defined by its intention than its literal sense.[4] Barthean myth is invariably constructed from signs which have been already constructed out of a previous alliance of a signifier and a signified. An example, an especially innocuous one, is given by the reading of a Latin sentence, '*quia ego nominor leo*', in a Latin grammar:

> *There is something ambiguous about this statement: on the one hand, the words in it do have a simple meaning: because my name is lion. And on the other, the sentence is evidently there in order to signify something else to me. Inasmuch as it is addressed to me, a pupil in the second form, it tells me clearly: I am a grammatical example meant to illustrate the rule about the agreement of the predicate. I am even forced to realize that the sentence in no way signifies its meaning to me, that it tries very little to tell me*

---

[4] This book, felicitously translated by Annette Lavers, consists of a number of 'mythologies' followed by the long essay, 'Myth Today'. It is from this latter that this reference and the following quotation come.

something about the lion and what sort of name he has; its true and fundamental signification is to impose itself on me as the presence of a certain agreement of the predicate. I conclude that I am faced with a particular, greater, semiological system, since it is co-extensive with the language: there is, indeed, a signifier, but this signifier is itself formed by a sum of signs, it is in itself a first semiological system (my name is lion*). Thereafter, the formal pattern is correctly unfolded: there is a signified (I am a grammatical example) and there is a global signification, which is none other than the correlation of the signifier and the signified; for neither the naming of the lion nor the grammatical example is given separately. (Barthes, 1972: 115–116)*

The parallels with our legend are pronounced. On the one hand, it too is loaded with simple meanings: *where on the map you find a red square, on the ground you will find a point of interest.* But as we have seen, the legend little commits itself to the unfurling of these meanings, even compared to the map image on which each is actually named – 'Singletary Lake Group Camp' or 'World Golf Hall of Fame'. The appearance of the red square on the legend thus adds nothing to our ability to understand the map. Instead it imposes itself on us as an assertion that *North Carolina has points of interest,* speaks *through* the map *about* the state. Yet, as in Barthes' example, this assertion about North Carolina is constructed out of, stacked on top of, the simpler significance of the red square on the legend, namely, to be identified with the words, 'Points of Interest'.

We thus have a two-tiered semiological system in which the simpler is appropriated by the more complex. Barthes has represented this relationship in the following way (Figure 14.5).

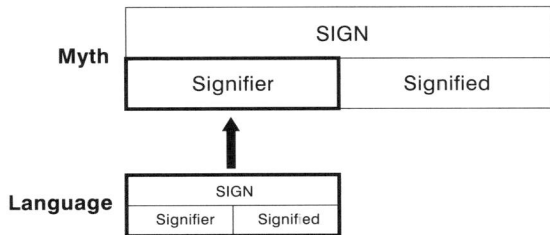

Figure 14.5   Signified and signifier are conjoined in the sign, the whole of which is seized by myth to be the signifier in its second-order semiological system. Barthes cautions that the spatialization here of the pattern of myth is only a metaphor. (Redrawn from the diagram, page 115 of Barthes, 1972.)

In our case, at the level of language we have as signifier the various marks that appear on the legend: the red square, the black dashed line, the bright blue asterisk; and as signified the respective phrases: 'Points of Interest', 'Ferry' and '24 Hour Hospital Emergency Service'. Taken together, the marks and phrases are *signs,* things which *in their sign function* are no longer usefully taken for themselves (there is no red square 350 yards on a side at Singletary Lake) but as indicative of or as pointing toward something else (a point of interest called Singletary Lake Group Camp). Collectively, these signs comprise the legend, *but this in turn is a signifier in another semiological system cantilevered out from the first.* At this level of myth we have as signified some version of what it might mean to be in North Carolina, some idea of its attractiveness (at least to a specifiable consumer), a concept signed also in the photos decorating the other side of the map, in the governor's message, in the 'Motorist's Prayer', a concept we could call 'North Carolinaness'. The signifier is, of course, the legend appropriated from the level of language by this myth to be its sign. Insidiously, this myth is not required to declare itself in language: this is its power. At the moment of reception, it evaporates: the legend is only a legend after all. One sees only its neutrality, its innocence. *What else could it be? It is after all a highway map!*

Indeed. *And so it is.* It is precisely this ambiguity that enables myth to work without being seen. Hidden on top of a primary semiological system, it resists transformation into symbols. As a legend or a map or a photograph, it retains always the fullness, the presence, of the primary semiological system to which it is endlessly capable of retreating. What viewed obliquely appears an advertising slogan, confronted directly is the blandest of legends, so that the slogan, still ringing in one's ears, is apprehended as no more than the natural echo of the facts of the map. It is in this way that North Carolinaness comes to be accepted as an attribute of the terrain instead of being seen as the promotional posture of state government it actually is. This constitutes, in Barthes' phrase, 'the naturalization of the cultural':

> *This is why myth is experienced as innocent speech: not because its intentions are hidden – if they were hidden they could not be efficacious – but because they are naturalized. In fact, what allows the reader to consume myth innocently is that he does not see it as a semiological system but as an inductive one. Where there is only an equivalence, he sees a kind of causal process: the signifier and the signified have, in his eyes, a natural relationship. This confusion can be expressed otherwise: any semiological system is a system of values; now the myth consumer takes the signification for a system of facts: myth is read as a factual system, whereas it is but a semiological system. (Barthes, 1972: 131)*

Not seen as a semiological system: this is the heart of the matter. Of all the systems so not seen, is there one more invisible than the cartographic? The most fundamental cartographic claim is *to be a system of facts,* and its history has most often been written as the story of its ability to present those facts with ever increasing accuracy. That this system can be corrupted everyone acknowledges: none are more vehement in their exposure of the 'propaganda map' than cartographers, but having denounced this usage they feel but the freer in passing off their own products as untainted by the very values which alone constitute the structure of a semiological system. It may no longer appear that an official state highway map is quite such a system of facts as it might previously have been supposed; but this is essentially a consequence of our presentation. Outside of this context, a highway map is accepted as inevitable, as about as natural a thing as can be imagined. Its presence in glove compartments, gas station racks and the backs of kitchen drawers is taken for granted. Yet as we have shown, even so innocent a part of the map *as the legend* carries an exhausting burden of myth, to say nothing of the prayer, governor's message, photographs and other paraphernalia cosseting the map image itself.

Nor does the map image escape the grasp of myth. On the contrary, it is more mythic precisely to the degree that it succeeds in persuading us that it is a natural consequence of perceiving the world. A state highway map, for instance, is unavoidably a map *of the state:* that is, an instrument of state polity, an assertion of sovereignty. There was, for example, no need from the perspective of the driver to have coloured yellow the states contiguous to North Carolina on its highway map. There was no *real* need to have shown the border. It is not, after all, as though the laws regulating traffic changed much at the borders, though to the extent they do, the map is silent.[5] At this level of language the map, like the legend, *seems* to proffer vital information; but it's an impression hard to sustain: *there is too little information to make what's provided useful.* Like the legend, the map in this regard makes no sense. From the perspective of myth, however, this delineation of the state's borders is of the essence. Though many will see in this only the most dispassionate neutrality (what could be more natural than the inclusion of the state's borders on its highway map?), there is nothing innocent about the map's affirmation of North Carolina's dominion over the land in white. Not only has effective territorial control long been dependent on effective mapping, but it is amongst other things the repetitive impact of the image of the territory mapped that lends credence to the claims of control (and hence the extensive logogrammatic application of the state's outline to seals, badges and emblems). Who would question the pretensions, the right to existence, the reality of North Carolina? Look! There it is on the map! The 1.6 million copies of the 1986–1987 edition of this map constitute 1.6 million assertions of the state's sovereignty, assertions which, however, at the moment of being noticed have the ability to fade back into the map where their appearance is taken entirely for granted, overlooked because expected, naturally a part of the surface.

Which is myth's way: the map is always there to deny that the significations piled on top of it are there at all. It is only a map after all, and the pretence is that it is innocent, a servant of the eye that sees things as they really are. But outside the world of speech, outside the world of maps, states carry on a precarious existence: little of nature, they are much of maps, for to map a state is to assert its territorial expression, to leave it off to deny its existence. Only when it is admitted that a state unrecognized (unmapped) is scarcely a state, that it is the determination (choice) of people to acknowledge (map) it that endows with substance an assertion of statehood, or not to acknowledge (map) it that relieves it of significance, is it possible to comprehend the anger directed at maps that acknowledge the independent existence of Bophuthatswana, Transkei, Ciskei and Venda; that deny the independent existence of Taiwan; or that, for that matter, run county borders through Indian Reservations, such as those of Swain and Jackson through the Cherokee Indian Qualla Boundary on the North Carolina highway map. It is not that the map is right or wrong (it is not a question of accuracy), but that it takes a

---

[5] This is even more obvious at the county level: it would be genuinely helpful to distinguish counties prohibiting the sale of alcoholic beverages from those selling beer and wine and mixed drinks. But in fact. the carefully delineated counties are not distinguished in any way. Then why show them? It is not a question that can be answered at the level of language. Only on the level of myth is their presence explicable, where North Carolina (and any other state), defender of states' rights (as it has to be), can be seen to dissolve in turn into its constituent counties, their boundaries an unscreened application of the yellow used to demarcate the sovereignties surrounding North Carolina, leaking, as it were, into the state via these county edges.

stand while pretending to be neutral on an issue over which people are divided.[6] Nor is it that those angered have confused the map with the terrain, but that they recognize what cartographers are at such pains to deny, that, like it or not, willingly or unwillingly, because *au fond* maps constitute a semiological system (that is, a system of values), they are ever vulnerable to seizure or invasion by myth. They are consequently, in all ways less like the windows through which we view the world and more like those windows of appearance from which pontiffs and potentates demonstrate their suzerainty, not because cartographers necessarily want it this way, but because given the manner in which systems of signs operate, they have no choice.

Paradoxically, it is an absence of choice founded on choice alone, for to choose is to reveal a value, and a map is a consequence of choices amongst choices. That the choice of mapping Bophuthatswana as an independent nation reveals a political attitude is something many will readily concede. But all choices are political, and it is no less revealing to choose to map *highways,* for this also is a value. That it would be difficult to produce a state highway map without highways is admitted, but there is no injunction on the state to map its roads anymore than there is for it to map the locations of deaths attributable to motor vehicles, or the density of cancer-linked emissions from internal combustion engines, or the extent of noise pollution associated with automotive traffic. It would be satisfying to live in a state that produced 1.6 million copies of such maps and distributed them free of cost to travellers, tourists, immigrants and industrial location specialists, but states find it more expedient to publish maps of highways. North Carolina *does* publish the *North Carolina Public Transportation Guide* – a highway map-like document displaying intercity bus, train and ferry routes – but it printed 15 000 copies of the most recent edition, less than a hundredth as many maps as it printed of its highways.[7] Not an advertisement, the public transportation map was produced without the assistance of the Department of Commerce. Could this be why, unlike the highway map amongst whose blond hikers, swimmers, golfers and white-water enthusiasts no blacks appear, blacks figure so prominently on the public transportation map? Here blacks buy intercity bus tickets, get on city buses, and in wheelchairs get assisted into specially equipped vans. The reek of special assistance is like sweat: 'Many of you have requested information on how to make your trip without using a private automobile.

---

[6] The issue reduces the editors of *The Times Atlas of the World, Seventh Comprehensive Edition* (Times Books, 1985) to stuttering incomprehensibility: 'In recent years much political significance has been attached to the manner in which international boundaries are depicted and the way names are spelled in atlases. The position *of The Times* as publishers of this and all other atlases has been stated repeatedly and unequivocally. To attempt to judge the rights and wrongs of territorial disputes is beyond the function of the publishers of an atlas . . . In its atlases *The Times* aims to show the territorial situation obtaining at the time of publication without regard to the *de jure* situation in contentious areas or rival claims of contending parties. The aim has always been to inform, to strive for accuracy and to be as up to date as possible. . ..' (viii) and much more gobbledygook of this tone. What can it mean? They do not attempt to judge, but they do attempt to show the situation obtaining: and how do they determine this without judging? With its pompous self-serving attitude this comes close to a perfect example of doublespeak, avowing everything while saying nothing much at all.

[7] North Carolina publishes the edition size and cost per copy on all public documents. The 1986 – available *Public Transportation Guide* – the map's second edition – carries a 1985 date. Curiously, while the governor's wife's photograph graces the highway map, it is missing from the guide, where he stands alone.

Because of these requests. . .' but there is nothing of this tone on the highway map. There was never any *need* to have requested a highway map: it, after all, *is a natural function of the state.* Everything conspires to this end of naturalizing the highway map (even the map of public transportation), of making the decision to produce such a map seem less a decision and more a gesture of instinct, of making its cultural, its historical, its political imperatives transparent: you see through them, and there is only the map, innocent, of nature, of the world as she really is.

14.3 Codes

It is, of course, an illusion: *there is nothing natural about a map.* It is a cultural artefact, an accumulation of choices made amongst choices every one of which reveals a value: not the world, but a slice of a piece of the world; not nature but a slant on it; not innocent, but loaded with intentions and purposes; not directly, but through a glass; not straight, but mediated by words and other signs; not, in a word, as it is, but in *code.* And of course it's in code: *all* meaning, *all* significance derives from codes, *all* intelligibility depends on them. For those who found their codes in the breakfast cereal box – little cardboard wheels arbitrarily linking letters and numbers – this generalization of the idea may occasion some disquiet. It shouldn't. When you wear a tie to work, you're dressing in code. When you frown, you're expressing in code. When you open a door for a lady – or wait for a man to open a door for you – you're gallanting in code. When you type or scribble, you're writing in code. Human languages are probably the most elaborate and complex codes we're familiar with – and the dictionary just a big clumsy breakfast cereal toy – but there are sublinguistic codes of incredible sophistication (those danced by Ginger Rogers and Fred Astaire) and supralinguistic codes of deep subtlety (such as the conventions underwriting the structure of James Joyce's *Ulysses*). Usually a number of different codes are used simultaneously (this is a text). Fred and Ginger were placed in settings, dressed, wore their hair a certain way, gestured, spoke and sang as well as danced and all was coded. The code of conventions structuring *Ulysses* cannot be encountered outside the code of English in which it is embedded. There is even a code of codes: a mime, for example, is forbidden the code of words, and in general the arts are distinguished by a code whose elements are other codes. It has long been a hallmark of cartography that it speaks in art as well as science.

More technically a code can be said to be an assignment scheme (or rule) coupling or apportioning items or elements from a conveyed system (the signified) to a conveying system (the signifier). The highway code is paradigmatic of the way this works. On the one side are intentions (she intends to turn), promises (Holly Springs will be encountered three miles down this road) and commands (not to pass, to stop, to go). On the other side are gestures (a hand stuck straight out the driver's window), words and numbers ('Holly Springs/3 miles'), and lights and lines (a red traffic light, a solid yellow line down the middle of the road). The intentions, promises and commands are elements of the system conveyed: signifieds (content). The gestures, words, numbers, lines and lights are elements of the system conveying (expression). The code (the rule – in this case, the Law) *assigns* the latter to the former, couples them. In so doing, it creates a *sign.*

An important distinction is being made here. The sign is *not* in the gestures or the lights, the words or the numbers: it is *not* the signifier. Nor is the sign in the intentions, promises or commands: it is *not* the signified. The sign exists solely, utterly and exclusively in its correlation (established by the code, the rule, by custom, by the law). There is nothing, for instance, inevitable (necessary) in the relationship between a driver sticking his arm straight out the left window and his intention to turn left (and in fact it has been largely supplanted by the flashing of lights on the left side of the car), any more than there is between a driver pointing to heaven and his intention to turn right (though doubtless there was some historical contingency that made it customary). They might, however, quite readily change places (may have already in some parts of the world), so that a left arm stuck straight out a left window signalled an intention to turn right and one stuck straight up signalled an intention to turn left: it would make no difference from the perspective of communication, for the meaning is in the code, and the new code could be as readily mastered as the old. Signs, in other words, are the creatures of codes with the loss of which they are rendered – like fat – into their constituent components, disembodied signifieds separated from insignificant signifiers. It is the codification in which the sign adheres, nothing else. Or, as Umberto Eco (1976: 48–49) puts it:

> *A sign is always an element of an* expression plane *conventionally correlated to one (or several) elements of a* content plane. *Every time there is a correlation of this kind, recognized by a human society, there is a sign. Only in this sense is it possible to accept Saussure's definition according to which a sign is the correspondence between a signifier and a signified. This assumption entails some consequences:* **a** a sign is not a physical entity, *the physical entity being at most the concrete occurrence of the expressive pertinent element;* **b** a sign is not a fixed semiotic entity *but rather the meeting ground for independent elements (coming from two different systems of two different planes and meeting on the basis of a coding correlation).*

Because signs neither have physical existence (unlike the signifier) nor permanence, they are frequently referred to as *sign functions,* or in Eco's (1976: 49) words:

> *Properly speaking there are not signs, but only* sign functions . . . *A sign function is realized when two* functives *(expression and content) enter into a mutual correlation; the same functive can also enter into another correlation, thus becoming a different functive and therefore giving rise to a new sign function. Thus signs are the provisional result of coding rules which establish* transitory *correlations of elements, each of these elements being entitled to enter – under given coded circumstances – into another correlation and thus form a new sign.*

This is not a game of words. The vocabulary is not important (not to us). What *is* important is the notion that signs – or sign functions or symbols: what they are called *does not matter* – that signs, to repeat, are realized *only* when coding rules bring into correlation two elements or items (or functives) from two domains or systems (the one signifying, of expression; the other signified, of content); and that *whenever* there is such a correlation, there is a sign. You may call this resulting sign an icon. You may call it a pictogram. You may call it a word. You may call it an index. You may call it a symbol. You may call it a piece of sculpture. You may call it a sentence. You may call it a map. You may call it New York City. In every case, whatever else it is, it is, *in its sign function*, also a sign, that is, a creature of a code.

No signs without codes. It must be insisted upon. That is, no self-explanatory signs. No signs which so resemble their referents as to self-evidently refer to them. They are inevitably arbitrary (inevitably reveal a value). Writing about the way Saussure and Peirce occasionally came to similar conclusions from different assumptions, Jonathan Culler (1981: 24) says:

> *Saussure, taking the linguistic sign as the norm, argues that all signs are arbitrary, involving a purely conventional association of conventionally delimited signifiers and signifieds; and he extends this principle to domains such as etiquette, arguing that however natural or motivated signs may seem to those who use them, they are always determined by social rule, semiotic convention. Peirce, on the contrary, begins with a distinction between arbitrary signs, which he calls 'symbols', and two sorts of motivated signs, indices and icons, but in his work on the latter he reaches a conclusion similar to Saussure's. Whether we are dealing with maps, paintings, or diagrams, 'every material image is largely conventional in its mode of representation'. We can only claim that a map actually resembles what it represents if we take for granted and pass over in silence numerous complicated conventions. Icons seem to be based on natural resemblance, but in fact they are determined by semiotic convention.*

Once the superordinate role of the convention (the rule, the code) is accepted it becomes easy to explain how what 'self-evidently' resembles a river on a map equally 'self-evidently' resembles veins on a diagram of the circulatory system, without invoking complicated principles of metaphor (not that these might not have been operant in the genesis of the sign). It is not that the reader thinks, 'Oh, yes, the deoxygenated blood is relatively bluer than that in the arteries, *and* under a clear blue sky the surface of rivers often seems blue; *and* both veins and arteries carry [whatever "carry" means] liquids in a branching [see "tree"] network [see "net", see "weaving"], sooo, let's see, that means . . .' This is not how it happens at all. What happens is that the reader finds himself in an entirely distinct coded circumstance *all at once*. At the level of language the diagram of the circulatory system is decoded without reference to the codes of the map, and vice versa. There is certainly no question of *resemblance* with respect to which Barthes (1981, 100–102) notes that it would be in any case a resemblance *to an identity* (the *identity* of the river, the *identity* of the vein), an identity 'imprecise, even imaginary, to the point where I can continue to speak of "likeness" without ever having seen the model', as those do who justify this sign for veins because 'they look like veins' without ever having seen a vein (without having seen a hepatic vein, without having seen an inferior vena cava), or the sign for a river (the Colorado) because 'it looks like a river' (the Thames? the Cuyahoga?) without having seen it (without having seen where the Colorado trickles all but dry into the Gulf of California).

It is not a matter of resemblance: the blue line is a blue line. It is the code that does the work, not the signifier. If there is involved an iconicism it is always at the level of the structure of the system (it is analogic not metaphoric). It is less the blueness of deoxygenation that says 'veins' than the simultaneous redness of the arteries, their characteristic jointure at the extremities, and their perfect parallelism; it is less the blue-between-black lines that says 'river' than its characteristic form, its characteristic relationship to other forms (other rivers, mountains, roads, towns and oceans); so that 'veins' can as easily be read in black or grey, and 'rivers' in diagrams of drainage basins and maps of flood insurance purchase. To say that it is the code that does the work, not the signifier, is just another way of saying that it is the code that makes the sign, not the mark.

So it is the *codes* upon which one must fasten if the map is to be decoded (or if a map is to be encoded). We think it possible to distinguish ten of these (there are doubtless others), which either the map exploits, or by virtue of which the map is exploited. Neither class is independent of the other, and no map fails to be inscribed in (at least) these ten codes. Those which the map exploits we term *codes of intrasignification*. These operate, so to speak, within the map: at the level of language. Those by virtue of which the map is exploited we term *codes of extrasignification*. These operate, so to speak, outside the map: at the level of myth.

Amongst the codes of intrasignification five at least are inescapable, the *iconic*, the *linguistic*, the *tectonic*, the *temporal* and the *presentational*. Under the heading *iconic* we subsume the code of 'things' ('events'), with whose relative location the map is enrapt: the streets of Genoa, rates of death by cancer, exports of French wine, the losses suffered in Napoleon's Russian campaign, airways, subways, the buildings of Manhattan, levels of air pollutants over six counties in Southern California, the rivers, roads, counties, airports, cities and towns of North Carolina. The iconic is the code of the inventory, of the world's fragmentation: into urban hierarchies, into hypsometric layers, into wet and dry.

The *linguistic* is the code of the names: the *Via Corsica*, the *Corso Aurelio Saffi*; trachea, bronchus and lung cancer, white males, age-adjusted rate by county, 1950–1969; *France, Amérique du Nord; Moscou, Polotzk*; DME chan 82 St John vsj 113.5; Cortland St World Tr Ctr N RR PATH; the Graybar Building, the Seagram; Orange County, Reactive Hydrocarbons; Cape Fear River, us 421, Pasquotank, Cherry Pt., Winston-Salem, Hickory. The linguistic is the code of classification, of ownership: identifying, naming, assigning.

The relationships of these things in space is given in the *tectonic* codes: in the *scalar* – in the number of miles (or feet) encoded in every inch – and in the *topological* – in the planimetry of cities, the stereometry of mountain ranges, the projective geometry of continents, the topographometry of the field traverse, the simple topology of the sketch map giving directions to the cocktail party. The tectonic is the code of finding, it is the code of getting there: it is the code of getting.

Because there is no connection, no communication, except in time, the codes of filiation are *temporal,* codes of duration, codes of tense. The *durative* establishes the scale, the map's *durée*, its 'thickness': as the map of rates of death from cancer, 1950–1969, is 'thicker' than the 1978–79 North Carolina highway map, which is 'thicker' than the map of reactive hydrocarbons, 6 to 9 a.m., July 22, 1979. The durative reveals (or hides or is mute about) lapses in cosynchronicity. The *tense* says when: some maps are in the past tense ('The World of Alexander the Great'), others in the future tense ('Tomorrow's Highways'), but most maps plump for the present ('State of the World Today'), or, if they can possibly get away with it, the aorist: no duration at all (no thickness), out of chronology (not lost – just out of it): free of time. These attain to myth at the very level of language.

Each of these codes – iconic, linguistic, tectonic and temporal – is embodied in signs with all the physicality of the concrete instantiation of the expressive pertinent element. On the page, on the sheet of paper, on the illuminated display with its flashing lights, these concrete instantiations are ordered, arranged, organized by the *presentational* code: they are *presented*. Title, legend box, map image, text, illustrations, inset map images, scale, instructions, charts, apologies, diagrams, photos, explanations, arrows, decorations, colour scheme and type faces are all chosen, layered, structured to achieve speech: coherent, articulate discourse. It is a question of the architecture of the picture plane, what's in the centre and what's at the edge, what's in fluorescent pink and what's in the blue of Williamsburg, whether the paper crackles with (apparent) age or sluffs off repeated foldings like a rubber sheet, whether the map image predominates or the text takes over. It is never, even at the lowest level, a question merely of escaping the stigmas of paranomia and aphrasia, dysphemia and idiolalia, dyslogia and cacology. From the very beginning it is a matter of fluency and eloquence, and soon enough of vigour and force of expression, of rhetoric, of polemic, for wherever it may begin, the code of presentation soon enough carries the map out of the domain of intrasignification into that of extrasignification, into that of the culture that insists upon its existence, that nurtures it, that consumes it.

Amongst the codes of extrasignification five again are inescapable, the *thematic*, the *topic*, the *historical*, the *rhetorical*, and the *utilitarian*. All operate at the level of myth, all make off with the map for their own purposes (as they made the map), all distort its meaning (its meaning at the level of language), subvert it to their own. If the presentational code permits the map to achieve a level of discourse, the *thematic* code establishes its subject: *on what shall the map discourse? What shall it argue?* Though it is precisely the thematic code that has dictated their appearance on the map, from the perspective of the reader the theme is experienced as a latency inherent in the 'things' iconically encoded in the map: roads, for instance, it is a map of roads and highways, it asserts the significance of roads and highways (if only by picturing them, if only by foregrounding them), its theme is Automobility (the legitimacy of Automobility). Or it is a general reference map, a map of hydrography and relief carved into political units and plastered with railroads and towns, a map, that is, of a landscape smothered by humanity, tamed, subdued (the red railroads – sometimes black – inevitably reminiscent of the bonds by means of which the Lilliputians restrained Gulliver), its theme is Nature Subdued.

And precisely as the thematic code runs off with the icons, so the *topic* code (with a long *o* from *topos*, place, as in *topography*, not *topicality*) runs off with the space established by the tectonic code, turns it from space to place, bounds it (binds it), gives it a name, sets it off from other space, asserts its existence: *this place is*.

Ditto the *historical* code. Only it works on the time established in the map by the temporal code. Are there bounding dates to the map's *durée?* Then the historical code appropriates them to an era, assigns it a name, incorporates it in a vision of history. So an archaeological map of Central America acquires the title, 'Before 1500/Pre-Columbian Glory', one of nineteenth century plantation crops, political units, selected urban places, cart roads, railroads and battles the title, '1821–1900/Time of Independence', yet another of similar subjects (though with the addition of a sign for refugee centres) the caption, '1945–Present/Upheaval and Uncertainty'.[8] There is no time that cannot be reduced to these sequacious causal schemata, absorbed into these platitudes, made comfortable and safe because grasped, understood.

If the thematic code sets the subject for the discourse, if the topic and historical codes secure the place and time, it is the *rhetorical* code that sets the tone, that, having consumed the presentational code, most completely orients the map in its culture (in its set of values), pointing in the very act of pointing somewhere else (to the globe) to itself, to its maker, to the culture that produced it, to the place and time and omphalos of that culture – the more dramatically as the aspect of the globe toward which it points is alien, is exotic, that is can have its title set in a type that mimics bamboo (Figure 14.6). It is a code of jingoisms, a code that beats its chest like Tarzan, a code of the sort of subtle chauvinisms that encourages the *National Geographic* to call a road a 'road' on its map of the Central Plains, 1803–1845, but to call it a 'cart road' on its map of Central America, 1821–1900.[9] But, after all, it *is* an 'American' map, that is, a map that reflects the genius of the *North* Americans, or at least those north of the Rio Grande (for according to the *National Geographic* the ancient Maya had but 'trade routes' and even the Camino Real is just a 'trail'); and, if only because it *is* the mapping culture, the mapping culture stands at stage centre, with all the others in the wings. For the rhetorical code the mere existence of the map is a sign of its higher culture, its sophistication: it is rhetorical *au fond*, and for this reason no map can eschew it. It is like clothing: even not to wear it is to be caught in the net of meanings woven by the code of fashion. To attempt to shed the rhetorical code is but to shout the more stridently through it: it is the very disregard for the subtler aspects of the code of presentation that so completely characterizes the publisher of *The Nuclear War Atlas* as 'socially conscious'[10]; it is nothing other than their violations of good taste that allow us to read the editors of *The State of the World Atlas* as angry.[11] Their qualified refusal of the power of the rhetorical code amounts to a bold proclamation of their rhetorical stance (cartographic nudism, cartographic streaking, cartographic punk), the very opposite of the position occupied by the United States Geological Survey, which obscures its stance beneath a rhetorically orchestrated denial of rhetoric (Brooks Brothers' shirts, clean classic clothing). The rigorous dispassion of a

---

[8] These examples come from the verso of 'Central America', published as a supplement to the *National Geographic* (April 1986: 466A).

[9] The Central America map is as cited above. That of the Central Plains comes from the verso of 'Central Plains', published as a supplement to *the National Geographic* (September 1985: 352A).

[10] *The Nuclear War Atlas,* a two by four foot sheet with 28 two-colour maps recto and text verso is published by The Society for Human Exploration, Victoriaville, Quebec. Our copy is undated.

[11] Kidron and Segal (1981), *The State of the World Atlas* (Simon and Schuster, New York). There is a second edition (1984), *The New State of the World Atlas* (Simon and Schuster, New York).

Figure 14.6 A television weatherman points to a map. At the same time, it points back to him, establishing and emphasising his modernity, sophistication, and thus his reliability. In turn, this flatters our sense of self-esteem for having selected this station over others. This is a map all but consumed by its rhetorical function.

topographic survey sheet is seductive precisely in the degree to which no sign of seduction is apparent: the message of Nature Subdued is the more powerful because it seems to be spoken not by the map (*it* appears to say nothing, appears to *allow* the world to speak) but by Nature itself. Here the map dresses itself in the style of Science. Elsewhere it will dress in the style of Art. Or in the style of the Advertisement. Or in the Vernacular (the North Carolina Highway map). The rhetorical code appropriates to its map the style most advantageous to the myth it intends to propagate. None is untouchable. All have been used.

As the map itself is finally used, picked up bodily by the *utilitarian* code to be carted off for any purpose myth might serve. A professor of curriculum and instruction, commenting on the availability of state highway maps for secondary classroom use, remarks, 'It has the governor's picture on it. You can get as many as you want'. It is here that the academic model of the map with its scanning eyes and graduated circle-comparing minds breaks down most completely. It has no room for the real uses of most maps, which are to possess and to claim, to legitimate and to name. Which great king, which emperor, which great republic has failed to signal its coming of age by the mapping of its domains? Whatever the pragmatic considerations (they are, after all, maps that speak also at the level of language), it has inevitably also been an act of conspicuous consumption, a sign of contemporaneity as well as wealth and power, a symbolic manifestation of the rights of possession. *These* are the uses of maps as certainly as it is the most important function of maps in geographic journals to certify the geographic legitimacy of the articles they decorate. The anthropology of cartography is an urgent project: what *are* all those maps actually used for? Signs, badges, tokens, emblems, billboards, gestures, leases, deeds, wallpaper, pretty picture: and do not say *not this one* – not a topographic survey sheet – for as surely as you do it will turn out to be that one with the most heinous agenda. Or that may be putting it too strongly. And yet this is how A.S. Hewitt, the man who in 1879 wrote the Geological Survey's enabling legislation, puts it in the epigram to the Survey's centennial history: 'What is there in this richly *endowed* land *of ours which may be dug*, or gathered, *or harvested*, and made part of the wealth of America and of the world, and how and where does it lie?' (Thompson, 1979: v). Whatever else this might be, it is not a language of disinterested curiosity, it is a language of exploitation. Dressed in their button-down white shirts and suitable ties it is the language spoken by the survey sheets as well, in their metered regularity (so many sheets per unit area), in their sensible no-nonsense layout, in their methodical tiling, their obsessive coverage. 'To catalogue', Barthes (1980: 27) notes, 'is not merely to ascertain, as it appears at first glance, but also to appropriate'. How are survey sheets different from maps of military targets?

14.4 Intrasignification

Clearly, the map is comprehended in two ways (Figure 14.7). As a medium of language (in the broadest sense) it serves as a visual analogue of phenomena, attributes and spatial relations: a model on which we may act, in lieu or anticipation of experience, to compare or contrast, mensurate or appraise, analyse or predict. It seems to inform, with unimpeachable dispassion, of the objects and events of the world. As myth, however, it refers to itself and to its makers, and to a world seen quite subjectively through their

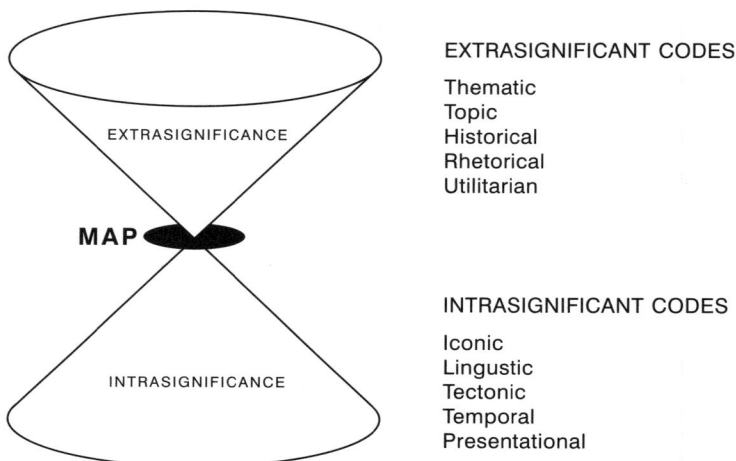

EXTRASIGNIFICANT CODES

Thematic
Topic
Historical
Rhetorical
Utilitarian

INTRASIGNIFICANT CODES

Iconic
Lingustic
Tectonic
Temporal
Presentational

Figure 14.7 The map as a focusing device between the domains of extrasignification and intrasignification: the map gathers up the constituent signs governed by the codes of intrasignification so that they will be able to act as signifiers in the sign functions governed by the codes of extrasignification – which specified them in the first place.

eyes. It trades in values and ambitions; it is politicized. Signing functions that serve the former set of purposes we have termed intrasignificant; those which serve the latter, extrasignificant. Whereas intrasignification consists of an array of sign functions indigenous to the map and which, taken jointly, constitute the map *assign,* extrasignification appropriates the complete map and deploys it *as expression* in a broader semiotic context. The map acts as a focusing device between these two planes of signification, as a lens that gathers up its internal or constituent signs and offers them up collectively *as a map.* But what effers from the map is not substantially different from what is afferent upon it – these have simply been repositioned in the semiological function – and, while extrasignification exploits the map in its entirety, we have seen how the initiatives of myth extend to even the most fundamental and apparently sovereign aspects of intrasignification, and are ultimately rooted in them. These aspects require our further attention.

The map is the product of a spectrum of codes that materialize its visual representations, orientate it in space and in time, and bind it together in some acceptable form. The actions of these codes are, if not entirely independent, reasonably distinct. *Iconic codes* govern the manner in which graphic expressions correspond with geographic items, concrete or abstract, and their attendant attributes. A *linguistic code* (occasionally two or several) is extended to the map to regulate the equivalence of typographic expressions and, via the norms of written language, a universe of terminology and nomenclature. As the space of the map is configured by *tectonic codes* – transformational procedures prescribing its topological and scalar relations to the space of the globe – *temporal codes* configure the time of the map in relation to the stream of events and observations from which it derives. The diversity of expressions that constitute the map are organized and orchestrated through a *presentational code* that fuses them into a coherent cartographic discourse.

14.4.1 Iconic Codes

Iconicity is the indispensable quality of the map. It is the source and principle of the map's analogy to objects, places, relations and events. In its capacity as geographic icon, the map subsumes a remarkable variety of visual representations and the codes, both general and specific, that underwrite them; yet the degree of iconicity evident in the map as a whole is not uniformly echoed amongst its constituents. The dot that represents a town is not iconic in the same way as the intricately shaped area representing a city; the blue line representing a river is not iconic in the same sense as the blue line representing a county road or, for that matter, a shoreline. Pursued far enough, every icon is seen as the product of two procedures: a symbolic (substitutive) operation that provides the basis of its *representative potential*, and a scheme of arrangement that yields its *specific and individual form*. The balance struck between these has frequently been the canon by which we judge representations as symbolic (of the town, for example) or iconic (of the city); and while this distinction will not be abandoned here, it will be applied with extreme care. No symbol is totally arbitrary unless it can be stripped entirely of connotation (an unlikely and undesirable prospect) and no icon is motivated free of convention because *representation without convention is not possible*. We can only say that some representations are more explicitly iconic or symbolic in function; and that media of cultural exchange – maps in particular – serve as proving grounds where iconic representations gradually acquire symbolic status through a process of reiteration and cultural distension.

The iconicity of Hermann Bollmann's 'New York Picture Map'[12] is so powerful that its representational conventions virtually disappear from view (Figure 14.8). On inspection, the picture plane melts away and our attention falls into a landscape of tangible urban forms: streets, sidewalks, roofs, facades, doors, windows. It seems so literal, so transparent to interpretation, so *natural* that it is difficult to accept as a highly conventionalized and essentially symbolic representation. Yet without our conventions of pictorial rendering this arresting image would be opaque and meaningless.[13] Make no mistake. Iconicity, as Bhattacharya (1984) has explained, is the product of a spatial transcription; and its derived form is an arrangement of marks in relation to one another and to the space they occupy.[14] The icon is motivated not by a monolithic precedent form but by the formal and necessarily spatial *arrangement* it would transcribe on the page, and it can only materialize through a transcriptive procedure. This procedure, in Bollmann's map, turns out to be extraordinarily elaborate: involving 67 000 photographs taken with specially-designed cameras, an axonometric projection spread in two dimensions by a calculated widening of streets, and, according to the map's jacket, 'several unique devices which remain his secret'. It emerges from a tradition of representation that is distinctly Western and intensively codified, and it speaks through a familiar (to us) regime of symbolic principles: lines demark intersections of planes and boundaries between solid and void; certain organizations of lines denote rectilinear volumes; recurring tonal patterns denote illuminated forms. To describe iconicity as a simple matter of visual likeness (as if this *could be* a simple matter), or as a formal correspondence between expression and referent, is to mystify its explanation and divorce it entirely from cultural enterprise. Iconicity derives from our ability to transcribe arrangements in space and mark them out in

---

[12] The *New York Picture Map* was created by Hermann Bollman for Pictorial Maps Incorporated, New York. In our copy, the recto carries Bollmann's rendering of midtown Manhattan in five colours, and the verso a two-colour planimetric map of the city of New York. Approximately 34 by 43 inches (86 × 100 cm), the map sheet folds to fit a jacket that includes forty-eight pages of text. It is not dated.

[13] Gregory (1973: 160–176) identifies personal experience and the geometry of environment as key ingredients of our ability to decode perspective transcriptions. He remarks that: 'In connection with the non-Western people, it is perhaps worth adding that they make little or nothing of drawings or photographs of familiar objects, and this was also true of the blind man made to see. It is likely that perspective cues are made use of only after considerable experience, when they are related to touch, and that it is only then that appropriate perspective cues give rise to distortions of size in flat figures'. In 'Pictorial Perception and Culture' (*Scientific American*, November 1972: 82–88), Jan Deregowski summarizes the application of Hudson's pictorial perception tests to a variety of African tribal and linguistic groups: 'The results from African tribal subjects were unequivocal: both children and adults found it difficult to perceive depth in the pictorial material. The difficulty varied in extent but appeared to persist through most educational and social levels'.

[14] This, and several of the references that follow, pertain to a special issue titled *The Semiotics of the Visual: On Defining the Field*, edited by Mihai Nadin. *Semiotica* is a Mouton publication, and is available in the USA and Canada through Walter de Gruyter, Inc., Hawthorne, New York.

conventional symbols – in other words, to *map* them. This ability is as fully realized in the drawing by da Vinci as in the Swiss topographic map, where the natural landscape – like Bollmann's urban landscape – is portrayed as a complex and continuous icon, bathed in light and rendered with the consummate authority of an iconism as richly meaningful for its audience as for its maker (Figure 14.9).

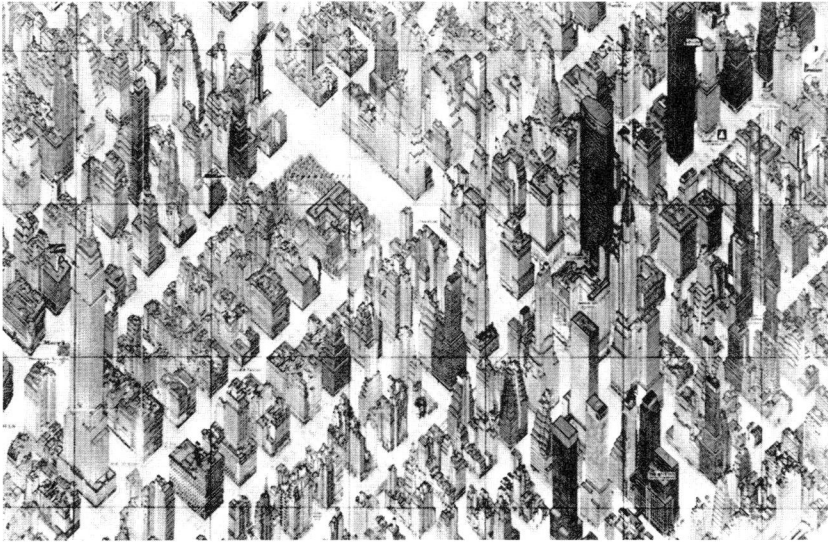

Figure 14.8 A portion of Bollmann's Manhattan. This compelling icon is an elaborate synthesis of Western representational conventions. (The map has been out of print in the United States for some time and the status of its US copyright is unclear. We understand that Bollmann has published similar maps of European cities through Bollmann Bildkarten-Verlag KG, Braunschweig, Germany.)

The map of population distribution produced by the US Bureau of the Census[15] has some of this same pretence. Substitute night for day, luminosity for reflectivity, and city form for architectural or geomorphic form, and we have an equally credible – if more remotely viewed – icon of human settlement. But the symbolism of this map is more explicit, and less uniform; in fact, it embraces several distinctly different representative principles. Urbanized areas, like Bollmann's office towers and Imhof's mountains, *enter* the map as geographic icons, shaped by the space of the features themselves transcribed onto the graphic plane. Isolated cities and towns, however, enter as geometrically pure squares and circles regardless of their geographic shape; they have undergone an abstraction conventionalizing their form and enacting their status as symbols.[16] Beyond

---

[15] Map GE-70, No. 1, *Population Distribution, Urban and Rural in the United States: 1970 (nighttime view)*, Bureau of the Census, US Department of Commerce, Washington, DC. The reproduction provided here is taken from a reduced portion published in *Maps for America*, Thompson (1979:13). Original scale is 1:7 500 000.

[16] The distinction being drawn here is essentially the same as that of Schlichtmann (1985). He differentiates 'plan information' from 'plan-free information' on the basis of the former's inclusion of location, and content items contingent thereon (i.e. transcribed shape and extent).

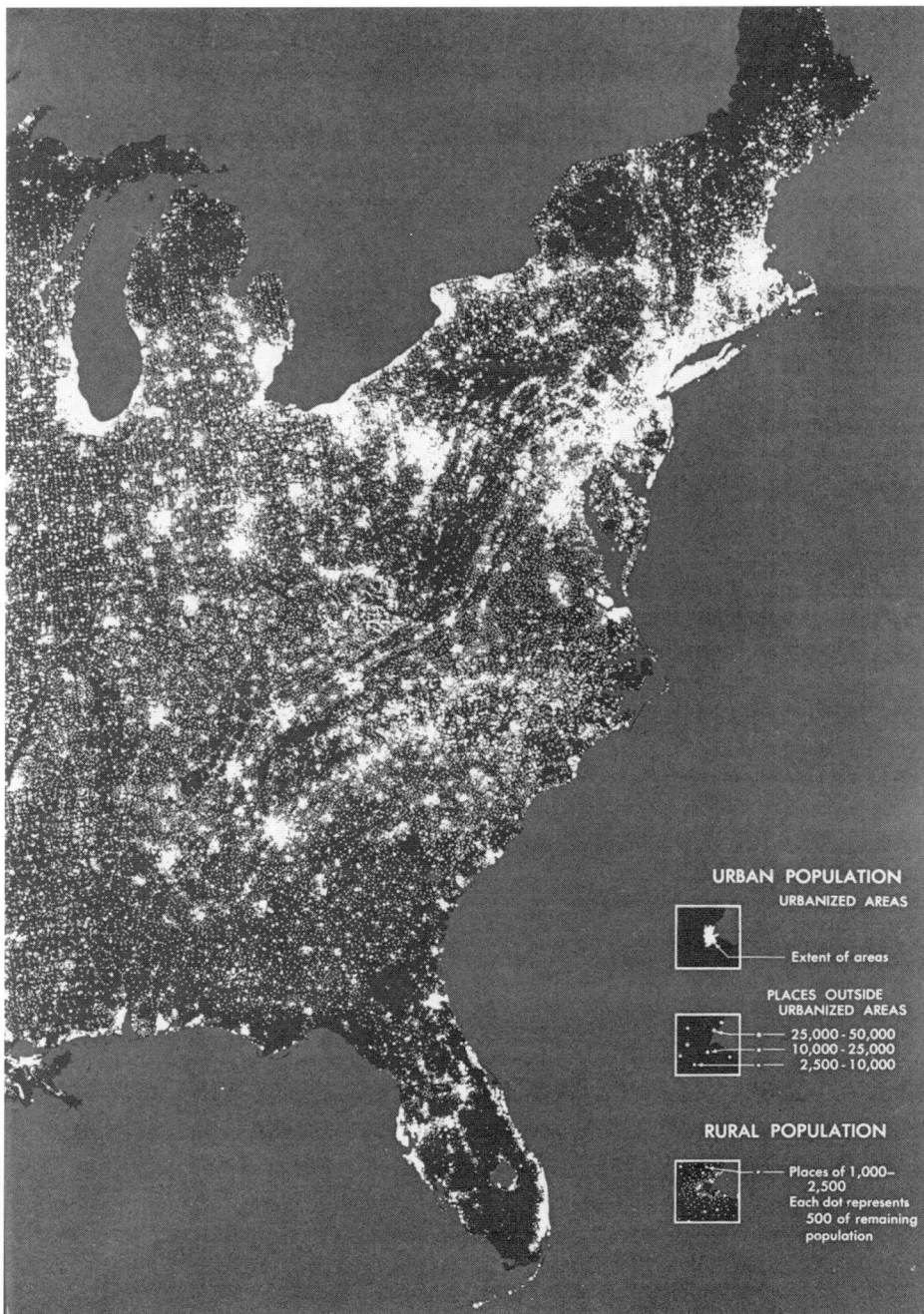

Figure 14.9 From a lexicon of graphic symbols, a geographic icon. While significant in itself, each mark, like a point of colour in a Seurat painting, is subservient to the impression of the whole (Thompson, 1979).

and between these, symbols are disengaged from exact spatial correspondence and referred to features which are in themselves abstractions. In the first instance, form is given as the consequence of the feature's spatial extension and the topological transformation that implants it on the page. Symbolism remains characteristic: white is city, dark blue is water (or foreign terrain), black is neither. In the second instance a formal symbolism is activated: white *square* is city or white *circle* is city. In the third instance, symbols are fixed not only in form but in value as well, and they acquire a limited but necessary mobility within a scheme that treats them not as localized occurrences (in which case they have no literal meaning) but as elements of a comprehensive system to be interpreted *en masse*. This map is truly a *tour de force*, an exemplar of cartographic representation deploying an arsenal of significant strategies from the most abstract and conventionalized to the most geographically constrained and overtly iconic. While we might expect, from this description, a baffling and practically indecipherable stew of signs, what we have instead is a remarkably legible and coherent representation, one which correlates strongly with a photographic representation of the same phenomena.[17] Profound differences of symbolic principle merge, almost seamlessly, in an icon that eschews the formal consequences of their application and takes their distribution as the basis of its own.

Signs formed, rather than just characterized, independently of geographic space are free to engage in formal metaphor. A lighthouse is signed with an ornamented triangle or an outlined circle and a complement of rays, a mine with an occluded dot or an emblematically crossed pick and shovel. Extracted from map context, these signs are icons in their own right – but icons of what? The triangular lighthouse sign and the circular mine sign are ostensible abstractions of their phenomenal counterparts and, regardless of their degree of abstraction, they remain icons insofar as they maintain a structural correspondence with them. But the circle and rays sign is iconic only in respect to the light, not the lighthouse, and it represents by virtue of a part-for-whole substitution. The pick and shovel sign (with no regard for technological currency) represents mining rather than mine by substituting artefact for process. These last two examples are conventional metaphors,[18] parallels to which abound in maps. They differ from the icons of urban form and symbols of city size in not referring literally to the phenomena they represent. They anticipate interpretation by singling out connotations and presenting them as surrogate icons. Icon is proffered, and taken, as symbol.

In signs, which *are* geographically conformal, metaphor operates through characteristic. Green symbolizes trees, and blue water, in our maps with the same conviction they did in the childhood drawings that emplaced these metaphors in our vocabulary. Never mind drought, Autumn, and acid rain, and never mind the cubic miles of eroded

---

[17] Compare, for example, the satellite image reproduced on pages 28 and 29 of the *Atlas of North America: Space Age Portrait of a Continent*, National Geographic Society, Washington, DC, 1985, or that on page 54 of Southworth and Southworth (1982).

[18] The term 'metaphor' is used here in the most general sense of representation through a surrogate interpretant. Johns (1984), distinguishes between metonymy (whole-for-part metaphor) and synechdoche (part-for-whole metaphor). Some authors invert this terminology. Within written language, distinctions among metaphoric types are numerous; but their applications to graphic signs are largely unexplored and of questionable utility.

silt that choke our rivers. In the map, our forests glow with the robust verdure of a perpetual Spring afternoon and even the Mississippi shines with a pristine Caribbean blue. These metaphors proclaim the map as ideal, or at least hyperbole: at once an analogue of our environment and an avenue for cultural fantasy about it. False coloration is hardly restricted to remotely-sensed imagery; it is characteristic of all our maps, which it dresses in the most reassuring tones.

The iconic code of the map is a complex mix of more specific codes – potentially any established or even *ad hoc* code of graphic representation, provided it either is or can be conventionalized. The map seems to have assimilated the entire history of visual communication, maintaining an immense pool of representational techniques and methodologies from which it draws freely, with little preference or prejudice, and which it augments through continual invention and recombination. While this inventory is far too extensive to be catalogued here, we can summarize the object of its application. The map is an icon, a visual analogue of a geographic landscape. It is the product of a number of deliberate, repetitive, symbolic gestures, carefully arranged and explicitly or implicitly referred to elements of a content taxonomy. Formal items – the discrete elements of iconic coding – may be shaped within the space of the map, in which case their symbolism and metaphorical potentials are characteristic, or preformed and imposed on the map, activating formal symbolism and formal metaphor as well. The diversity of cartographic expression far surpasses that of written language or any other medium of practical exchange; but map signs are only as diverse as our abilities to interpret them and their formation is as firmly prescribed by the confines of our own visual culture, the array of conventions that dictate how we may equate marks and meanings. The iconic code of the map is the sum of its various conventions of graphic representation; the comprehensive icon of map image is the synthesis of their actions.

14.4.2 Linguistic Codes

It is difficult to imagine a map without language. However separate the evolution of iconic and linguistic representation, the map has, for millennia, embraced both. External to the map image, language assumes its familiar textual forms: identifying, explaining, elaborating, crediting, cautioning. Its main role, though, lies within the map image and in its interpretive template, the map legend. Like graphic marks, typographic marks sign the content of the map, on different yet complementary grounds.

In the legend, semantic connections are made between classes of graphic images or image attributes and linguistic representations on the phenomena to which they refer. In this capacity, the legend acts as interpreter between the unique semiological system of individual map and the culturally universal system of language; so that on seeing a red circle, for example, we may hear the words 'Welcome Center' (even if we're not entirely sure what that means). If it is legitimate to say that maps are read, then they are read in this respect. In translating graphic expression to linguistic expression we make the map literate, and its meanings subject to literary representation and manipulation. It seems our compulsion and need to do so.

Within the map image, linguistic signs address not only what things are called ('Lake') but also what they are named ('Superior'). Thus identification is a matter of both designation and nomenclature. Much of our geographic nomenclature carries a residuum of designation, as in 'Union City', 'Youngstown', 'Louisville', 'Pittsburgh'; but with respect to natural features it is practically obligatory. One word, 'river' for instance, may occur hundreds of times within a single map image. The cartographer who would erase this redundancy, however, finds that rivers are no longer distinguishable from creeks, nor lakes from reservoirs. Here language is not just naming features, but illuminating content distinctions which have, for whatever reason, escaped iconic coding.

If the function of language in maps were simply toponymic, we could assume that the linguistic signifiers themselves, if recognizably formed and correctly arranged, would be fixed in meaning. This is clearly not the case. Within the map image, elements of visible language serve as counterparts to iconic signs, overlapping their content and spatial domains and echoing their iconic properties. In the map image, entire words and arrangements of words are given iconic license, generating a field of linguistic signs best likened to concrete poetry. Letters expand in size, increase in weight, or assume majuscule form to denote higher degrees of importance. Stylistic, geometric and chromatic variations signal broad semantic divisions. Textual syntax is largely abandoned as words are stretched and contorted and word groups rearranged to fit the space of their iconic equivalents. Clearly this code invokes more than the disposition of phonetic archetypes[19] (Figure 14.10).

It's not that the map rejects the groundrules of textualized language; if it did, it would quickly degenerate to a vehicle for newspeak or nonsense. Even seemingly absurd statements like 'Lac Champlain Lake' and 'Rio Grande River' are grammatically functional in a bilingual or multilingual culture. What this code gains in the cartographic context is nearly unrestricted access to the means of iconic coding. Amongst attempts to produce maps entirely from linguistic signs, the more successful have been cognisant of these means[20]; and in even the most familiar maps the field of topographic signs, taken on its own, visualizes the geographic landscape in much the same way as the field of graphic signs. The map is simultaneously language and image. As word lends icon access to the semantic field of its culture, icon invites word to realize its expressive potentials in the visual field. The result is the dual signification virtually synonymous with maps, and the complementary exchange of meaning that it engenders. The map image provides a context in which the semantics of the linguistic code are extended to embrace a variety of latent iconic potentials[21]; to the same end, it imposes a secondary syntax which shapes entire linguistic signifiers into local icons.

---

[19] Bartz (1969) summarizes the iconic ('analogous') characteristics of letterforms in the cartographic context as those referring to location (point location, linear and areal extent, shape and orientation of feature), quality, quantity and value (relative importance).

[20] Southworth and Southworth (1982: 189) reproduce two examples; Lynch (1976: 158–9 and dust jacket) reproduces another.

[21] Viglionese (1985) foregrounds these potentials in a series of analyses attentive to the pre-phonographic origins of linguistic expression and the cultural bases of iconicity.

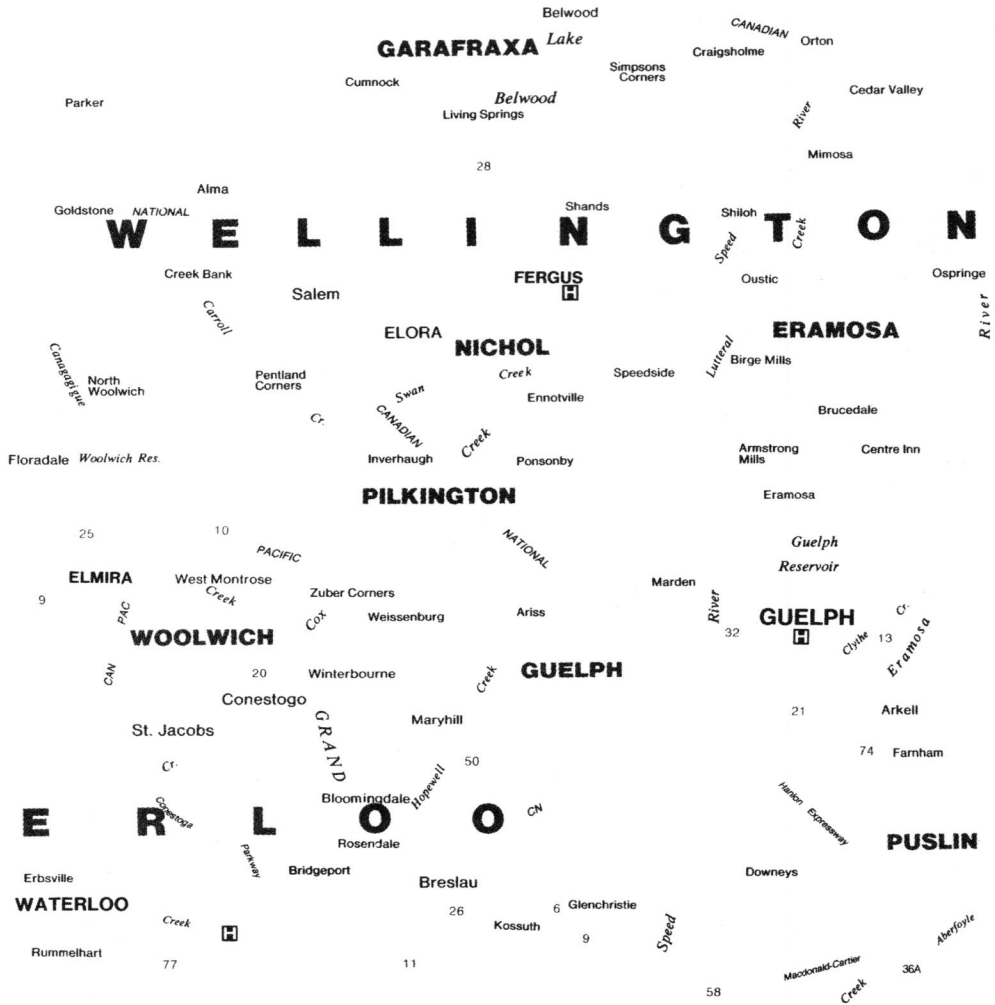

Figure 14.10 A field of linguistic map signs. Even without internal distinctions of colour, its iconicity is immediately apparent in contrast to the surrounding text. (From artwork by Gerald Boulet for 'Midwestern Ontario/Outdoor Recreation', published by the Kitchener–Waterloo Record, 1982.)

14.4.3 Tectonic Codes

Before approaching this subject, we should refresh our understanding of codes. A code is an interpretive framework, a set of conventions or rules, which permits the equivalence of expression (a graphic or typographic mark) and content (forest, population of less than 1000 persons, or multilane limited-access highway). In effect, a code legislates how something may be construed as signifying, as representing, something else. In this respect signs are encoded in formation and decoded in interpretation; and it is only through the mediation of a code that signification is possible.

Each map employs a tectonic code, a code of construction, which configures graphic space in a particular relation to geodesic space. This code effects a *topological* transformation from spheroid to plane in sign production and plane to spheroid in interpretation. It has a *scalar* function as well, logically separable from the topological but not practically independent of it. While the role of this code as representative principle is evident, its content and expression are less so, because both of these functives are abstract space. The tectonic code governs a sign function, which has as its content a topology and as the product of its action a correlative topology. If cartographic projections and scales have not been widely recognized as codes, it is not because they are difficult to formulate (reducible to concise mathematical expressions they are much more easily formulated than the codes of iconic and linguistic representation) but because they do not in themselves produce material imagery. They offer space for space, abstraction for abstraction, and their work is not visible until it is subjected to iconic coding. The mesh of graticule lines cradling the map image is not the tectonic code itself, but an icon of the topology acted upon by this code. Nor is it obligatory to render this topology: frequently it is manifest only in the shape and disposition of features and, when it is visualized, it serves primarily as a referencing system to implement the literalization or numeralization of space (Figure 14.11).

This code traffics in spatial meanings, and the messages it allows us to extract from the map are messages of distance, direction and extent. It shapes and scales the graphic plane in such a way that these messages emerge, veridically or erroneously, from the map image. While iconic and linguistic codes access the semantic field of geographic knowledge, the tectonic code provides their syntactical superstructure; this is the code through which we *signify not* what, but *where*. In moulding the map image, the tectonic code allows it to refer to the space which we occupy and experience; and inevitably it is laden with our preconceptions about that space. It is hardly surprising to find the map projection at the centre of political controversy, pretending as it does to validate our cultural centrism and objectify our territorial aims. It has these potentials because it allows us to view the world as we choose – as much or as little of it as we like, from whatever vantage point we like, and with whatever distortions we like – and, even though we know better, it projects an aura of ubiquity and authenticity. It can do so because we recognize it as the only thing exact – if in the most limited sense – in a practice that propagandizes exactitude as if this were the reason for its existence.

14.4.4 Temporal Codes

'Every map is out-of-date before it's printed'. This adage is a staple of the cartographic office. It is customarily dragged out for the benefit of the novice, held up as a fact of life (like death or taxes), and then put aside as an inevitable consequence of the complexities of the mapping process. If it is ever meant seriously, then that's as a barb at the sluggishness of the mapping bureaucracy – every member of the bureaucracy except, of course, the cartographer. But for the most part it evokes laughter or sentient smiles

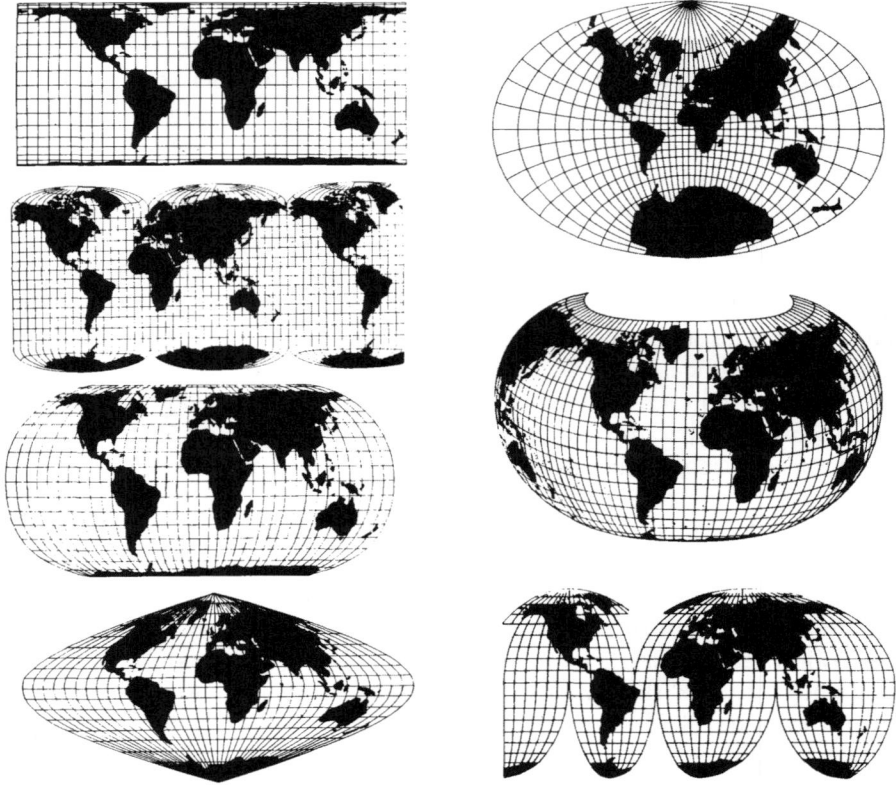

Figure 14.11 Icons of geodesic space, transcribed through a variety of tectonic codes. While scale and viewpoint maintain a general constancy, extreme regional distortions arise as the consequence of topological transformation. The cartographer's choice is not based on a chimerical concept of objectivity, but on the degree to which these distortions support the underlying proposition of the map (Bertin, 1983).

rather than angst (let's not get too wound up over it; we said out-of-date, not obsolete), and it's really not the sort of thing that cartographers lose sleep over. It just makes them a little uneasy.

Somehow we've gotten the idea that maps have nothing to do with time. We'll indicate a date of publication, and perhaps a timeframe for data collection, but that's about as far as it goes – and these gestures have more to do with the status of the map as a document than with any issue *of map time*. We shrug that one off, if a bit nervously, because we've learned to make maps in the terms they can resolve. Anything that changes fast enough to render the map genuinely obsolete before it can reach its audience doesn't belong in the map in the first place. The map is opaque to these things; it filters them out. That's partly a function of scale: maps are macroscalar and macroscopic and, after all, we *are* mapping mountains and not the pebbles inching down their slopes. But the things we're increasingly interested in mapping don't have this short-term permanence at any scale; they're more in the nature of *behaviours* than geographic fixtures. These interests may inspire new map forms, but they haven't forced us yet to admit that maps embody time as surely as – in fact because – they embody space. Most of us continue to think of the map as either a snapshot – in time but not of it; something with time evaporated out of it – or as akin to a three-hour exposure of Grand Central Station, in which actions, events and processes disappear and all that register are *objects of permanence*. We may be acutely aware of emplacing time in the photograph, and even of permanence as the arbitrary consequence of this act, but we don't generally extend these understandings to the map. There time retains the character of a hidden dimension, a cartographic *Twilight Zone*. But the map does encode time, and to the same degree that it encodes space, and it invokes a temporal code that empowers it to signify in the temporal dimension. That the action of this code on temporal attributes should be explained by the action of two subcodes which parallel those acting on spatial attributes is hardly surprising. The map employs a code *of tense,* concerning its temporal topology, and a code of *duration,* which concerns its temporal scale.

Tense is the direction in which the map points, the direction of its reference in time. It refers to past, to present (or a past so immediate as to be taken as present), or future – relative, of course, to its own temporal position. So we have maps in the past tense ('East Asia at the time of the Ch'ing Dynasty'), maps in the present tense (the '1986–1987 North Carolina Transportation Map'), and maps in the future tense (of tomorrow's weather, or a simulation of nuclear winter). We also have temporal *postures,* like that of the fantastic map (of 'Middle Earth', 'Dune' or 'Slobbovia') that has a present and past separate – but not entirely detached – from our own, or the allegorical map ('The Map of Matrimony', 'The Gospel Temperance Railroad Map', 'The Road to Hell') (Post, 1973) that proclaims itself atemporal or eternal and, in doing so, makes a stance of tense that more closely resembles the *aorist* of Greek than any English form. As maps slide into the past they become *past* maps ('antique' is a term reserved for past maps of some virtue or special appeal) where they continue to refer to *their* pasts, presents and imagined futures. The posture of the facsimile and the counterfeit is one of position rather than reference, the facsimile admitting (if only in a whisper) of its true temporal position.

The distinction between present and past is always difficult. A map positioned in the last century is obviously *past* – or is it? The physiographic map of 1886 is past by virtue of its cultural references – its references to the state of physiographic knowledge or the state of graphic representation in 1886 – not by virtue of its content, which we still insist we can scale into immutability. Erwin Raisz's physiographic maps, interleaved amongst the pages of the modern atlas, appear transported there from another time – and they are – but we take them all the same as maps *of the present.*[22] Without a more stable yardstick, the passage of cartographic time is marked off in editions. For the atlas it is accelerated by political and developmental pace and braked by the constraints of map production; for the topographic map it's modulated by the intensity of localized activity; and with

---

[22] We refer here to the maps occupying pages 80–81 and 148–149 of *Goode's World Atlas,* sixteenth edition, Rand McNally, 1982.

the digital database it's fixed in a perpetual virtual present.[23] Meanwhile, the current incarnation of the USGS quadrangle sheet *expresses* temporal distance, the distance between the present map and its predecessor, with a violent purple denoting *these* things as having happened between then (whenever then was) and now. Cherished globes have been sacrificed to garage sales and flea markets, the megabuck atlas is becoming an art investment, and we even have a class of disposable maps (with a lifespan roughly equal to that of a newspaper) characterized not so much by its funk as its anticipated, and almost immediate, obsolescence. We are increasingly conscious of the distance between present tense and past tense; and while it's still remarkably elastic, it is – as everyone tells us – shrinking fast.

The durative code of the map operates on the scalar aspect of time. As spatial scale is a relation between the space of the map and the space of the world, temporal scale is a relation between the time of the map and the time of the world. To understand this, we have to see the map as having thickness in time. Take, for example, an electronic map of traffic density in downtown Raleigh. Let's say that, in one minute, it plays out on a colour graphics terminal the events of an entire day. This map has a *temporal scale* that is the ratio of one interval (a minute) to another (twenty-four hours), or 1:1440. It compresses time in the same manner that it compresses space. Of course, that was a convenient example. Consider instead the newcomer to Raleigh mapping out his environment from a bus window. It's Saturday afternoon and he's just boarded the South Saunders bus at the central transfer point on Martin Street.

4:51 It will be four minutes before the bus leaves. Outside a few dozen people sit around on benches talking, reading newspapers, or just waiting, enjoying the Spring sun slanting between the banks and commercial buildings lining the Fayetteville Street Mall. In one direction the Mall slides down to the glassed and steel-trussed Convention Centre. At the other end, three blocks away, the turquoise dome of the State Capital bulges over its massive oaks. The view in both directions is fragmented by the Mall's decor: saplings, floral planters, a scattering of sculptures, a clock mounted on a mirrored kiosk. There are seven other passengers on the bus now, one of them thrusting his hand relentlessly into a box of candied popcorn. The next seat bears five knife slits, and here and there a nom de plume stands out in the faded graffiti: 'Catbird', 'The Non Stop Crew', 'Woogie Tee'.

4:55 The bus rolls from the curb, stops abruptly as another nudges in front of it, then groans away. The street is compressed by grey and beige walls rising a half dozen stories from the sidewalk. At eye level the bus reflects dimly in the plate glass of old shop fronts. Everything is in shadow.

---

[23] One might reflect here on the currency of data drawn from geographic information systems, the difference in time between their point of acquisition and point of use, and the liability potentially incurred. Given the naive tendency of most users to accept any electronically-coded information as current, the onus is clearly on the purveyor of information to inform the user to the contrary. Political bubble-bursting notwithstanding, this is a responsibility that the system manager ignores at his own peril: unearthing a telephone cable is one thing; cracking open an oil tanker is quite another.

4:57 A right turn onto Blount Street. To the left, ageing warehouses catch the sunlight head on. One of them announces its renovation. The next block's been levelled on both sides and, to the right, a sea of asphalt and windshields foregrounds the city's nucleus of office towers. Several blocks of shotgun shacks, verandas crowded with laundry lines and painted metal chairs, then the expanse of South Street slashed clear around Memorial Auditorium, an imposing chunk of institutionalized Art Deco.

4:59 The bus dips beneath the Shaw University pedestrian bridge, careens right onto Smithfield, and stops beside a tiny parkette of juniper. Here Wilmington and Salisbury streets merge into Highway 50 and zip off in six grass-trimmed lanes of new pavement toward the Garner suburbs. As cars burst past in both directions, the driver weighs his odds.

5:01 Past the commuters' raceway, the bus rattles over a set of railway tracks and the backside of Memorial Auditorium jumps across the right windows. Swinging left onto old Fayetteville Street, it stops below a cascade of terraces capped by an archetypal red brick elementary school. Directly across the street, a project sprawls out sheathed with brown wood siding and decorated in spray-bomb cursive. One person leaves the bus and two teenage girls hoist a stroller through the front doors.

5:03 To the right a fresh canopy of leaves spreads over the weathered monuments of Mount Hope Cemetery, and to the left the project gives way to squared-off little homes. The bus wheels right onto Maywood and the small homes persevere, gradually brightening. On the neighbourhood basketball court, a girl in a pink jumpsuit buries a fifteen-footer.

5:06 The bus lurches across a graded swath of red soil that imprints the future widening of South Saunders Street, and brakes to a halt opposite Earp's Seafood. It turns right onto South Saunders, then left at Carroll's Used Tires, then right again onto Fuller. A stretch of tidy compact houses ends suddenly at Lake Wheeler Road. A tire swing (one of Carroll's?) hangs outside the near window. Several passengers disembark here; one boards and is recognized. 'How ya doin?" 'Awright!'

5:08 The bus cuts right onto Lake Wheeler Road and descends a long grade To the left a high chain link fence tracks its descent, staking out the boundary of Dorothea Dix Hospital. To the right a precipitous slope tumbles into a clutter of rooftops and ahead Raleigh's best downtown panorama spreads over the windshield. At the foot of the grade, the road dovetails back into South Saunders where a column of plaster hens files across the eaves of RB's Chicken'n'Ribs.

5:11 Passing the entrance to the Dorothea Dix grounds, the bus stops in front of Heritage Park (another housing project but far more ambitious than the one on Fayetteville Street). Three riders step out cradling their afternoon purchases, and a right turn onto South Street aims just off the downtown core. Another descent, bottoming out below a closely set pair of railway trestles, then a quick rise and a confusion of lanes. With Memorial Auditorium a block ahead the bus pivots left onto McDowell.

5:13 On the left, a parking lot then a Chevy dealership. On the right another parking lot, then another, then another. Cars everywhere. No people, just cars, waiting. The downtown towers against the right window and then disappears behind a four storey parking deck. A cluster of satellite dishes crowd together on an office rooftop.

5:15 At the corner of McDowell and Martin the green expanse of Nash Square spreads out over the driver's left shoulder. A handful of people wander, without apparent intention, across the park. Turning right, the bus squeezes between the walls of Martin Street, gets lucky at the Salisbury traffic light, and then slips against the curb. The doors hiss open. It's still 79 degrees outside but in the shadows it feels cool.

If the bus didn't return to Martin Street, there would be nothing especially spatial about this experience; it unfolds *in time* as a sequence of impressions, and its spatial quality remains latent until it reconnects with its point of origin and becomes a *closed traverse* (Figure 14.12). At that point, everything witnessed becomes synchronous and the previously confounded immigrant exclaims, 'I know where I am!' (implying that, to some degree, 'I know where I've been'). Space has been surrounded and captured (unlike the tenuously connected scenes lingering along its perimeter, beyond the grasp of its closure); time has collapsed into space. It is still present in the map, but *as space*.[24] In Minard's *Carte Figurative* of Napoleon's Russian campaign ((reproduced, with some fanfare, in Tufte, 1983: 41, 176), time is literally distance, marked out by the rhythm of falling boots and shrinking roll calls. Less dramatically, but more explicitly, the 'Driving Distance Chart' at the back of the AAA road atlas[25] recognizes each segment as simultaneously a spatial interval (255 miles) and a temporal interval (5 hours and 20 minutes). Curiously – or perhaps predictably – it also tries to subvert its identity as a map, even proclaiming itself a 'chart' (read '*not* a map'), but it still looks like a map and it still functions as one.

We can pretend that the dimensions of the map are entirely synchronic, that it has no diachronic quality except as a specimen of technical or methodological evolution; but every cartographer who has grafted a new road onto an old, or dropped the still warm symbols of his latest research onto the cool plate of a twenty year old base map, should know better. The potential for anachronism is enormous; and sometimes it runs amok, as in the map that drags our earliest continental explorers across a fabric of forty-eight American states or ten Canadian provinces ('Native states? What native states?!'). Time is always present in the map because it is inseparable from space. They are alternative and complementary distillations, projections of a space/time of a higher dimensional

---

[24] Carlstein (1982: 38–64) argues convincingly for a 'time–space' framework of geographic description, employing the Hägerstrand time–geographic model and its system of graphic notation. We have used this system here to construct Figure 14.12.

[25] The example at hand concludes the *North American Road Atlas* published by the American Automobile Association, Falls Church, Virginia, 1984.

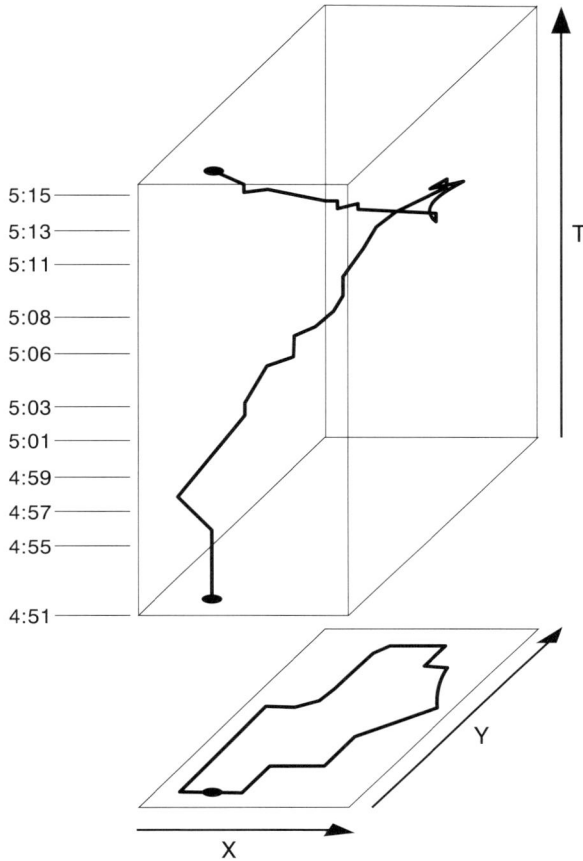

Figure 14.12 A spatio-temporal map of the bus trip, and a planar projection in which the temporal dimension has been collapsed to zero thickness. Space emerges as the product of synchronization (temporal flattening) and the closure of movement.

order. We cannot have a map without thickness in time unless we can have a map without extension in space; we cannot squeeze time out of the map, only into it.

14.4.5 Presentational Codes

The time of the map, the space of the map, the phenomena materialized in this framework, and the roster of terms and toponyms cast into it are not the map. Expressed through a complex of iconic and linguistic marking schemes, they become the content of the map image; but the map, as we have already pointed out, is much more than this solitary image orphaned on its audience's doorstep. The map image is accompanied by a crowd of signs: titles, dates, legends, keys, scale statements, graphs, diagrams, tables, pictures, photographs, more map images, emblems, texts, references, footnotes, potentially any device of visual expression. The map gathers up this *potpourri* of signs and makes of it a coherent and purposeful proposition. How these signs come together is the province of a presentational code, which takes as content the relationships amongst messages resident in the map and offers as expression a structured, ordered, articulated and affective display: a legitimate discourse.

The more apparent aspects of this code are intrasignificant. It acts on the structure of the map, dividing and proportioning the space of the page, staking out the prospective geometry of blocks, columns, channels and margins. It proceeds from the primacy of the rectangle, echoing our Euclidean systemisation of environment (objects, rooms, buildings, streets, cities), use (trims, folds, stacks, racks, packages, pigeonholes) and reading itself. Within this latent superstructure the ingredients of the map are laid out, ordered by a positional scheme fixing relations of sign to sign and sign to ground and imposing on the map a *program,* a discursive strategy. Discourse is articulated through emphasis (large or small, prominent or subdued) and elaboration (the relative complexity of signs, the intricacy of their meaning).

But the presentational code works beyond schemes of graphic organization. As it acts on the map as a whole, its effects are *manifest in the whole map*; and some of these are aimed clearly toward extrasignification. The map has a discursive tone: soft/loud, even/dynamic, complacent/agitated, polite/aggressive, soothing/abrasive. The majority of 'good' maps position themselves on the left side of these oppositions, more conscious of the demands of professional decorum than sensitive to those of their subject matter – or perhaps their intent is to pacify by shading even the most urgent and disturbing themes into Muzak (the reverse is equally incongruous: some of the most thematically mundane maps bludgeon their viewers with symbols that weigh on the page like musket balls). The map also reflects on itself. It asserts its status amongst maps as mean or lavish, frugal or conspicuous, in its consumption of resources: the scale of its effort, the virtuosity of its craft, its opulence of colour, material sensuality, the abundance of surface left unprinted, its sheer size. These gestures are all the more obvious in the atlas, where they can pile up into an object of palpable thickness and weight. So at one extreme we have the Park Avenue hedonism of the *World Geo-Graphic Atlas,*[26] bound by a cloth-wrapped and gold-imprinted cover a quarter of an inch thick and framed by striking end papers that sprawl over nearly five square feet. At the other extreme we have the grim imperative of *The Nuclear War Atlas*[27]: an anti-atlas taking the form of a Marxist tabloid, a document one could well imagine run off after hours on a hand-cranked press and thrust at nervous yuppies on street corners, or nailed to a senator's door. Government maps are especially status conscious, announcing the cost of their printing or the percentage of recycled pulp in their stock in an effort to disarm the bellicose taxpayer. The map also proclaims its alignment: its professional camp (a Cartographer's

---

[26] *The World Geo-Graphic Atlas: A Composite of Man's Environment,* edited and designed by Herbert Bayer, was produced in 1953 for the Container Corporation of America. Described in the foreword as 'an effort to contribute modestly to the realms of education and good taste', it is, as a gesture of corporate good will or a device of corporate promotion (take your pick), an exceptionally lavish and ambitious volume.

[27] *The Nuclear War Atlas,* The Society for Human Exploration, Victoriaville, Quebec. While hardly likely to inspire professional envy among most cartographers, this atlas assumes the form appropriate to its purpose. It would be difficult to imagine as an expensive coffee-table book except, perhaps, as a device of the blackest humour.

map as opposed to a Designer's map), its institutional allegiance (a National Geo-graphic map as opposed to a Bartholomew, a Rand McNally as opposed to a AAA), and occasionally the method and aesthetic of its author (a Bollmann map of Manhattan as opposed to an Anderson). It has a projective aspect as well: it's prepared for a particular audience. It is manufactured for the urbane or the profane, the casual or the attentive, for those at ease with maps or for the cartophobic, for the executive or the mercenary, the well-to-do or the student, the sighted or the blind. It speaks in *their* language: in clinical ascetic, in hot-colour High-tech, in journalistic cartoon, in Country and Western, or suburban rec-room (Figure 14.13).

The presentational code of the map can't be explained as a simple set of rules for graphic organization, especially without defining *whose* rules. Its action is not limited to the structural aspects of presentation or confined to affairs of visual priority and reading sequence (not at least until computers produce maps *for* computers). The map isn't a debating club exercise; it's set firmly in the real world, where the abstractions of structure, order, and articulation cannot be cut away from issues of aesthetics or even belief – any more than the grammar of this text can be separated from its meaning or the attitudes and values of its authors.

14.5 Sign Functions

Maps are about relationships. In even the least ambitious maps, simple presences are absorbed in multilayered relationships integrating and disintegrating sign functions, packaging and repackaging meanings. The map is a highly complex supersign,[28] a sign composed of lesser signs, or, more accurately, a synthesis of signs; and these are supersigns in their own right, systems of signs of more specific or individual function. It's not that the map conveys meanings so much as *unfolds* them *through a cycle of interpretation* in which it is continually torn down and rebuilt; and, to be truthful, this is not really the map's work but that of its user, who creates a wealth of meaning by selecting and subdividing, combining and recombining its terms in an effort to comprehend and understand. But, however elaborate, this is not an unbounded

---

[28] This term is more widely accepted among graphic designers than among linguisticians. Ockerse and Van Dijk (1979: 363) describe the supersign as: 'a sign which allows for a complex simultaneity of possible interpretants'. In 'De-Sign/Super-Sign', Ockerse (1984: 251–252) elaborates on: 'The problem of defining the so-called "super-sign". This means to provide a rational system for communication wherein the sum forms the major mode of signification. The participating elements within this complex whole contribute bits of information. The whole is actually a sign made up of other signs; more precisely, the supersign is a sign system. This system is intended to include all signs that operate within the system or that can/will influence the system: the bits, their structural relations, the sum representations created by the juxtapositions of micro- and macro-elements (bits to bits, bits to groups, groups to groups, groups to the whole, the whole to others etc.). Involved are potential layers and levels of information (in terms of importance, denotative and connotative references) for the reader/viewer. The supersign is like a text; but its potential is even intertextual, a characteristic of signs. In fact, the supersign concept even provides a system that invites the reader/viewer to become an active participant in a generative process.' It will become apparent that, in our analysis, the term 'system' has a more specific meaning than that intended by Ockerse; but this does not indicate disagreement over the nature or function of the supersign.

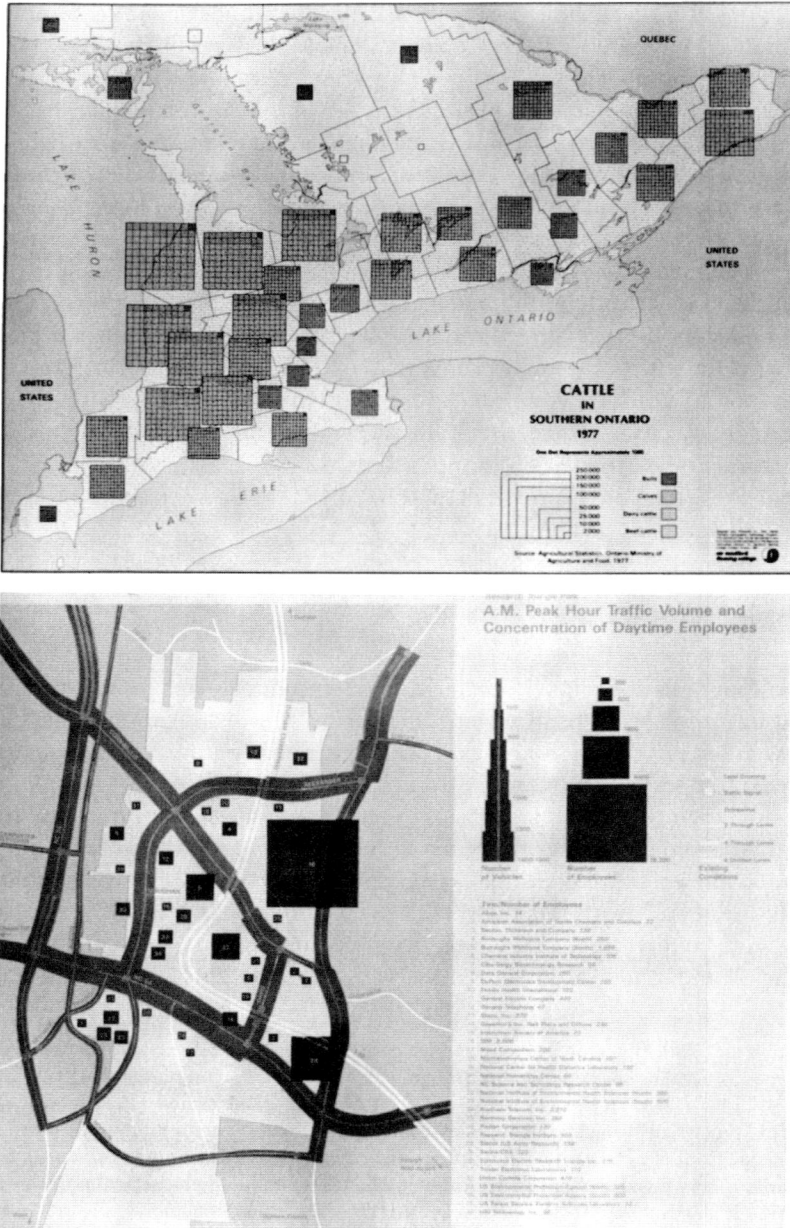

Figure 14.13   In two students' maps, differences of professional alignment are asserted through subtleties of presentation: the structure of the page, relations of image to edge and graphic to linguistic sign aggregates, typographic style and format, (*top:* by Blair Watke, Sir Sandford Fleming College, 1981; *bottom:* by Patricia Gwaltney, School of Design, North Carolina State University, 1985.)

process. Inevitably, it has a lower bound, the most particular sign function that resists decomposition into constituent signs, and an upper bound, the integral supersign of the entire map that accesses the realm of extrasignification; and between these extremes it is stratified. Twofold stratifications have been repeatedly proposed,[29] and widely accepted, but these don't go far enough. If we intend to explain how the map generates and structures the signing processes by virtue of which it is a map, then we need at least four strata or levels of signification: which we'll call *elemental, systemic, synthetic,* and *presentational.*

At the elemental level, visual occurrences (marks) are linked with geographic occurrences (features) in the set of germinal sign functions announced, if incompletely, by the map legend. At the systemic level, signs (supersigns) are composed of similar elements, forming systems of features and corresponding systems of marks. At the synthetic level (super-supersign?) dissimilar systems enter into an alliance in which they offer meaning to one another and collude in the genesis of an embracing geographic icon. We have at this point a map image; but we don't have a *map* without at least title and legend and, more typically, a host of supportive signs assuming textual, pictorial, diagrammatic and even cartographic forms. Presentation is the level at which the map image is integrated with and positioned in relation to relevant signs in other significant domains, and with which we have finally – or primarily – a complete and legitimized map. We will not take the position that maps are assembled from constituents (perceptually composed) nor that they are dismantled into constituents (perceptually decomposed), but will assume that the map is entered at any level of signification and that interpretation proceeds in either direction, by integration or disintegration, toward map or toward mark.[30] But not necessarily in a straight line. It may be tempting to regard these levels of signification – partly because of the order of their discussion, partly because of the connotations attached to terms like 'synthesis' or 'decomposition', and partly because of logical predispostion – as stages in a sequential process which, set in motion, moves inexorably toward a condition of greatest or least integration. That is not

---

[29] Head (1984) stresses two levels of interpretation, citing the following: Petchenik (1979); Olson (1976); and Bertin (1979). Among these, however, it turns out that only Petchenik's analysis is entirely restricted to two levels ('being-in-place' and 'knowing-about-space'): Olson's 'Level One' and 'Level Two' are supplemented by a 'Level Three' that is curiously distinct in its attention to meanings; and Bertin (1983: 141, 151), acknowledges a variety of 'intermediate' levels between the 'elementary' and the 'overall'. Schlichtmann (1985: 25, 27–28) identifies three levels of signification – 'minimal signs, macrosigns and texts' – which seem to differ more in extent than degree of synthesis. While none of these analyses recognizes a presentational, or discursive, level of signification, our terms are probably in closest agreement with Schlichtmann's.

[30] Our concern here is not the neurological processing of stimuli, but the *interpretation* of visual signs. The map user, regardless of – and oblivious to – physiological means, is obviously capable of both composing and decomposing complex signs; one of these abilities is of little use without the other. There seems to be a tendency among cartographers to regard perception as an exclusively constructive – even additive – process, encouraged perhaps by an affinity for mechanistic perceptual models that, for the most part, simply invert the biological metaphors of technological design (offering cameras for eyes, telecommunications systems for neural systems, or industrial robot vision for human cognition), and driven by a virtual obsession with the measurement of responses to largely decontextualized cartographic expressions. But the issue at hand is one of interpretive strategy: a strategy that operates on the organization of meanings, and the construction and deconstruction of *meaningful structures.* Its application is bidirectional and comprehensive.

our view. These interpretive levels are *simultaneous states* and, although the map may occupy only one of these states at one instant for one observer, they are all equally accessible through a process of perceptual transformation – that is, a restructuring or refiguring of the map.

14.5.1 Elemental Signs

Elemental map signs, by definition, cannot be decomposed to yield lesser signs referring to *distinct geographic entities.* They are the least significant units which have specific reference to features, concrete (Omaha) or abstract (1000 pigs), within the map image. Appraised in terms of the map's graphic signifiers, this criterion is easily confused; and we must keep in mind that a sign is not its expression, but the marriage of expression and content. The elemental map sign operates at the lower bound of the map's content taxonomy, and below this bound reside connotation and characteristic but nothing which can be construed as feature. Strict linguistic models of maps become hopelessly contorted over this issue if their analogies are pushed too far. *Q. – What is the graphic equivalent of a phoneme? A1. – There isn't one. A2. – It's a misguided question.* As we have seen, the map is an iconic medium that imposes its behaviour on language, not the other way around; and there is no reason to expect graphic signs to observe the rigidly contrived, and separately evolved, protocol of phonetic representation.

At the elemental level, graphic mark (a triangular dot, a blue line) is equated with feature (an occurrence of cobalt, a river). But the elemental sign is not of necessity, univocal. It is common practice in thematic cartography to invent map signs which (as elements) are polymorphic, polychromatic, polyscalar and, in consequence, polysemic; and, although each sign generated through such principles refers to one feature, it expresses simultaneously several of that feature's attributes (this subject is given thorough treatment by Bertin, 1983: 195–268 and 321–408). The elemental nature of map signs resides in the singularity of their geographic reference, not the simplicity of their meaning. Visual simplicity is no yardstick either; elemental signifiers are not restricted to visual primitives like dots and lines. They may just as easily assume more complex or more overtly iconic forms: a juxtaposition of flags signifies a border crossing, a bull's-eye a city, a string of dots and dashes a political boundary. In spite of their complexity these are elemental signs; they are not decomposed in interpretation: one flag signifies nothing without the other; the dot of the bull's-eye cannot be stripped of its enclosing circle; the patterned line cannot be reduced to Morse Code. None of these will dissolve into autonomous signs.

The autonomy of a sign, and therefore its elemental status, can only be assessed in view of the entire lexicon of the map that accommodates it. Take, for example, the signification of a church with the image of a square surmounted by a crucifix. If the square is also deployed *sans* crucifix to represent buildings in general, or if other signifiers can be exchanged for the crucifix to denote a variety of building types, then the square is an elemental expression and the crucifix (or anything else) appended to it is subelemental. The crucifix is, in effect, a qualifier. Its content is characteristic, not feature; and, regardless of its symbolic potency or self-sufficiency outside the map, in the map it has no *geographic* reference independent of the square that serves as its vehicle. This is an elemental *construct:* the syntactical product of two signs, one conjugated with another. Its expression is structurally divisible into two or more signifiers with both separate and joint meaning (building + Christianity = church). If, on the other hand, the square appears only in conjunction with the crucifix, it has no reference independent of their union and they must be jointly taken, not as construct, but as an undifferentiated element similar to the juxtaposed flags. This distinction is an important one because it indicates the presence or absence of an elemental syntax.

How are we to interpret two signifiers which apparently claim equal reference to the same feature, as both blue line and blue-tinted area do in the cartographically standard lake sign? We could regard these as coextensive signs manifest, in Klee's (1968: 18–21) terms, as medial and active conditions of the same visual plane.[31] This may be valid with respect to *possible* representations of lakes, but a map can only admit one such possibility to the exclusion of all others: we will not find one lake portrayed as outline, its neighbour as coloured area, and the next as both. Neither signifier is redundant in the map, which adopts both because, in that context, neither signifies in the other's absence. An alternative analysis, also from the Formalist perspective, would identify the lake sign as one visual element: formed by its outline and characterized by the colour blue (blue in this case has no form but is only an attribute of form). Taken as a basis for explaining how the sign functions, how it relates content and expression, this puts us in an absurd position. A lake is signified by a blue line which closes on itself; and, if within that figure we find a blue tint, then the lake is characterized as having water in it! Both of these postures – the former accepting line and area as simultaneous signifiers of the same signified, and the latter accepting only the line as denoting feature and denying formal status to the area it encloses – refuse to acknowledge what we already take for granted: that the blue line represents the shoreline of the lake and the blue tint the surface of the lake. Correctly or incorrectly, with naive or deliberate motive, this is how we interpret it and this is how we map it. Of course the shoreline feature, strictly speaking, does not exist except as a boundary between water and land or as a locus at which the depth of the water table reaches zero with respect to the land surface (whatever that is) – and Keates' (1982: 82) objection to the use of boundary signs in street plans applies here as well – but if we can accept contour lines, and other isolines, then we have certainly learned to accept the shoreline. The surface of the lake isn't any more concrete – it is just the boundary between water and air – and the fact that it's planar (we can water ski on it) rather than linear makes it no less an abstraction. In principle, we regard the land surface and the water table as roughly parallel planes (and as everywhere coextensive) and, where these planes intersect, we conventionally demark their intersection with a blue line and place a blue tint to one side of that line (preferably the wet side). What we have

[31] First published in 1925, and first translated in 1953, this, together with Kandinsky (1979), root the Formalist approach to visual design firmly in the curriculum and practice of the Bauhaus. Contemporary treatments of a general nature include Dondis (1973), Wong (1972), and, despite its title, Bertin's (1983) *Semiology of Graphics.* For decades, Formalism has dominated the methodology of cartographic design: its appearance in the modern textbook is effectively compulsory, and a bibliography of papers that construct 'design guides' from Formalist principles would be too extensive to present here. For a relatively concise, cartographically-oriented, review see Fisher (1982).

then are two abstractions, shoreline and water surface, that we are willing to grant status as features (and to map accordingly) while at the same time recognizing them as two of many aspects or *connotations* of the lake (or pond or ocean) feature. So we have another type of sign construct (shoreline + surface = lake), only this time both of its components are features. And it turns out that the blue line, in and of itself, does not represent the shoreline after all (although it may represent a river in the same map), but does so only in the presence of a blue tint on one side and none on the other: *as part of a sign construct*. While the language of the map is drawn from a store of culturally prescribed possibilities, its terms are specifically defined only in application, where the semantic field and syntactical procedures of the individual map form a unique dialect or *sémie* (Figure 14.14).

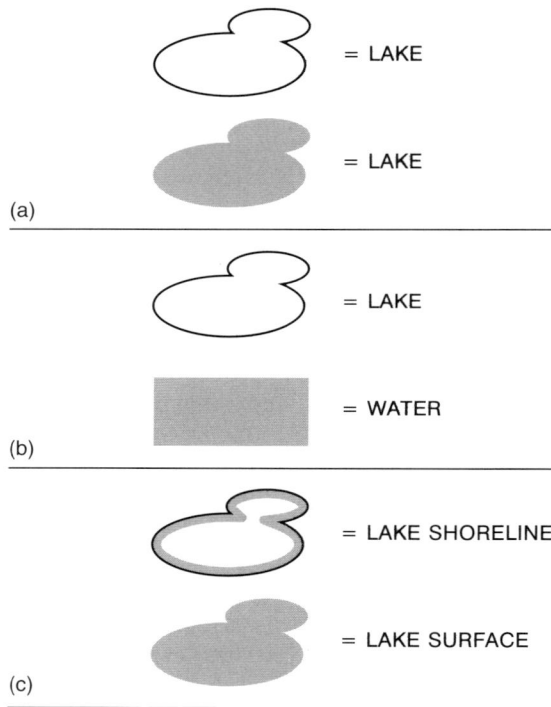

Figure 14.14 Alternative interpretations of the lake sign: (a) and (b) from a Formalist perspective, and (c) as a sign construct. The resemblance between the shoreline in (c) and pre-lithographic lake signs is anything but coincidental.

We have tried to demonstrate why we must insist that map signs be considered in terms of both expression *and content,* and to point out the inadequacy of a Formalist perspective that regards only signifiers and not signs; as well as to suggest the degree to which our conceptualization of phenomena structures, even dictates, the manner in which we represent them. Thus an elemental sign is a *sign of elemental meaning,* one which refers to an element of the landscape that, however artificial, we are not inclined to tear into constituent bits. With this premise it is possible to build systems of signs, and systemic meaning, from elements.

14.5.2 Sign Systems

By sign system we mean a set or family of similar elemental signs *extensive in the space of the map image:* a distribution of statistical units, a network of channels, a matrix of areal entities, a nesting of isolines. In this respect, we identify a road system, a river system, or a system of cities. It requires that we interpret many like signs as one sign, again a syntactical product but now one of geographic syntax. The systemic signifier is shaped by the disposition of its corresponding set of phenomena in geodesic space and by the topological transformation that brings this space to the surface of the page. It is also shaped by the way we define elements in the first place. If we were to map, say, the distribution of mountainous regions in the United States by taking as our criterion the (rather over-simplified) notion that all lands elevated 1500 meters or more qualify and that those of lesser elevation do not, we will find in our map a quite different sign system than if we had chosen 2000 meters as our benchmark. It isn't usually this innocent. What if we were mapping toxic levels of airborne pollutants? What the map says on this subject is determined by what standards, *whose* standards, we accept as a yardstick of toxicity. In content a system is, after all, a system of features – and features only exist when we recognize them as such (Figure 14.15).

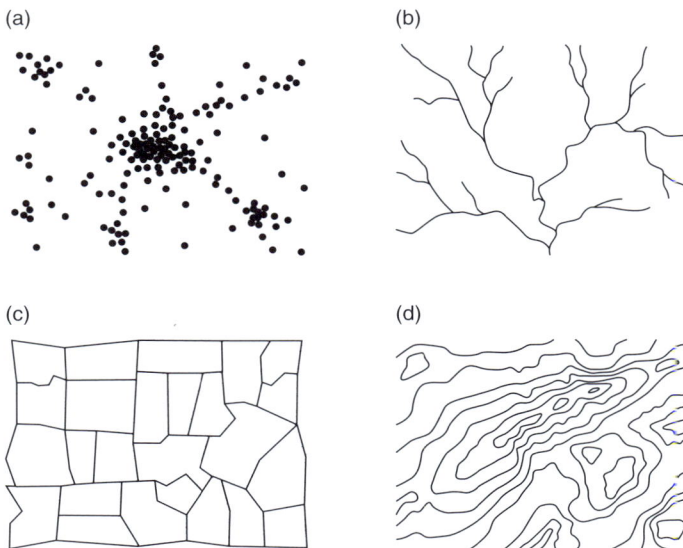

Figure 14.15　Typical cartographic sign systems: (a) a discrete distribution, (b) a network of signs, (c) a sign matrix, (d) nested signs. Regardless of implantation or graphic symbolism, each system structures the landscape in a distinctly different manner.

An arrangement of signifiers on the map constitutes a system only, of course, by virtue of our ability to perceptually organize its elements into something whole. At the systemic level, the bases of affinity amongst elements are those of implantation (yielding point, line or area systems) and those formal and chromatic attributes variously termed qualitative, nominal, distinguishing or differential. Not surprisingly, the latter are as effective amongst linguistic signs as amongst iconic signs, distinguishing hydrographic nomenclature, for example, by italic form or blue colour. What is surprising, however, is the degree of variation the systemic signifier will tolerate without falling to pieces. Our highway maps, almost to the last, serve up pavement in *a smörgasbord of* colours: red, blue, yellow, black, brown, whatever's in the printer's pantry. If the object is to represent a coherent highway *system,* then we could hardly do more to subvert its recognition. But that object is secondary to the marking out of politically-based *subsystems,* the sifting out of the relative accomplishments of federal, state and county treasuries. These maps can't just be written off as the products of illogical design of aesthetic insensitivity; they are graphic examples of how the extrasignificant functions of the map penetrate to its most practical and seemingly dispassionate design decisions.

The reason we can get away with this sort of thing is that, with the exception of scattered distributions, cartographic sign systems are typified by connectivity. Their elements link up, abut, cradle or nest within one another. They have anatomies. We recognize primarily their structure and use the characteristics of their elements mainly to highlight subsystems which would be otherwise undifferentiated, or to unstick systems of similar structure. That is to say, we attend more to the syntax of the system than the sematic import of its components. We don't distinguish blue highways from rivers because their signifiers are a little wider and little less sinuous, and we do so in spite of their most salient attributes of blueness and linearity, but because they are *structured differently as systems*, because they are manifestly *different landscapes.* The system is a landscape because, while the element simply *is* somewhere, the system *goes* somewhere; and, in doing so, it structures the space of the map.

14.5.3 Synthesis

There is no such thing as a monothematic map. Consider that emblem of thematic cartography: an array of graduated circles against the barest outline of subject area. This map image signifies at least the shoreline (usually elaborated beyond any conceivable utility), the water surface, the land surface and probably one or more proprietary boundaries (in which case we've differentiated several political states as well), and – almost forgot – whatever it is the graduated circles represent. Stripping off the circles leaves us with an absolute minimum of three sign systems, and typically twice that many, lurking behind the ostensibly servile trace of the pen. Sure, cartographers design maps for cartographers – as architects design buildings for architects and politicians make laws for politicians – but to call this monothematic is going too far. Can we really take that much for granted? Are we so thoroughly hypnotized that we can't even *see* the map?

Maps are About Relationships. In other words, they are about how one landscape – a landscape of roads, of rivers, of cities, government, sustenance, poison, the good life, of whatever – is positioned in relation to another. The map synthesizes these diverse landscapes, projecting them onto and into one another, with less than subtle hints that one is correlative to another or that *this* is an agent or effect of *that.* The map can't simply say that something is present (present in what?) or that it is distributed in a certain way (distributed in relation to what?); it's after the *big picture,* the kind of insight that only comes with an omnipresent viewpoint and the power to choose what inhabits the world. At this level the map image as a whole (whole in content if not necessarily in scope) is the supersign, and the various systems it resolves to are its constituent signs. And signs can only have meaning in relation to other signs. Merleau-Ponty (1964: 39) puts it this way:

> *What we have learned from Saussure is that, taken singly, signs do not signify anything, and that each one of them does not so much express a meaning as mark a divergence of meaning between itself and other signs. Since the same can be said for all other signs, we may conclude that language is made of differences without terms; or more exactly, that the terms of language are engendered only by the differences which appear amongst them. This is a difficult idea, because common sense tells us that if term A and term B do not have any meaning at all, it is hard to see how there could be a difference of meaning between them; and that if communication really did go from the whole of the speaker's language to the whole of the hearer's language, one would have to know the language in order to learn it. But the objection is of the same kind as Zeno's paradoxes; and as they are overcome by the act of movement, it is overcome by the use of speech.*

What is signified by any system in the last illustration? Nothing. If they were juxtaposed with a sign system that we could recognize, or furnished with a nomenclature that allowed us to supply that system, they could become signs, not by virtue of any abstract geographic reference but *in relation to* another sign system that holds meaning for the observer. If you have to resort to the map title to determine that *this* map of teenage suicides takes place in Los Angeles, then you're probably too far removed to care. What the map does (and this is its most important internal sign function) is permit systems to open and maintain a dialogue with one another. It is obvious why a road folds back on itself when we can see the slope it ascends, or why two roads parallel one another a stone's throw apart when we can see them on opposite banks of a river, or why an interstate cramps into a tense circle when we can see the city and its rush-hour torment. We know the behaviour of this system so well, in fact, that we can take it as an index[32] of other systems in the total absence of their direct representation. On the face of it, the map confirms these understandings; but they are understandings *that have already been created by maps.*

---

[32] This term is used in the sense intended by Peirce: to express a causal relation between object (steep slope, river, city) and interpretant (twisting road, parallel roads, circular highway segment). For Peirce, *icon, index,* and *symbol* constitute the second of three trichotomies which jointly define and elaborate taxonomy of signs. See Peirce (1955: 98–119) or Hartshorne and Weiss (1960: 134–173).

The *gestalt*[33] of each sign system is positioned against the semiotic ground of another sign system, or a subsynthesis of systems. The roads in the state highway map aren't grounded against an insignificant white surface; they're grounded against North Carolina or Illinois or Texas. What lies between the roads isn't aether (it isn't 40 lb. Springhill Offset either): it's tobacco and loblolly pine and patches of red dirt rolling over the Piedmont, or rugose mats of corn dotted with crows and John Deeres, or relentless miles of sand and prickly pear rippling in the heat. *There is nothing in the map that fails to signify.* Not even in a map of the Moon. So the flow of water is interpreted against the ground of land form, and vice versa; and the pattern of forestation is interpreted against the ground of both, as both and each are interpreted against it. In the synthesized map image, every sign system is potentially figure and every sign system is potentially ground (Figure 14.16). There is nothing inherently or irrevocably ground about even the land mass: try telling a truckload of surfers that the shoreline in the highway map is just a backdrop to the road system. They'll tell you that you have it all backwards.

The map image is a synthesis of spatially and temporally registered *gestelten,* each a synthesis in its own right; and to pretend that this whole is no more than the sum of its parts, or that we can do more than recommend a certain alignment of their priorities, is to reduce our concept of the map to that of a diagram. No degree of thematic constriction can silence the conversation amongst map signs. The map models the world as an interplay of systems and presents it to us as a multivoiced analogue, with harmonies and dissonances clearly discernible. Through the map we observe how systems respond to one another, and appraise the nature and degree of that response. We *explore the world through the map,* not as vicarious Amazon travellers hacking across the pages of *National Geographic,* but by remaking it in our own chosen terms and wringing as much meaning as we can out of what we've made.

14.5.4 Presentation

In presentation the map attains the level of discourse. Its discursive form may be as simple as a single map image rendered comprehensible by the presence of title, legend and scale; or as complex as those in *The New State of the World Atlas* (Figure 14.17) (Kidron and Segal, 1984) hurling multiple map images, diagrams, graphs, tables and texts at their audience in a raging polemic.[34] It may be as diverse as vacation triptiks, rotating cardboard star finders, perspex-slabbed shopping centre guides, chatty supermarket video displays, or place mats for Formica diner tables. Presentation is more

---

[33] The familiar example of the musical theme, which retains its identity despite transposition to another key or rescoring for a different ensemble of instruments, is remarkably evocative of the cartographic sign system that retains *its* identity throughout numerous topological and scalar transformations, spatial reorientations, and symbolic representations. Clearly, the recognizable whole, in both cases, is an artefact of structure rather than sensation – *a gestalt.*

[34] This atlas presents fifty-seven map plates, and corresponding micro-essays, addressing urgent (and frequently controversial) socio-political issues of global scope. Its overcrowded page layouts, animated symbolism, disturbing colours, pointed titles, and terse text form the ingredients of an acerbic discourse on the corruption and repression of the modern nation-state.

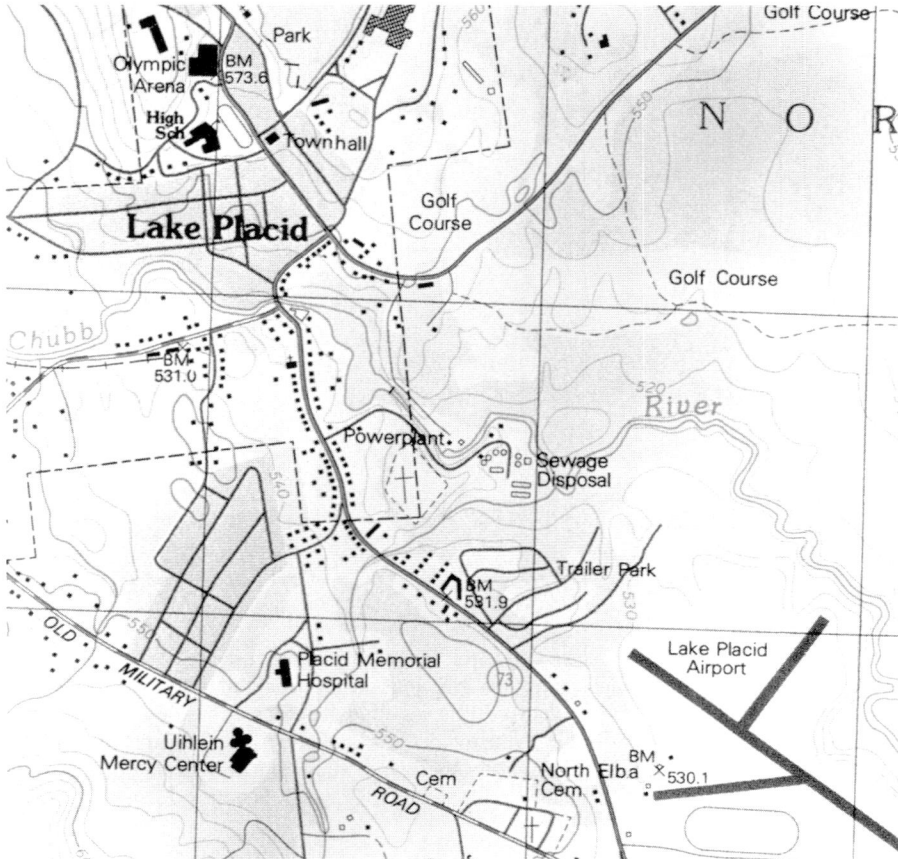

Figure 14.16 A synthesis of signs. Thematically diverse landscapes merge in a richly-coded supersign, exhaustively deconstructed and reconstructed by the map user in an effort to reveal topical and relational meanings (Thompson, 1979).

than placing the map image in the context of other signs; it's placing the map in the context of its audience.

Robert Scholes (1982: 144) identifies discourse, in the arena of literature, as:

. . . those aspects of a text which are appraisive, evaluative, persuasive, or rhetorical, as opposed to those which simply name, locate, and recount. We also speak of 'forms of discourse' as generic models for utterances of particular sorts. Both the sonnet and the medical prescription can be regarded as forms of discourse that are bound by rules which cover not only their verbal procedures but their social production and exchange as well.

And he notes (1982: 34) that the:

. . . coding of discourse is a formal strategy, a means of structuring that enables the maker of the discourse to communicate certain kinds of meaning.

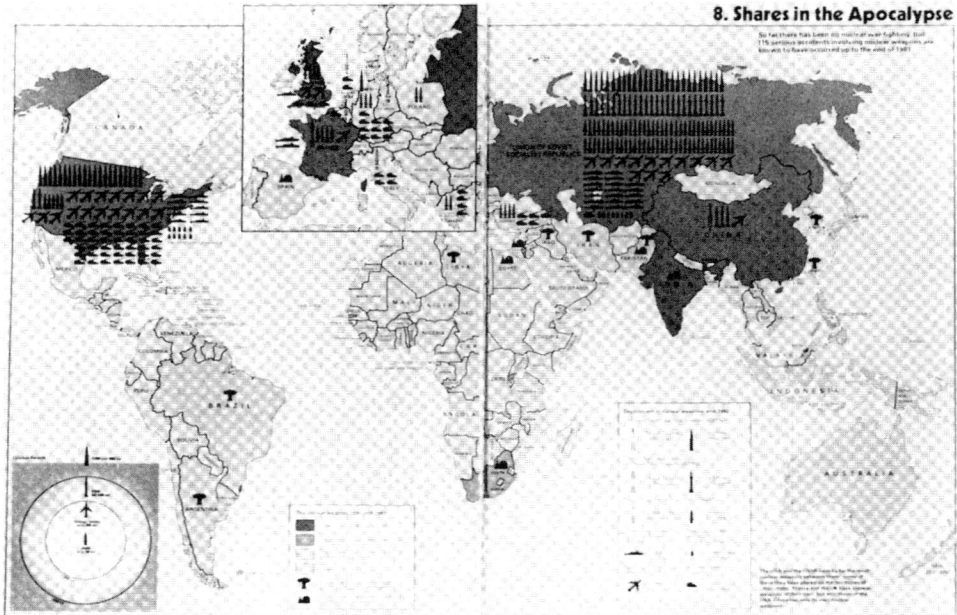

Figure 14.17   An exceptionally compact and intense presentation that impresses the urgency of its theme. The proliferation of weapons surrounds the non-nuclear island of Mongolia, overflows the borders of the United States, and even demands the rescaling of Europe. Virtually all of the northern hemisphere is coloured red and darkened by sinister pictograms. Textual and diagrammatic statements propel the map's message. (Plate 8. 'Shares in the Apocalypse', from 'The New State of the World Atlas' by Michael Kidron and Ronald Segal; published in the United States by Simon and Schuster, and in the United Kingdom by Pluto Press.)

Discourse is preceded by a code of presentation, and by the notion of an audience capable of applying that code to reach meaning *through* structure. For us, this means that the idea of 'percipient' must be extended to the entire culture of map makers and map users and include, as one of its most prominent aspects, their ability to generate and use strategic codes that permit maps to speak *about* the world rather than simply of it.

In bringing the map to this point we make it entirely accessible to the processes of extrasignification, and subject to their appropriation. It can be seized and carried off whole (necessarily whole) to serve the motives of mythic representation. The plan of the shopping centre, colour-coded, with shops topically and alphabetically organized and numerically keyed – a paradigm of logical graphic representation for the illogical masses – becomes an expression of the fact that 'We've got it all: trendy clothes, trendy shoes, books, records, tools, cameras, jewellery, fondue pots, exotic coffees, pizza and parking'. The diner placemat ceases to be a regional guide to places of interest and focal points of recreation (it was never meant as a gravy blotter or it wouldn't have been printed in the first place) and becomes the Chamber of Commerce's propaganda vehicle, complete with smiling checker-shirted fishermen tugging against smiling bass the size of Volkswagens. Which brings us back to where we started. The map is simultaneously an instrument of communication – intrasignification, given the benefit of doubt – and an instrument of persuasion – extrasignification and its propensity toward myth.

Presentation locates the map front and centre in all this action, at the vertex of both planes of signification. It's not a quirk of house style that populates the *National Geographic* map with maize-laden Cherokee or the state highway map with trees, bees, civil war artefacts and cavorting tourists. It's the deliberate activation of popular visual discourse. It's not just pragmatism or objectivity that dresses the topographic map with reliability diagrams and magnetic error diagrams and multiple referencing grids, or the thematic map with the trappings of F-scaled symbols and psychometrically divided greys. It's the urge to claim the map as a scientific instrument and accrue to it all the mute credibility and faith that this demands. Presentation, as the end and the beginning of the map, closes the loop of its design. It makes the map whole and, in doing so, prepares it for a role that begins where its avowed attention to symbolism, geodesic accuracy, visual priority and graphic organisation leaves off. It injects the map into its culture.

References

Barthes, R. (1972) *Mythologies*, Hill and Wang, New York.

Barthes, R. (1980) The plates of the Encyclopedia, in *New Critical Essays* (ed. R. Barthes), Hill and Wang, New York.

Barthes, R. (1981) *Camera Lucida*, Hill and Wang, New York.

Bartz, B.S. (1969) Type variation and the problem of cartographic type legibility - part one. *The Journal of Typographic Research*, **3**(2), 130–135.

Bertin, J. (1979) La test de base de la graphique. *Bulletin du Comité Français de Cartographie*, **79**, 3–18.

Bertin, J. (1983) *Semiology of Graphics*, University of Wisconsin Press, Madison, WI.

Bhattacharya, N. (1984) A picture and a thousand words. *Semiotica*, **52**(3/4), 213–246.

Carlstein, T. (1982) *Time Resources, Society and Ecology*, George Allen and Unwin, London.

Culler, J. (1981) *The Pursuit of Signs: Semiotics, Literature, Deconstruction*, Cornell University Press, Ithaca, NY.

Deregowski, J. (1972) Pictorial perception and culture. *Scientific American*, 82–88.

Dondis, D.A. (1973) *A Primer of Visual Literacy*, MIT Press, Cambridge, MA.

Eco, U. (1976) *A Theory of Semiotics*, Indiana University Press, Bloomington, IN.

Fisher, H.T. (1982) *Mapping-Information: The Graphic Display of Quantitative Information*, Abt Associates, Cambridge.

Gregory, R.L. (1973) *Eye and Brain: The Psychology of Seeing*, 2nd edn, McGraw-Hill, New York.

Hartshorne, C. and Weiss, P. (1960) *Collected Papers of Charles Sanders Peirce: Volume II, Elements of Logic*, Harvard University Press, Cambridge.

Head, C.G. (1984) The map as natural language: A paradigm for understanding. *Cartographica*, **21**(1), 1–32.

Johns, B. (1984) Visual metaphor: lost and found. *Semiotica*, **52**(3/4), 291–333.

Kandinsky, W. (1979) *Point and Line to Plane*, Dover, New York.

Keates, J.S. (1982) *Understanding Maps*, Longman, London.

Kidron, M. and Segal, R. (1981) *The State of the World Atlas*, Simon and Schuster, New York.

Kidron, M. and Segal, R. (1984) *The New State of the World Atlas*, 2nd edn, Simon and Schuster, New York.

Klee, K. (1968) *Pedagogical Sketchbook*, Faber and Faber, London.

Lynch, K. (1976) *Managing the Sense of a Region*, MIT Press, Cambridge, MA.

Merleau-Ponty, M. (1964) *Signs*, Northwestern University Press, Evanston, IL.

NGS (1985) *Atlas of North America: Space Age Portrait of a Continent*, National Geographic Society, Washington, DC.

Ockerse, T. (1984) De-sign/super-sign. *Semiotica*, **52**(3/4), 251–252.

Ockerse, T. and Van Dijk, H. (1979) Semiotics and graphic design education. *Visible Language*, **13**(4), 358–378.

Olson, J.M. (1976) A co-ordinated approach to map communication improvement. *American Cartographer*, **3**, 151–159.

Peirce, C.S. (1955) *Philosophical Writings of Peirce* (ed. J. Buchler), Dover, New York.

Petchenik, B.B. (1979) From place to space: The psychological achievement of thematic mapping. *The American Cartographer*, **6**, 5–12.

Post, J.B. (1973) *An Atlas of Fantasy*, Mirage Press, Baltimore, MD.

Rand McNally (1982) *Goode's World Atlas*, 16th edn, Rand McNally and Co., Chicago, IL.

Robinson, A. and Sale, R. (1969) *Elements of Cartography*, 3rd edn, John Wiley & Sons, Inc., New York.

Schlichtmann, H. (1985) Characteristic traits of the semiotic system map symbolism. *The Cartographic Journal*, **22**(1), 23–30.

Scholes, R. (1982) *Semiotics and Interpretation*, Yale University Press, New Haven, CT.

Southworth, M. and Southworth, S. (1982) *Maps: A Visual Survey and Design Guide*, Little, Brown, and Co., Boston, MA.

Times Books (1985) *The Times Atlas of the World*, Seventh Comprehensive Edition, Times Books, London.

Thompson, M. (1979) *Maps for America: Cartographic Products of the U.S. Geological Survey and Others*. US Government Printing Office, Washington, DC.

Tufte, E.R. (1983) *The Visual Display of Quantitative Information*, Graphics Press, Cheshire, CT.

Viglionese, P.C. (1985) The inner functioning of words: Iconicity in poetic language. *Visible Language*, **14**(3), 373–386

Wong, W. (1972) *Principles of Two-Dimensional Design*, Van Nostrand Reinhold, New York.

15

Reflection Essay: *Designs on Signs/Myth and Meaning in Maps*

Denis Wood[1] and John Fels[2]

[1] *Independent Scholar*

[2] *College of Natural Resources, North Carolina State University, USA*

Designs on Signs/Myth and Meaning in Maps has led a charmed if unusual existence. All but ignored on publication, its incorporation into a pair of Smithsonian exhibitions, and inclusion in the exhibitions' 'accompanying book', *The Power of Maps* (Wood and Fels, 1992), has given the paper a currency enjoyed by few *Cartographica* articles. Its title in this book, 'The Interest Is Embodied in the Map in Signs and Myths', captures the paper's focus on the interests motivating maps, as well as our conviction that these interests pervaded the map, penetrating to the very level of the marks out of which any map is built.

We wrote *Designs on Signs* during the academic year, 1985–1986. We wrote it in Boylan Heights, in Raleigh, North Carolina, USA, where Denis was on the faculty of the School of Design at North Carolina State University and John was spending a sabbatical year. We had met in the summer of 1984. John had spent the previous ten years developing and teaching the core Design curriculum in the Cartography Program at Sir Sandford Fleming College in Ontario, Canada, and he had short-listed several eastern universities as possible sites for his well-earned sabbatical. The key attraction at North Carolina State University was Denis, known to John through the work Denis had published in *Cartographica*, but the School of Design was well regarded at the time and its Visual Design Program also held some interest for John.

It is probably important to say that at the time we met Denis had only published a commentary and four book reviews in *Cartographica* (Wood, 1980a, 1982a, 1982b, 1983a, 1983b). Though he had long been romanced by maps, Denis was teaching design,

Classics in Cartography: Reflections on Influential Articles from Cartographica   Edited by Martin Dodge
© 2011 John Wiley & Sons, Ltd

landscape history and environmental psychology at North Carolina State University – he had been for ten years – and writing about maps was, at most, a sideline. In fact, by 1984 Denis had published far more about film – in *The Journal of Popular Film* (Wood, 1978a, 1979, 1980b, 1981a), *Film Quarterly* (Wood, 1981b), *Literature/Film Quarterly* (Wood, 1978b), and so on – than he had about maps; but the reviews he had published in *Cartographica* – passionate, often intemperate – had definitely struck a chord and not just with John. In 1985 Brian Harley, during his post-banquet speech at the Eleventh International Conference on the History of Cartography, was to say, 'Nor have we welcomed the criticism of outsiders. When a gust of fresh air blows in – as with the bracing polemics of Denis Wood ...'.

'Outsider ...' It must have seemed that way to people reading Denis' reviews who knew no more about him than his position at the School of Design, but Denis had completed his doctorate under cartographer George McCleary at Clark University, where fellow students had included Borden Dent, Karl Chang and Barbara Buttenfield; and since McCleary had received his doctorate from Arthur Robinson, Denis was anything but an outsider. It was John, despite teaching map design, who was the outsider, for John had come to his position at Sir Sandford Fleming with little more than an undergraduate degree in architecture from Washington University in St. Louis. But it was precisely John's Bauhaus-orientated designer's perspective that David Jupe found attractive.[1] Jupe was then pioneering the design-based mapmaking for which Sir Sandford Fleming would become famous,[2] and John's background suggested that he would approach map design from an explicitly designer's perspective, though one grounded in John's preceding five-year experience as a practicing cartographer in Ontario's Ministry of Natural Resources.

So we were well matched, an outsider who was an insider and an insider who was an outsider, but had John and his wife, Vicki, not been looking for a place to spend his sabbatical, it's unlikely we would have met. It's certainly unlikely we would have discovered our mutual interest in semiology, as Denis referred to it, or semiotics, as John preferred to call it, which emerged during lunch the day that John and Vicki visited the School of Design. Amongst other attractions, the prospect of spending a year discussing the semiotics of maps proved irresistible to John, and the following summer he and Vicki returned to Raleigh, renting a house a few blocks from Denis and his family. Shortly thereafter we began a series of weekly meetings to explore the relevance of semiology to maps.

We brought different but complementary commitments to these meetings. For the previous few years Denis had been reading everything by Roland Barthes he could lay his

---

[1] David Jupe had been Supervising Cartographer in the Cartography Section of the Ontario Department of Geology. In the late 1960s he joined the faculty of the School of Natural Resources at Sir Sandford Fleming College, where he founded the Cartography Program. John joined the faculty to fill in for Jupe on a 1976 sabbatical.

[2] Fleming students have repeatedly captured the Best Student Award in the American Congress on Surveys and Mapping (ACSM) Competition in Map Design, along with a majority of the Outstanding Achievement and Honourable Mention Awards granted. They have received eleven Canadian Cartographic Association President's Prizes, and won the CIG Intergraph Award for Computer Mapping five times. See their website for a complete list of honorees dating to 1981 (www.geomaticsatfleming.ca).

hands on, stimulated by a special issue that *Visible Language* had devoted to Barthes in 1977. Barthes in turn had stimulated Denis' reading of Saussure, Derrida, Foucault and Lacan. Denis' collaborator, the psychologist Robert Beck, was working along parallel lines and the two had already initiated the work which, shaped by their reading of Barthes' *S/Z* (Barthes, 1974), they would subsequently publish as *Home Rules* (Wood and Beck, 1994). But until John made the connection, Denis had not been thinking about the relevance of his reading to maps. John had been thinking about ways to systematize an approach to the visual design of maps for some time, thinking that had traversed the formalist traditions to which he'd been exposed as an architecture student where the writings of Paul Klee, Wassily Kandinsky and Johannes Itten were especially important. The 1983 translation into English of Jacques Bertin's *Semiology of Graphics* 'really caught me with my pants down (in the best possible way)', as John said, and he began searching for kindred thinkers closer to home. At first the search turned up little except for a few pieces by Hansgeorg Schlichtmann and Thomas Ockerse, but it was certainly to bear fruit in the meetings with Denis. Catching each other up, and exploring as our conversations dictated, we embarked on an intensive reading programme in semiotics (Morris, Peirce, Eco, Greimas, Barthes, Sebeok), linguistics (Saussure, Jakobson, Hjemlslev), structuralism (Lévi-Strauss, Lacan, Kristeva, Foucault), deconstructionism (Derrida) and phenomenology (Husserl, but especially Merleau-Ponty).

Aside from the excitement of the readings, the great attraction of our meetings was their openness. We had no agenda, no plan, no goal, except that of tackling the problem, as we saw it, of how maps worked, that is, how they did what they did. Our only starting points were a profound dissatisfaction with the way maps were treated in the professional – and popular – literatures, and a conviction that while maps did *not* comprise a language – contra Jim Blaut, Lech Ratajski, Jan Pravda, C. Grant Head and others – maps undoubtedly were composed of signs and did comprise some sort of sign system. Most often the meetings involved the deconstruction of a map or maps that we had brought with us, deconstructions that opened with questions like 'What's this map about?' and 'What's going on here?'; questions that worked their way through ever finer interrogations of increasingly magnified map marks; and questions that closed the hermeneutic circle with, 'So what's *really* going on here?' Weeks later we would return to maps we thought we had squeezed dry, only to wring ourselves through the process yet again, deepening, we hoped, our understanding with each pass. These semiological analyses, these 'close readings', became our fundamental method.

There really was no one map in which we were particularly interested, and the range we explored is perhaps best suggested by the maps we tackled years later in the meetings that would result in our *The Natures of Maps*: placemat maps, advertising maps, illustrative maps, school maps, maps produced by the full range of sciences, from every discipline, all sizes, and of every degree of seriousness. One map that soon appeared on the table was a highway map, the *1978–1979 North Carolina Transportation Map & Guide to Points of Interest* (Figure 14.1). It was an everyday map but one rich enough – with its inset maps, legend, mileage chart, safety tips, motorist's prayer, and so on – to sustain long discussion, soon enough intensive study, and ultimately a rich text. Though it literally did just come 'to hand when we were casting about for an example', our choice of this state highway map would come to be seen by some as an 'attack' (see below) not only on the map, but on the great state of North Carolina.

There came a point in our meetings when it was clear that we had made some kind of breakthrough, that our analysis of the map as myth, that is, as Barthean myth – which is to say a kind of 'speech' better defined by its intention than its literal sense – amounted to a total recasting of the terms in which maps had to be discussed.[3] Because myth was a sign system cantilevered from a simpler system of signs, it meant we had to tackle this basal sign system as well, and this necessitated our articulation of the codes maps exploited to unite signifieds and signifiers. Here we depended heavily on Eco's theory of semiotics, though in working through the ways in which elemental signs were combined into sign systems and greater syntheses we found ourselves thinking through the approaches of Ockerse, Van Dijk, Bertin, John's formalist avatars Kandinsky and Klee, Merleau-Ponty, Peirce, and others. Once we'd sketched the ten intrasignificant and extrasignificant codes we realized that we had something publishable and we began the drafting, redrafting and re-redrafting of what became *Designs on Signs/Myth and Meaning in Maps*. Denis took final responsibility for the opening 'myth' half of the paper, while John took the second 'meaning' section in hand, and though the paper is co-authored in every sense of the word, Denis wrote the first half, John the second.

15.1 From Paper to Exhibition

At over forty pages it turned out to be a bear of a paper, with more than 20 000 words and nearly two dozen illustrations. We harboured doubts that anyone would publish such a behemoth, and entertained the possibility that it might have to appear in two pieces, but Bernard Gutsell was the personification of enthusiasm and the piece appeared across fifty pages in *Cartographica*'s Autumn 1986 issue.

Publication was met with . . . resounding silence. Indeed, we're hard-pressed to recall anyone saying anything about it at all for several years at least. We can suggest a number of reasons for this reception. To begin with it's a big, fat, difficult paper. Even today, when much of it reads like dogma, there remain whole sections that people have yet to get their heads around. And, then, it was written in a conversational style wholly unfamiliar to readers used to academic prose. It could veer from the slangy to the esoteric within a single sentence and it was scarred by more than one of the ellipses critics loved to hate in Denis' prose. And who *were* these people foisting this long and radical piece about a highway map – a *single* highway map – on *Cartographica*'s readers? True, Denis *had* published those few reviews in *Cartographica* (and for those who were paying attention in *The Professional Geographer*, *The American Cartographer*, and US and Canadian map librarians' journals as well) *and* a piece about the evolution of hill signs. But *it* had appeared in *Prologue: The Journal of the National Archives*, and what

---

[3] Barthes' idea of the myth also remains a fruitful resource for dealing with map art. See Denis's treatment in 'A map is an image proclaiming its objective neutrality: A response to Mark Denil' (*Cartographic Perspectives*, 2007, **56**: 4–16); and the seventh chapter of his *Rethinking the Power of Maps* (Guilford, New York, 2010).

cartographer reads *that*? And John's students might have been *sweeping up* awards in the annual American Congress on Surveying and Mapping competitions – over the years Fleming students have captured over 80 of the ACSM awards – but his students' maps were not signed with his name and . . . what was Sir Sandford Fleming College anyway?

No, we were definitely outsiders and what we were trying to say was . . . *unheard of*. Considering the paper's reception it's important to remember that it was *the very first* salvo in the critical cartography wars. Brian Harley's 'Maps, Knowledge, and Power' didn't come out until the following year, his 'Deconstructing the Map' not until 1989 (reproduced as Chapter 17 of this volume). Robert Rundstrom's first paper reassessing mapping amongst First Nations peoples came out only in Rundstrom, 1990; John Pickles' 'Geography, GIS, and the Surveillant Society' only in Pickles, 1991. So, the ideas were new – wholly new to the cartographic community – the authors were nobodies, and the paper was difficult and overly long. It could easily have sunk with scarcely a trace.

But then, in the summer of 1990, Denis and his family were vacationing in New York. One morning he called the School of Design in Raleigh to check in and learned that Griselda Warr from the Cooper–Hewitt Museum in New York had called only moments earlier hoping to speak to him. Denis and his son, Randall, strolled across Central Park and presented themselves at the museum where Griselda and Lucy Fellowes wanted to talk about an exhibition of maps they were planning for 1992. Many meetings followed – many experts were consulted – but a year and a half later Denis became co-curator of what became *The Power of Maps*. His role was to shape the exhibition's structure, its thesis, its point; and he built this around the new thinking about the power of maps that he, John, Harley, David Woodward, Pickles, Rundstrom and others had been working out. The show featured more than 400 maps, and Denis turned *Designs on Signs/Myth and Meaning in Maps* into its centrepiece, the argument toward which it built, the 'polemical zenith' as Chuck Twardy called it (Twardy, 1992). In fact, *Designs on Signs* took over the fifth room where the headline bellowed, 'Whose agenda is in your glove compartment?' (Figure 15.1). In a vitrine in the wall below the headline was a glove compartment sawn from the dashboard of a car. Spilling from the open compartment was a slew of highway maps. Another vitrine was stuffed with North Carolina automotive memorabilia to drive home the paper's contention that the highway map's theme was the 'legitimacy of automobility'. A third vitrine displayed a collection of older highway maps, emphasizing the interest oil companies had in producing them. A frieze of North Carolina state license plates ran around the room. The most recent North Carolina state highway map, spread out on a table in the room's centre, was festooned with call-outs that spelled out the paper's argument. On another wall were the alternative maps, North Carolina's *Public Transportation Guide*, for example, that *Designs on Signs* had contrasted with the highway map to nail down the point that, anything but a functional response to a public demand, the map was first and foremost an *advertisement* for the state. These alternative maps, too, were festooned with callouts that helped the room embody *Designs on Signs*.

Strongly supported by American Express, the show was a huge success, attracting more than 60 000 visitors. Designed by Pentagram's Peter Harrison, the exhibit garnered no fewer than seven design awards that ranged from *Business Week*'s Silver

Figure 15.1 The complexities of the Designs on Signs article reduced to a slogan in the Washington, DC exhibition installation of The Power of Maps. (Source: authors.)

Award for Industrial Design Excellence to a Federal Design Achievement Award from the National Endowment for the Arts. The show was the subject of a study by the Smithsonian's Institutional Studies Office. Later published as 'Communication and Persuasion in a Didactic Exhibition: *The Power of Maps* Study', this demonstrated, in the words of later critic Ramona Fernandez (2001), that the exhibit 'was highly successful in transmitting its central abstractions. It is significant that this exhibit caused its visitors to think critically about maps as systems of knowledge constructed out of ideology. A detailed visitor study [over a thousand visitors participated] demonstrated that it, unlike many exhibits, was able to convince visitors of its thesis'. By all accounts the North Carolina room, that is, Designs on Signs, was key to making the exhibit's case, a point driven home by the press coverage. If the exhibit as a whole was central, the North Carolina room *hogged* attention. For example, the full page that *Newsweek* devoted to the show was headlined, 'Beware the Glove Compartment', and a nice piece of the highway map was splayed across the centre of the page. Rather than vanishing without a trace, the argument we had made in *Designs on Signs* was becoming notorious.

15.2 From Paper to Book

Amplifying the notoriety was *The Power of Maps*, the book. Very early in 1992, while work on the show was in full swing, Peter Wissoker, then at Guilford Press, asked Denis if he'd be interested in writing a book. Wissoker had originally approached Denis about editing a collection of the then recently deceased Brian Harley's writings, but when it transpired that prior to his death Harley had submitted a collection to Johns Hopkins – published a decade later as *The New Nature of Maps* – Wissoker broached the subject of a book of Denis' own, perhaps to parallel the exhibit (which was not planning a catalogue) and develop the show's themes to a degree not possible on the walls. The book, ultimately described in the show's press kit as its 'accompanying publication', was released the day the show opened and it did indeed parallel the show, room by chapter. Because Guilford was adamant about releasing the book at the show's opening, there was little time in which to write it, and Denis had to plunder things he'd already written, including, as the fifth chapter and paralleling the exhibit's fifth room, all of *Designs on Signs/Myth and Meaning in Maps*, adding John's name to the book's title page.

Like the show the book was a hit, becoming a selection of the History, Quality Paperback, and Book-of-the-Month clubs and a best seller for Guilford. It was widely reviewed in both the popular and professional press, was twice translated into Chinese (first in Taiwan, then in Beijing), and remains in print seventeen years after its release. As Jane Jacobs (2003) said a few years ago in an editorial in the *Transactions of the Institute of British Geographers*, '*The Power of Maps* was a minor sensation and has been widely reviewed, routinely used in teaching the history of geographical knowledge, and rarely goes without citation in scholarship on the geopolitics of maps'. Other than the book's general thesis that maps express particular views in support of specific interests, and present information selectively to shape our view of the world and our place in it, the most frequently cited material all comes out of the *Designs on Signs* chapter. Of signal importance remain the map's construction as myth and our articulation of the map codes.

The Power of Maps exhibition proved popular enough to remount two years later at the Smithsonian on the Mall in Washington. Though most of the original maps were replaced by others, the structure of the original show was replicated and the North Carolina room was essentially reproduced. 'Whose agenda is in your glove compartment?' teased the provocative header, this time in type twelve inches high, bright red on a white wall. A subhead in only slightly smaller type elaborated that: 'Even an ordinary map has hidden messages. Denis Wood, co-curator of this exhibition and a resident of North Carolina, shares his reading of maps from that state. You could do the same for your state'. A line below this in still smaller type wondered: 'Must driving be the only way? Between 1945 and 1990, 63 511 people died on North Carolina highways'. More than a rending of the veil, it was a call to action.

If journalists liked to play up the 'Beware the Glove Compartment' theme, they also liked to reassure. 'Are road maps really a government plot? Nah', ran the *Newsweek* subhead. The lurid headline reflected Americans' pervasive interest in hidden messages – in plots – while the dismissive question and the 'Nah', reflected their common-sense belief that a road map was just a road map. North Carolina papers were especially prone to handling the story this way. In full caps over a photo of the 'Whose agenda is in your glove compartment?' header from the Washington version of the show ran the title: 'Subliminal messages claimed to unfold from state highway map.' The related story was headlined, 'Propaganda in your glove box?' where the question mark alone made the point of *Newsweek*'s 'Nah'. But papers elsewhere harped on the theme too: 'Reading between the borderlines', ran the headline of the article in the *Chicago Tribune*, with subheads that read, 'Flexible truth', 'Maps reveal more about the mapmaker than about the terrain' and 'Not accurate for long'. 'Exhibit reveals global agendas', ran a subhead in the *Washington Times*.

Without exception the 'global agendas', 'flexible truth', 'propaganda' and 'subliminal messages' were construed as laid *on top of* an otherwise straightforward truth: at bottom maps remained reliable representations of reality, no matter that a superficial gloss might be able to *twist them* into serving *special interests*.[4] Indeed, attention turned on these *special* interests, since *common* interests tend not to be thought about as interests at all; and since *everybody* shares such interests, what could it matter that they're socially constructed? From this perspective, our reading of the state highway map came to little more than 'a celebration of [our] own pet peeves', as William Burpitt complained in a letter to the editor of Raleigh's *News and Observer* (Burpitt, 1994). From Burpitt's perspective the argument *we'd* advanced – in which it is interest alone that motivates mapmaking, *common interest especially* – crumbles, to be replaced by one in which, at worst, factual maps might be distorted by special interests. This argument was made with exceptional forcefulness – if also some confusion – by Helen Bunn (1994) in another letter to the *News and Observer*:

Liberals are indeed a strange breed. NC State University professor Denis Wood claims that North Carolina's highway (not cycle) map 'advanced a specific political agenda', while the map actually 'is intended to show the highway system and to promote our state', according to state cartographer Clarence Poe Cox, who overseas [sic] the production of the map.

Later in the article the truth comes out. Wood's criticism is 'based on the 1992 map issued by former Republican Governor Jim Martin's transportation department rather than on the newer version produced under Democratic Governor Jim Hunt'. It appears to me that Wood also has a specific political agenda. Is it now politically correct for the pot to call the kettle black?

Promoting North Carolina as a 'leisure paradise' in my view is not misrepresentation – it's a fact. From the mountains to the sea, North Carolina is the greatest state in the greatest nation on Earth. It is evident that Wood feels no loyalty toward North Carolina, so it must not be his home state.

Where are Jesse Helms and Lauch Faircloth? Readers should have called on them to demand the removal of this partisan exhibit from the Smithsonian's Ripley Centre.

---

[4] To promote a different *special interest-* or *counter-interest-* Fels has produced the 'North Carolina Watersheds' map, which portrays the topography, hydrography, physiographic regions and major watersheds of North Carolina, but *no roads or highways whatsoever* (Fels 2000). His stated intention was 'to encourage a different notion of geographic "connectedness", one not based on automobiles and pavement but on the intrinsic connections of North Carolina's natural systems'. Any cartographer can play at this game, if that's what one thinks this is. We reproduced the map in Wood and Fels 2008 (p. 174).

Only *special*, indeed only *partisan* interest is objectionable, but *it* suffices to justify the removal of the exhibition! In being attacked by letters to the editor, *Designs on Signs/ Myth and Meaning in Maps* had come a long way from being an unreadable article in a marginal professional journal.

It has a way to go too. The Designs on Signs chapter of *The Power of Maps*, updated and split into two, will be the only piece of *The Power of Maps* carried forward into the book's second edition. Why? Because no matter how we have come to think about maps, no matter the future of maps themselves, they will always be embodied in signs; and, for this reason alone, semiological analysis will always be necessary, not just interesting but *essential*. Our analysis of the propositional logic of the map – as laid out in our recent *The Natures of Maps* (Wood and Fels, 2008) – assumes, simply takes for granted, the semiological arguments of Designs on Signs. Though in *The Natures of Maps* we may dissolve the map's surface into the atomic propositions we call postings, these postings *have to be realized in signs*. If we devote less than one of the book's 230 pages to 'Map Logic and Its Semiotic Expression', this is only because we'd already laid that part of the argument out, you know, back in 1986 . . . across fifty pages of *Cartographica*.

Further Reading

Barthes, R. (1972) *Mythologies*, Hill and Wang, New York. (Barthes wrote extensively on semiotic theory, but in these popular essays he used it to highlight the political dimensions of advertisements, film, and other everyday things, revealing their character as myth.)

Bertin, J. (1983) *Semiology of Graphics: Diagrams, Networks, Maps*, University of Wisconsin Press, Madison, WI. (Bertin introduced semiotics to mapmakers with this seminal text which remains, despite its age, indispensable.)

Casti, E. (2005) Towards a theory of interpretation: Cartographic semiosis. *Cartographica*, **40** (3), 1–16. (Casti argues that a semiotic approach de-emphasizes the map as mediant to emphasize it as agent.)

Eco, U. (1976) *A Theory of Semiotics*, Indiana University Press, Bloomington, IN. (Eco's semiotics differs in crucial ways from both Barthes' and Bertin's. There are as many semiotics as there are semioticians.)

Wood, D. and Fels, J. (1992) *The Power of Maps*, Guilford Press, New York. (Embedding Designs on Signs in this book made clear the power of a semiotic analysis to reveal the mythic character of a wide range of maps, especially when allied to other forms of analysis.)

Wood, D. and Fels, J. (2008) *The Natures of Maps*, University of Chicago Press, Chicago. (Here the semiotic analysis of Designs on Signs has been wed to a non-representational, indeed propositional characterization of the map.)

Wood, D., Fels, J. and Krygier, J. (2010) *Rethinking the Power of Maps*, Guilford Press, New York. (Here a truly propositional map logic is founded on the semiotic basis laid in Designs on Signs.)

References

Barthes, R. (1974) *S/Z*, Hill and Wang, New York.

Bertin, J. (1983) *Semiology of Graphics*, University of Wisconsin Press, Madison, WI.

Bunn, H. (1994) Partisan (map) politics. *News and Observer*, 22 January, 13A.

Burpitt, W. (1994) A professor's pet peeves. *News and Observer*, 22 January, 13A.

Fels, J. (2000) *North Carolina Watersheds*, 3rd ed., John Fels Cartographics, Raleigh, NC.

Fernandez, R. (2001) *Imagining Literacy: Rhizomes of Knowledge in American Culture and Literature*, University of Texas Press, Austin, TX.

Harley, B. (1988) Maps, knowledge, and power, in *The Iconography of Landscape: Essays on the Symbolic Representation, Design, and Use of Past Environments* (eds. D. Cosgrove and S. Daniels), Cambridge University Press, Cambridge, pp. 277–312.

Jacobs, J. (2003) Editorial: home rules. *Transactions of the Institute of British Geographers NS*, **28**, 259–263.

Ockerse, T. (1984) De-sign/super-sign. *Semiotica*, **52**(3/4), 247–272.

Pickles, J. (1991) Geography, GIS, and the surveillant society. *Papers and Proceedings of Applied Geography Conferences*, 14, 80–91.

Rundstrom, R. (1990) A cultural interpretation of Inuit map accuracy. *Geographical Review*, **80** (2), 155–168.

Schlichtmann, H. (1985) Characteristic traits of the semiotic system 'map symbolism'. *The Cartographic Journal*, **22**(1), 23–30.

Twardy, C. (1992) Denis Wood: Escaping the tyranny of the map. *Inform*, **3**(4), 6–7.

Wood, D. (1972) On mental maps. *The Canadian Cartographer*, **9**(2), 149–150.

Wood, D. (1977) Now and then: Comparisons of ordinary Americans' symbol conventions with those of past cartographers. *Prologue: The Journal of the National Archives*, **9**(3), 151–161.

Wood, D. (1978a) The stars in our hearts: Critical commentary on George Lucas' *Star Wars*. *The Journal of Popular Film and Television*, **6**(3), 262–279.

Wood, D. (1978b) Growing up among the stars: More on George Lucas' *Star Wars*. *Literature/ Film Quarterly*, **6**(4), 327–341.

Wood, D. (1979) As sand on the beach: Commentary on Matthew Robbins and Hal Barwood's *Corvette Summer*. *The Journal of Popular Film and Television*, **7**(2), 130–145.

Wood, D. (1980a) *The History of Topographical Maps* by P. D. A. Harvey. *Cartographica*, **17** (3), 130–133.

Wood, D. (1980b) All the words we cannot say: Critical commentary on Michael Cimino's *The Deer Hunter*. *The Journal of Popular Film and Television*, **7**(4), 366–382.

Wood, D. (1981a) The bodies we keep tripping over: Critical commentary on Sam Fuller's *The Big Red One*. *Journal of Popular Film and Television*, **9**(1), 2–12.

Wood, D. (1981b) The empire's new clothes. *Film Quarterly*, **34**(3), 10–16.

Wood, D. (1982a) *Concepts in the History of Cartography* by M. J. Blakemore and J. B. Harley. *Cartographica*, **19**(1), 73–75.

Wood, D. (1982b) *The Mapmakers* by John Noble Wilford. *Cartographica*, **19**(3/4), 127–131.

Wood, D. (1982c) *Concepts in the History of Cartography* by M. J. Blakemore and J. B. Harley. *The American Cartographer*, **9**(1), 91–93.

Wood, D. (1983a) *Early Thematic Mapping in the History of Cartography* by Arthur Robinson. *Cartographica*, **20**(3), 109–112.

Wood, D. (1983b) *The Visual Display of Quantitative Information* by Edward R. Tufte. *Cartographica*, **20**(4), 104–107.

Wood, D. (1991) The humanization of cartography. Bulletin – Geography and Map Division, Special Libraries Association, March.

Wood, D. (1994) *Home Rules*, Johns Hopkins University Press, Baltimore, MD.

Wood, D. and Fels, J. (1992) *The Power of Maps*, Guilford, New York.

Wood, D. and Fels, J. (2008) *The Natures of Maps*, University of Chicago Press, Chicago.

Wood, D., Fels, J., and Krygier, J., *Rethinking the Power of Maps*, Guilford Press, New York.

SECTION THREE
POLITICS AND SOCIETY

16

Deconstructing the Map

J.B. Harley[1]

Abstract

The paper draws on ideas in postmodern thinking to redefine the nature of maps as representations of power. The traditional rules of cartography – long rooted in a scientific epistemology of the map as an objective form of knowledge – will firstly be reviewed as an object of deconstruction. Secondly, a deconstructionist argument will explore the textuality of maps, including their metaphorical and rhetorical nature. Thirdly, the paper will examine the dimensions both of external power and of the omnipresence of internal power in the cartographic representation of place.

A map says to you, 'Read me carefully, follow me closely, doubt me not.' It says, 'I am the earth in the palm of your hand. Without me, you are alone and lost.'

And indeed you are. Were all the maps in this world destroyed and vanished under the direction of some malevolent hand, each man would be blind again, each city be made a stranger to the next, each landmark become a meaningless signpost pointing to nothing.

Yet, looking at it, feeling it, running a finger along its lines, it is a cold thing, a map, humourless and dull, born of calipers and a draughtsman's board. That coastline there, that ragged scrawl of scarlet ink, shows neither sand nor sea nor rock; it speaks of no mariner, blundering full sail in wakeless seas, to bequeath, on sheepskin or a slab of wood, a priceless scribble to posterity. This brown blot that marks a mountain has, for the casual eye, no other significance, though twenty men, or ten, or only one, may have squandered life to climb it. Here is a valley, there a swamp, and there a desert; and here is a river that some curious and courageous soul, like a pencil in the hand of God, first traced with bleeding feet. (Beryl Markham, 1983)

[1] Originally published: 1989, *Cartographica*, **26**(2), 1–20.At the time of publication: Harley was Professor of Geography at University of Wisconsin in Milwaukee and Director of the Office of Map History in the American Geographical Society Collection.

Classics in Cartography: Reflections on Influential Articles from Cartographica   Edited by Martin Dodge
© 2011 John Wiley & Sons, Ltd

The pace of conceptual exploration in the history of cartography – searching for alternative ways of understanding maps – is slow. Some would say that its achievements are largely cosmetic. Applying conceptions of literary history to the history of cartography, it would appear that we are still working largely in either a 'premodern' or a 'modern' rather than in a 'postmodern' climate of thought.[2] A list of individual explorations would, it is true, contain some that sound impressive. Our students can now be directed to writings that draw on the ideas of information theory, linguistics, semiotics, structuralism, phenomenology, developmental theory, hermeneutics, iconology, Marxism and ideology. We can point to the names in our footnotes of (amongst others) Cassirer, Gombrich, Piaget, Panofsky, Kuhn, Barthes and Eco. Yet despite these symptoms of change, we are still, willingly or unwillingly, the prisoners of our own past.

My basic argument in this essay is that we should encourage an epistemological shift in the way we interpret the nature of cartography. For historians of cartography, I believe a major roadblock to understanding is that we still accept uncritically the broad consensus, with relatively few dissenting voices, of what *cartographers* tell us maps are supposed to be. In particular, we often tend to work from the premise that mappers engage in an unquestionably 'scientific' or 'objective' form of knowledge creation. Of course, cartographers believe they have to say this to remain credible but historians do not have that obligation. It is better for us to begin from the premise that cartography is seldom what cartographers say it is.

As they embrace computer-assisted methods and Geographical Information Systems, the scientist rhetoric of map makers is becoming more strident. The 'culture of technics' is everywhere rampant. We are told that the journal now named *The American Cartographer* will become *Cartography and Geographical Information Systems*. Or, in a strangely ambivalent gesture toward the nature of maps, the British Cartographic Society proposes that there should be two definitions of cartography, 'one for professional cartographers and the other for the public at large'. A definition 'for use in communication with the general public' would be 'Cartography is the art, science and technology of making maps"; that for 'practicing cartographers' would be 'Cartography is the science and technology of analysing and interpreting geographic relationships, and communicating the results by means of maps'.[3] Many may find it surprising that 'art' no longer exists in 'professional' cartography. In the present context, however, these signs of ontological schizophrenia can also be read as reflecting an urgent need to rethink the nature of maps from different perspectives. The question arises as to whether the notion of a progressive science is a myth partly created by cartographers in the course of their own professional development. I suggest that it has been accepted too uncritically by a wider public and by other scholars who work with maps.[4] For those concerned with the history of maps it is especially timely that we challenge the cartographer's assumptions. Indeed, if

---

[2] For these distinctions see Eagleton (1983); for an account situated closer to the direct concerns of cartography see Ferraris (1988: 12–24).

[3] Reported in *Cartographic Perspectives: Bulletin of the North American Cartographic Information Society*, 1989, **1**(1): 4.

[4] Others have made the same point: see, especially, the trenchantly deconstructive turn of the essay by Wood and Fels (1986).

the history of cartography is to grow as an interdisciplinary subject amongst the humanities and social sciences, new ideas are essential.

The question becomes how do we as historians of cartography escape from the normative models of cartography? How do we allow new ideas to come in? How do we begin to write a cartographic history as genuinely revisionist as Louis Marin's (1988) 'The King and his Geometer' (in the context of a seventeenth century map of Paris) or William Boelhower's (1984) 'The Culture of the Map' (in the context of sixteenth century world maps showing America for the first time; pages 41–53) (see also Boelhower, 1988)? These are two studies informed by postmodernism. In this essay I also adopt a strategy aimed at the deconstruction of the map.

The notion of deconstruction[5] is also a password for the postmodern enterprise. Deconstructionist strategies can now be found not only in philosophy but also in localized disciplines, especially in literature, and in other subjects such as architecture, planning (Knox, 1988; Gregory, 1987) and, more recently, geography (Dear, 1988). I shall specifically use a deconstructionist tactic to break the assumed link between reality and representation which has dominated cartographic thinking, has led it in the pathway of 'normal science' since the Enlightenment, and has also provided a ready-made and 'taken for granted' epistemology for the history of cartography. The objective is to suggest that an alternative epistemology, rooted in social theory rather than in scientific positivism, is more appropriate to the history of cartography. It will be shown that even 'scientific' maps are a product not only of 'the rules of the order of geometry and reason' but also of the 'norms and values of the order of social . . . tradition' (Marin, 1988: 173). Our task is to search for the social forces that have structured cartography and to locate the presence of power – and its effects – in all map knowledge.

The ideas in this particular essay owe most to writings by Foucault and Derrida. My approach is deliberately eclectic because in some respects the theoretical positions of these two authors are incompatible. Foucault anchors texts in socio-political realities and constructs systems for organizing knowledge of the kind that Derrida loves to dismantle.[6] But even so, by combining different ideas on a new terrain, it may be possible to devise a scheme of social theory with which we can begin to interrogate the hidden agendas of cartography. Such a scheme offers no 'solution' to an historical interpretation of the cartographic record, nor a precise method or set of techniques, but as a broad strategy it may help to locate some of the fundamental forces that have driven map making in both European and non-European societies. From Foucault's writings, the key revelation has been the omnipresence of power in all knowledge, even though that power is invisible or implied, including the particular knowledge encoded in maps and atlases. Derrida's notion of the rhetoricity of all texts has been no less a challenge.[7] It demands a search for metaphor and rhetoric in maps where previously scholars had

---

[5] Deriving from the writings of Jacques Derrida: for exposition see the translator's Preface Derrida (1976: ix-lxxxvii); Norris (1982, 1987).

[6] As an introduction I have found to be particularly useful Said (1978); also the chapters 'Jacques Derrida' by Hoy (1985) and 'Michel Foucault' by Philp (1985).

[7] On the other hand, I do not adopt some of the more extreme positions attributed to Derrida. For example, it would be unacceptable for a social history of cartography to adopt the view that nothing lies outside the text.

found only measurement and topography. Its central question is reminiscent of Korzybski's (1948: 58, 247, 498, 750–751) much older dictum 'The map is not the territory' but deconstruction goes further to bring the issue of how the map represents place into much sharper focus.

Deconstruction urges us to read between the lines of the map – 'in the margins of the text' – and through its tropes to discover the silences and contradictions that challenge the apparent honesty of the image. We begin to learn that cartographic facts are only facts within a specific cultural perspective. We start to understand how maps, like art, far from being 'a transparent opening to the world', are but 'a particular human way . . . of looking at the world' (Blocker, 1979: 43).

In pursuing this strategy I shall develop three threads of argument. Firstly, I shall examine the discourse of cartography in the light of some of Foucault's ideas about the play of rules within discursive formations. Secondly, drawing on one of Derrida's central positions, I will examine the textuality of maps and, in particular, their rhetorical dimension. Thirdly, returning to Foucault, I will consider how maps work in society as a form of power-knowledge.

16.1 The Rules of Cartography

One of Foucault's primary units of analysis is the discourse. A discourse has been defined as 'a system of possibility for knowledge' (Philp, 1985: 69). Foucault's method was to ask, it has been said,

> *what rules permit certain statements to be made; what rules order these statements; what rules permit us to identify some statements as true and others as false; what rules allow the construction of a map, model or classificatory system . . . what rules are revealed when an object of discourse is modified or transformed . . . Whenever sets of rules of these kinds can be identified, we are dealing with a discursive formation or discourse. (Philp, 1985: 69)*

The key question for us then becomes, 'What type of rules have governed the development of cartography?' Cartography I define as a body of theoretical and practical knowledge that map makers employ to construct maps as a distinct mode of visual representation. The question is, of course, historically specific: the rules of cartography vary in different societies. Here I refer particularly to two distinctive sets of rules that underlie and dominate the history of Western cartography since the seventeenth century.[8] One set may be defined as governing the technical production of maps and is made explicit in the cartographic treatises and writings of the period.[9]

---

[8] 'Western cartography' is defined as the types of survey mapping first fully visible in the European Enlightenment and which then spread to other areas of the world as part of European overseas expansion.
[9] The history of these technical rules has been extensively written about in the history of cartography, though not in terms of their social implications nor in Foucault's sense of discourse: see, for example, the later chapters of Crone (1953, 1978).

The other set relates to the cultural production of maps. These must be understood in a broader historical context than either scientific procedure or technique. They are, moreover, rules that are usually ignored by cartographers so that they form a hidden aspect of their discourse.

The first set of cartographic rules can thus be defined in terms of a scientific epistemology. From at least the seventeenth century onward, European map makers and map users have increasingly promoted a standard scientific model of knowledge and cognition. The object of mapping is to produce a 'correct' relational model of the terrain. Its assumptions are that the objects in the world to be mapped are real and objective, and that they enjoy an existence independent of the cartographer; that their reality can be expressed in mathematical terms; that systematic observation and measurement offer the only route to cartographic truth; and that this truth can be independently verified.[10] The procedures of both surveying and map construction came to share strategies similar to those in science in general: cartography also documents a history of more precise instrumentation and measurement; increasingly complex classifications of its knowledge and a proliferation of signs for its representation; and, especially from the nineteenth century onward, the growth of institutions and a 'professional' literature designed to monitor the application and propagation of the rules (for evidence see Wolter, 1975a, 1975b). Moreover, although cartographers have continued to pay lip service to the 'art and science' of map making,[11] art, as we have seen, is being edged off the map. It has often been accorded a cosmetic rather than a central role in cartographic communication (Morris, 1982). Even philosophers of visual communication – such as Arnheim (1986: 194–202), Eco (1976: 245–257), Gombrich (1975) and Goodman (1968: 170–171; 228–230) – have tended to categorise maps as a type of congruent diagram – as analogs, models, or 'equivalents' creating a similitude of reality – and, in essence, different from art or painting. A 'scientific' cartography (so it was believed) would be untainted by social factors. Even today many cartographers are puzzled by the suggestion that political and sociological theory could throw light on their practices. They will probably shudder at the mention of deconstruction.

The acceptance of the map as 'a mirror of nature' (to employ Richard Rorty's (1979) phrase) also results in a number of other characteristics of cartographic discourse even where these are not made explicit. Most striking is the belief in progress: that, by the application of science, ever more precise representations of reality can be produced. The methods of cartography have delivered a 'true, probable, progressive, or highly confirmed knowledge' (Laudan, 1977: 2). This mimetic bondage has led to a tendency not only to look down on the maps of the past (with a dismissive scientific chauvinism) but also to regard the maps of other non-Western or early cultures (where the rules of map making were different) as inferior to European maps.[12] Similarly, the

---

[10] For a discussion of these characteristics in relation to science in general see Campbell (1973); also Woolgar (1988, especially Chapter 1), and Hooykaas (1987) for a more specifically historical context.

[11] See, for example, the definition of cartography in Meynen (1973: 1,3): or, more recently, Wallis and Robinson (1987: xi), where cartography 'includes the study of maps as scientific documents and works of art'.

[12] For a discussion of these tendencies in the historiography of early maps see Harley (1988a, 1988b).

primary effect of the scientific rules was to create a 'standard' – a successful version of 'normal science'[13] – that enabled cartographers to build a wall around their citadel of the 'true' map. Its central bastions were measurement and standardization and beyond there was a 'not cartography' land where lurked an army of inaccurate, heretical, subjective, valuative and ideologically distorted images. Cartographers developed a 'sense of the other' in relation to nonconforming maps. Even maps such as those produced by journalists, where different rules and modes of expressiveness might be appropriate, are evaluated by many cartographers according to standards of 'objectivity', 'accuracy' and 'truthfulness'. In this respect, the underlying attitude of many cartographers is revealed in a recent book of essays on *Cartographie dans les medias* (Gauthier, 1988). One of its reviewers has noted how many authors attempt to exorcise from

> the realm of cartography any graphic representation that is not a simple planimetric image, and to then classify all other maps as 'decorative graphics masquerading as maps' where the 'bending of cartographic rules' has taken place ... most journalistic maps are flawed because they are inaccurate, misleading or biased. (Andrews, 1989)

Or, in Britain, we are told, there was set up a 'Media Map Watch' in 1984. 'Several hundred interested members [of cartographic and geographic societies] submitted several thousand maps and diagrams for analysis that revealed [according to the rules] numerous common deficiencies, errors and inaccuracies along with misleading standards' (Balchin, 1988: 33–48). In this example of cartographic vigilantism the 'ethic of accuracy' is being defended with some ideological fervour. The language of exclusion is that of a string of 'natural' opposites: 'true and false'; 'objective and subjective'; 'literal and symbolic' and so on. The best maps are those with an 'authoritative image of self-evident factuality' (Lupton, 1986: 53).

In cases where the scientific rules are invisible in the map we can still trace their play in attempting to normalize the discourse. The cartographer's 'black box' has to be defended and its social origins suppressed. The hysteria amongst leading cartographers at the popularity of the Peters' projection,[14] or the recent expressions of piety amongst Western European and North American map makers following the Russian admission that they had falsified their topographic maps to confuse the enemy, give us a glimpse of how the game is played according to these rules. What are we to make of the 1988 newspaper headlines such as 'Russians Caught Mapping' (*Ottawa Citizen*), 'Soviets Admit Map Paranoia' (*Wisconsin Journal*) or (in the *New York Times*) 'In West, Mapmakers Hail "Truth"' and 'The rascals finally realized the truth and were able to tell

---

[13] In the much-debated sense of Kuhn (1962). For challenges and discussions, see Lakatos and Musgrave (1970).

[14] Arno Peters (1983) *The New Cartography* (New York: Friendship Press). The responses included Loxton (1985), *Cartographic Journal* (1985), Robinson (1985), Porter and Voxland (1986) and, for a more balanced view, Snyder (1988).

it, a geographer at the Defense Department said'?[15] The implication is that Western maps are value free. According to the spokesman, our maps are not ideological documents, and the condemnation of Russian falsification is as much an echo of Cold War rhetoric as it is a credible cartographic criticism.

This timely example also serves to introduce my second contention, that the scientific rules of mapping are, in any case, influenced by a quite different set of rules, those governing the cultural production of the map. To discover these rules, we have to read between the lines of technical procedures or of the map's topographic content. They are related to values, such as those of ethnicity, politics, religion or social class, and they are also embedded in the map producing society at large. Cartographic discourse operates a double silence toward this aspect of the possibilities for map knowledge. In the map itself, social structures are often disguised beneath an abstract, instrumental space, or incarcerated in the coordinates of computer mapping. And in the technical literature of cartography they are also ignored, notwithstanding the fact that they may be as important as surveying, compilation or design in producing the statements that cartography makes about the world and its landscapes. Such an interplay of social and technical rules is a universal feature of cartographic knowledge. In maps it produces the 'order' of its features and the 'hierarchies of its practices' (Foucault, 1973: xx). In Foucault's sense the rules may enable us to define an *episteme* and to trace an archaeology of that knowledge through time (Foucault, 1973: xxii).

Two examples of how such rules are manifest in maps will be given to illustrate their force in structuring cartographic representation. The first is the well-known adherence to the 'rule of ethnocentricity' in the construction of world maps. This has led many historical societies to place their own territories at the centre of their cosmographies or world maps. While it may be dangerous to assume universality, and there are exceptions, such a rule is as evident in cosmic diagrams of pre-Columbian North American Indians as it is in the maps of ancient Babylonia, Greece or China, or in the mediaeval maps of the Islamic world or Christian Europe.[16] Yet what is also significant in applying Foucault's critique of knowledge to cartography is that the history of the ethnocentric rule does not march in step with the 'scientific' history of map making. Thus, the scientific Renaissance in Europe gave modern cartography coordinate systems, Euclid, scale maps and accurate measurement, but it also helped to confirm a new myth of Europe's ideological centrality through projections such as those of Mercator (Peters, 1983). Or again, in our own century, a tradition of the exclusivity of America was enhanced before World War II by placing it in its own hemisphere ('our hemisphere') on the world map.[17] Throughout the history of cartography ideological

[15] 'Soviet aide admits maps were faked for 50 years' and 'In West, mapmakers hail truth', *The New York Times*, 3 September 1988; 'Soviets admit map paranoia', *Wisconsin State Journal* Saturday, 3 September 1988; 'Soviets caught mapping!' *The Ottawa Citizen* Saturday, 3 September 1988; 'Faked Russian maps gave the Germans fits', *The New York Times*, 11 September 1988; and 'National geo-glasnost?' *The Christian Science Monitor*, 12 September 1988.
[16] Many commentators have noted this tendency. See, for example, Tuan (1974), Chapter 4, 'Ethnocentrism, symmetry, and space', 30–44. On ancient and medieval European maps in this respect see Harley and Woodward (1987). On the maps of Islam and China see Harley and Woodward (1992).
[17] For the wider history of this 'rule' see Whitaker (1954); also Whittemore Boggs (1945), Henrikson (1975).

'Holy Lands' are frequently centred on maps. Such centricity, a kind of 'subliminal geometry' (Harley, 1988b: 289–290), adds geopolitical force and meaning to representation. It is also arguable that such world maps have, in turn, helped to codify, to legitimate and to promote the world views which are prevalent in different periods and places.[18]

A second example is how the 'rules of the social order' appear to insert themselves into the smaller codes and spaces of cartographic transcription. The history of European cartography since the seventeenth century provides many examples of this tendency. Pick a printed or manuscript map from the drawer almost at random and what stands out is the unfailing way its text is as much a commentary on the social structure of a particular nation or place as it is on its topography. The map maker is often as busy recording the contours of feudalism, the shape of a religious hierarchy or the steps in the tiers of social class,[19] as the topography of the physical and human landscape.

Why maps can be so convincing in this respect is that the rules of society and the rules of measurement are mutually reinforcing in the same image. Writing of the map of Paris, surveyed in 1652 by Jacques Gomboust, the King's engineer, Louis Marin points to 'this sly strategy of simulation-dissimulation':

> *The knowledge and science of representation, to demonstrate the truth that its subject declares plainly, flow nonetheless in a social and political hierarchy. The proofs of its 'theoretical' truth had to be given, they are the recognizable signs; but the economy of these signs in their disposition on the cartographic plane no longer obeys the rules of the order of geometry and reason but, rather, the norms and values of the order of social and religious tradition. Only the churches and important mansions benefit from natural signs and from the visible rapport they maintain with what they represent. Townhouses and private homes, precisely because they are private and not public, will have the right only to the general and common representation of an arbitrary and institutional sign, the poorest, the most elementary (but maybe, by virtue of this, principal) of geometric elements; the point identically reproduced in bulk. (Marin, 1988: 173)*

Once again, much like 'the rule of ethnocentrism', this hierarchicalization of space is not a conscious act of cartographic representation. Rather, it is taken for granted in a society that the place of the king is more important than the place of a lesser baron, that a castle is more important than a peasant's house, that the town of an archbishop is more important than that of a minor prelate, or that the estate of a landed gentleman is more worthy of emphasis than that of a plain farmer. Cartography deploys its vocabulary accordingly so that it embodies a systematic social inequality. The distinctions of class

---

[18] The link between actual mapping, as the principal source of our world vision, and *mentalité* still has to be thoroughly explored. For some contemporary links see Henrikson (1987). For a report on research that attempts to measure this influence in the cognitive maps of individuals in different areas of the world see Saarinen (1987).

[19] For a general discussion see Harley (1988a, 1988b: 292–294); in my essay on 'Power and legitimation in the English geographical atlases of the eighteenth century' (Harley, 1997), these 'rules of the social order' are discussed in the maps of one historical society.

and power are engineered, reified and legitimated in the map by means of cartographic signs. The rule seems to be 'the more powerful, the more prominent'. To those who have strength in the world shall be added strength in the map. Using all the tricks of the cartographic trade – size of symbol, thickness of line, height of lettering, hatching and shading, the addition of colour – we can trace this reinforcing tendency in innumerable European maps. We can begin to see how maps, like art, become a mechanism 'for defining social relationships, sustaining social rules and strengthening social values' (Geertz, 1983: 99).

In the case of both these examples of rules, the point I am making is that the rules operate both within and beyond the orderly structures of classification and measurement. They go beyond the stated purposes of cartography. Much of the power of the map, as a representation of social geography, is that it operates behind a masque of a seemingly neutral science. It hides and denies its social dimensions at the same time as it legitimates. Yet whichever way we look at it the rules of society will surface. They have ensured that maps are at least as much an image of the social order as they are a measurement of the phenomenal world of objects.

16.2 Deconstruction and the Cartographic Text

To move inward from the question of cartographic rules – the social context within which map knowledge is fashioned – we have to turn to the cartographic text itself. The word 'text' is deliberately chosen. It is now generally accepted that the model of text can have a much wider application than to literary texts alone. To non-book texts such as musical compositions and architectural structures we can confidently add the graphic texts we call maps.[20] It has been said that 'what constitutes a text is not the presence of linguistic elements but the act of construction' so that maps, as 'constructions employing a conventional sign system' (McKenzie, 1986: 35), become texts. With Barthes we could say they 'presuppose a signifying consciousness' that it is our business to uncover (Barthes, 1973: 110). 'Text' is certainly a better metaphor for maps than the mirror of nature. Maps are a cultural text. By accepting their textuality we are able to embrace a number of different interpretative possibilities. Instead of just the transparency of clarity we can discover the pregnancy of the opaque. To fact we can add myth, and instead of innocence we may expect duplicity. Rather than working with a formal science of communication, or even a sequence of loosely related technical processes, our concern is redirected to a history and anthropology of the image, and we learn to recognize the narrative qualities of cartographic representation[21] as well as its claim to provide a synchronous picture of the world. All this, moreover, is likely to lead to a

---

[20] This is cogently argued by McKenzie (1986: especially 34–39), where he discusses the textuality of maps. Robinson and Petchenik (1976: 43), reject the metaphor of map as language: they state that 'the two systems, map and language are essentially incompatible', basing their belief on the familiar grounds of literality that language is verbal, that images do not have a vocabulary, that there is no grammar, and the temporal sequence of a syntax is lacking. Rather than isolating the differences, however, it now seems more constructive to stress the *similarities* between map and text.

[21] The narrative qualities of cartography are introduced by Wood (1987).

rejection of the neutrality of maps, as we come to define their intentions rather than the literal face of representation, and as we begin to accept the social consequences of cartographic practices. I am not suggesting that the direction of textual enquiry offers a simple set of techniques for reading either contemporary or historical maps. In some cases we will have to conclude that there are many aspects of their meaning that are undecidable.[22]

Deconstruction, as discourse analysis in general, demands a closer and deeper reading of the cartographic text than has been the general practice in either cartography or the history of cartography. It may be regarded as a search for alternative meanings. 'To deconstruct', it is argued,

> *is to reinscribe and resituate meanings, events and objects within broader movements and structures; it is, so to speak, to reverse the imposing tapestry in order to expose in all its unglamorously dishevelled tangle the threads constituting the well-heeled image it presents to the world. (Eagleton, 1986: 80, quoted in Soja, 1989: 12)*

The published map also has a 'well-heeled image' and our reading has to go beyond the assessment of geometric accuracy, beyond the fixing of location and beyond the recognition of topographical patterns and geographies. Such interpretation begins from the premise that the map text may contain 'unperceived contradictions or duplicitous tensions' (Hoy, 1985: 540) that undermine the surface layer of standard objectivity. Maps are slippery customers. In the words of W.J.T. Mitchell, writing of languages and images in general, we may need to regard them more as 'enigmas, problems to be explained, prison houses which lock the understanding away from the world'. We should regard them 'as the sort of sign that presents a deceptive appearance of naturalness and transparency concealing an opaque, distorting, arbitrary mechanism of representation' (Mitchell, 1986: 8). Throughout the history of modern cartography in the West, for example, there have been numerous instances of where maps have been falsified, of where they have been censored or kept secret, or of where they have surreptitiously contradicted the rules of their proclaimed scientific status (Harley, 1988c).

As in the case of these practices, map deconstruction would focus on aspects of maps that many interpreters have glossed over. Writing of 'Derrida's most typical deconstructive moves', Christopher Norris (1987: 19) notes that

> *deconstruction is the vigilant seeking-out of those 'aporias', blindspots or moments of self-contradiction where a text involuntarily betrays the tension between rhetoric and logic, between what it manifestly means to say and what it is nonetheless constrained to mean. To 'deconstruct' a piece of writing is therefore to operate a kind of strategic reversal, seizing on precisely those unregarded details (casual metaphors, footnotes, incidental turns of argument) which are always, and necessarily, passed over by interpreters of a more orthodox persuasion. For it is here, in the margins of the*

---

[22] The undecidability of textual meaning is a central position in Derrida's criticism of philosophy: see the discussion by Hoy (1985: 54–58).

text – the 'margins', that is, as defined by a powerful normative consensus – that
deconstruction discovers those same unsettling forces at work.

A good example of how we could deconstruct an early map – by beginning with what
have hitherto been regarded as its 'casual metaphors' and 'footnotes' – is provided by
recent studies reinterpreting the status of decorative art on the European maps of the
seventeenth and eighteenth centuries. Rather than being inconsequential marginalia, the
emblems in cartouches and decorative title pages can be regarded as *bask to the way* they
convey their cultural meaning,[23] and they help to demolish the claim of cartography to
produce an impartial graphic science. But the possibility of such a revision is not limited
to historic 'decorative' maps. A recent essay by Wood and Fels (1986) on the Official State
Highway Map of North Carolina indicates a much wider applicability for a deconstructive strategy by beginning in the 'margins' of the contemporary map. They also treat the
map as a text and, drawing on the ideas of Roland Barthes (1973: 103–159) of myth as a
semiological system, develop a forceful social critique of cartography which though
structuralist in its approach is deconstructionist in its outcome. They begin, deliberately,
with the margins of the map, or rather with the subject matter that is printed on its verso:

One side is taken up by an inventory of North Carolina points of interest – illustrated
with photos of, among other things, a scimitar horned oryx (resident in the state zoo), a
Cherokee woman making beaded jewelry, a ski lift, a sand dune (but no cities) – a ferry
schedule, a message of welcome from the then governor, and a motorist's prayer ('Our
heavenly Father, we ask this day a particular blessing as we take the wheel of our car...').
On the other side, North Carolina, hemmed in by the margins of pale yellow South
Carolinas and Virginias, Georgias and Tennessees, and washed by a pale blue Atlantic, is
represented as a meshwork of red, black, blue, green and yellow lines on a white
background, thickened at the intersections by roundels of black or blotches of pink....To
the left of... [the] title is a sketch of the fluttering state flag. To the right is a sketch of a
cardinal (state bird) on a branch of flowering dogwood (state flower) surmounting a
buzzing honey bee arrested in midflight (state insect). (Wood and Fels, 1986: 54)

What is the meaning of these emblems? Are they merely a pleasant ornament for the
traveller or can they inform us about the social production of such state highway maps?
A deconstructionist might claim that such meanings are undecidable, but it is also clear
that the State Highway Map of North Carolina is making other dialogical assertions
behind its masque of innocence and transparency. I am not suggesting that these
elements hinder the traveller getting from point A to B, but that there is a second text
within the map. No map is devoid of an intertextual dimension and, in this case too, the
discovery of intertextuality enables us to scan the image as more than a neutral picture of
a road network.[24] Its 'users' are not only the ordinary motorists but also the State of

---

[23] Most recently, Clarke (1988); also Harley (1984; 1988b: especially 296–299; 1984; 1985).

[24] On the intertextuality of all discourses – with pointers for the analysis of cartography – see Todorov (1984:
60–74); also Bakhtin (1981). I owe these references to Dr. Cordell Yee, History of Cartography Project,
University of Wisconsin at Madison.

North Carolina that has appropriated its publication (distributed in millions of copies) as a promotional device. The map has become an instrument of State policy and an instrument of sovereignty (Wood and Fels, 1986: 63). At the same time, it is more than an affirmation of North Carolina's dominion over its territory. It also constructs a mythic geography, a landscape full of 'points of interest', with incantations of loyalty to state emblems and to the values of a Christian piety. The hierarchy of towns and the visually dominating highways that connect them have become the legitimate natural order of the world. The map finally insists 'that roads really *are* what North Carolina's all about' (Wood and Fels, 1986: 60). The map idolises our love affair with the automobile. The myth is believable.

A cartographer's stock response to this deconstructionist argument might well be to cry 'foul'. The argument would run like this: 'Well after all it's a state highway map. It's designed to be at once popular and useful. We expect it to exaggerate the road network and to show points of interest to motorists. It is a derived rather than a basic map'.[25] It is not a scientific map. The appeal to the ultimate scientific map is always the cartographers' last line of defence when seeking to deny the social relations that permeate their technology.

It is at this point that Derrida's strategy can help us to extend such an interpretation to all maps, scientific or non-scientific, basic or derived. Just as in the deconstruction of philosophy Derrida was able to show 'how the supposedly literal level is intensively metaphorical' (Hoy, 1985: 44), so too we can show how cartographic 'fact' is also symbol. In 'plain' scientific maps, science itself becomes the metaphor. Such maps contain a dimension of 'symbolic realism' which is no less a statement of political authority and control than a coat-of-arms or a portrait of a queen placed at the head of an earlier decorative map. The metaphor has changed. The map has attempted to purge itself of ambiguity and alternative possibility.[26] Accuracy and austerity of design are now the new talismans of authority culminating in our own age with computer mapping. We can trace this process very clearly in the history of Enlightenment mapping in Europe. The topography as shown in maps, increasingly detailed and planimetrically accurate, has become a metaphor for a utilitarian philosophy and its will to power. Cartography inscribes this cultural model upon the paper and we can examine it in many scales and types of maps. Precision of instrument and technique merely serves to reinforce the image, with its encrustation of myth, as a selective perspective on the world. Thus maps of local estates in the European *ancien regime*, though derived from instrumental survey, were a metaphor for a social structure based on landed property. County and regional maps, though founded on scientific triangulation, were an articulation of local values and rights. Maps of the European states, though constructed along arcs of the meridian, served still as a symbolic shorthand for a complex of nationalist ideas. And world maps, though increasingly drawn on mathematically

---

[25] The 'basic' and 'derived' division, like that of 'general purpose' and 'thematic', is one of the axiomatic distinctions often drawn by cartographers. Deconstruction, however, by making explicit the play of forces such as intention, myth, silence and power in maps, will tend to dissolve such an opposition for interpretive purposes except in the very practical sense that one map is often copied or derived from another.
[26] I derive this thought from Eagleton (1983: 135), writing of the ideas of Roland Barthes.

defined projections, nevertheless gave a spiralling twist to the manifest destiny of European overseas conquest and colonisation. (These examples are from Harley, 1988b: 300.) In each of these examples we can trace the contours of metaphor in a scientific map. This in turn enhances our understanding of how the text works as an instrument operating on social reality. In deconstructionist theory the play of rhetoric is closely linked to that of metaphor. In concluding this section of the essay I will argue that notwithstanding 'scientific' cartography's efforts to convert culture into nature, and to 'naturalize' social reality (Eagleton, 1983: 135–136), it has remained an inherently rhetorical discourse. Another of the lessons of Derrida's criticism of philosophy is 'that modes of rhetorical analysis, hitherto applied mainly to literary texts, are in fact indispensable for reading *any* kind of discourse' (Norris, 1982: 19). There is nothing revolutionary in the idea that cartography is an art of persuasive communication. It is now commonplace to write about the rhetoric of the human sciences in the classical sense of the word rhetoric (McCloskey, 1985; Nelson *et al.*, 1987). Even cartographers – as well as their critics – are beginning to allude to the notion of a rhetorical cartography but what is still lacking is a rhetorical close-reading of maps.[27]

The issue in contention is not whether some maps are rhetorical, or whether other maps are partly rhetorical, but the extent to which rhetoric is a universal aspect of all cartographic texts. Thus, for some cartographers the notion of 'rhetoric' would remain a pejorative term. It would be an 'empty rhetoric' which was unsubstantiated in the scientific content of a map. 'Rhetoric' would be used to refer to the 'excesses' of propaganda mapping or advertising cartography or an attempt would be made to confine it to an 'artistic' or aesthetic element in maps as opposed to their scientific core. My position is to accept that rhetoric is part of the way all texts work and that all maps are rhetorical texts. Again we ought to dismantle the arbitrary dualism between 'propaganda' and 'true', and between modes of 'artistic' and 'scientific' representation as they are found in maps. All maps strive to frame their message in the context of an audience. All maps state an argument about the world and they are propositional in nature. All maps employ the common devices of rhetoric such as invocations of authority (*especially* in 'scientific' maps[28]) and appeals to a potential readership through the use of colours, decoration, typography, dedications or written justifications of their method.[29] Rhetoric may be concealed but it is always present, for there is no description without performance.

The steps in making a map – selection, omission, simplification, classification, the creation of hierarchies and 'symbolization' – are all inherently rhetorical. In their intentions as much as in their applications they signify subjective human purposes

---

[27] For a notable exception see Wood and Fels (1986). An interesting example of cartographic rhetoric in historical atlases is described in Goffart (1988).

[28] In Wood and Fels (1986: 99), the examples are given for topographical maps of reliability diagrams, multiple referencing grids, and magnetic error diagrams; on thematic maps 'the trappings of F-scaled symbols and psychometrically divided greys' are a similar form of rhetorical assertion.

[29] The 'letter' incorporated into Gomboust's map of Paris, as discussed by Marin (1988: 169–174), provides an apposite example.

rather than reciprocating the workings of some 'fundamental law of cartographic generalization'.[30] Indeed, the freedom of rhetorical manoeuvre in cartography is considerable: the map maker merely omits those features of the world that lie outside the purpose of the immediate discourse. There have been no limits to the varieties of maps that have been developed historically in response to different purposes of argument, aiming at different rhetorical goals, and embodying different assumptions about what is sound cartographic practice. The style of maps is neither fixed in the past nor is it today. It has been said that 'The rhetorical code appropriates to its map the style most advantageous to the myth it intends to propagate' (Wood and Fels, 1986: 71). Instead of thinking in terms of rhetorical versus non-rhetorical maps, it may be more helpful to think in terms of a theory of cartographic rhetoric which accommodated this fundamental aspect of representation in all types of cartographic text. Thus, I am not concerned to privilege rhetoric over science, but to dissolve the illusory distinction between the two in reading the social purposes as well as the content of maps.

16.3 Maps and the Exercise of Power

For the final stage in the argument I return to Foucault. In doing so I am mindful of Foucault's criticism of Derrida that he attempted 'to restrict interpretation to a purely syntactic and textual level' (Hoy, 1985: 60; for further discussion see Norris, 1987: 213–220), a world where political realities no longer exist. Foucault, on the other hand, sought to uncover 'the social practices that the text itself both reflects and employs' and to 'reconstruct the technical and material framework in which it arose' (Hoy, 1985: 60). Though deconstruction is useful in helping to change the epistemological climate, and in encouraging a rhetorical reading of cartography, my final concern is with its social and political dimensions, and with understanding how the map works in society as a form of power-knowledge. This closes the circle to a context-dependent form of cartographic history.

We have already seen how it is possible to view cartography as a discourse – a system which provides a set of rules for the representation of knowledge embodied in the images we define as maps and atlases. It is not difficult to find for maps – especially those produced and manipulated by the state – a niche in the 'power/knowledge matrix of the modern order' (Philp, 1985: 76). Especially where maps are ordered by government (or are derived from such maps), it can be seen how they extend and reinforce the legal statutes, territorial imperatives and values stemming from the exercise of political power. Yet to understand how power works through cartographic discourse and the effects of that power in society further dissection is needed. A simple model of domination and subversion is inadequate and I propose to draw a distinction between *external* and *internal* power in cartography. This ultimately derives from Foucault's ideas about power-knowledge, but this particular formulation is owed to Joseph Rouse's (1987) recent book on *Knowledge and Power,* where a theory of the internal power of science is, in turn, based on his reading of Foucault.

---

[30] This is still given credence in some textbooks: see, for example, Robinson *et al.* (1984: 127).

The most familiar sense of power in cartography is that of power *external* to maps and mapping. This serves to link maps to the centres of political power. Power is exerted *on* cartography. Behind most cartographers there is a patron; in innumerable instances the makers of cartographic texts were responding to external needs. Power is also exercised *with* cartography. Monarchs, ministers, state institutions, the Church, have all initiated programmes of mapping for their own ends. In modern Western society maps quickly became crucial to the maintenance of state power – to its boundaries, to its commerce, to its internal administration, to control of populations and to its military strength. Mapping soon became the business of the state: cartography is early nationalized. The state guards its knowledge carefully: maps have been universally censored, kept secret and falsified. In all these cases maps are linked to what Foucault called the exercise of 'juridical power' (Foucault, 1980: 88; see also Rouse, 1987: 209–210). The map becomes a 'juridical territory': it facilitates surveillance and control. Maps are still used to control our lives in innumerable ways. A map-less society, though we may take the map for granted, would now be politically unimaginable. All this is power *with* the help of maps. It is an external power, often centralized and exercised bureaucratically, imposed from above and manifest in particular acts or phases of deliberate policy.

I come now to the important distinction. What is also central to the effects of maps in society is what may be defined as the power *internal* to cartography. The focus of enquiry therefore shifts from the place of cartography in a juridical system of power to the political effects of what cartographers do when they make maps. Cartographers manufacture power: they create a spatial panopticon. It is a power embedded in the map text. We can talk about the power of the map just as we already talk about the power of the word or about the book as a force for change. In this sense maps have politics. (I adapt this idea from Winner, 1980: 121–136.) It is a power that intersects and is embedded in knowledge. It is universal. Foucault writes of

> *The omnipresence of power: not because it has the privilege of consolidating everything under its invincible unity, but because it is produced from one moment to the next, at every point, or rather in every relation from one point to another. Power is everywhere; not because it embraces everything, but because it comes from everywhere. (Foucault, 1978: 93)*

Power comes from the map and it traverses the way maps are made. The key to this internal power is thus cartographic process. By this I mean the way maps are compiled and the categories of information selected; the way they are generalized, a set of rules for the abstraction of the landscape; the way the elements in the landscape are formed into hierarchies; and the way various rhetorical styles that also reproduce power are employed to represent the landscape. To catalogue the world is to appropriate it,[31] so that all these technical processes represent acts of control over its image which extend beyond the professed uses of cartography. The world is disciplined. The world is normalized. We

---

[31] Adapting Barthes (1980: 27), who writes much like Foucault, 'To catalogue is not merely to ascertain, as it appears at first glance, but also to appropriate'. Quoted in Wood and Fels (1986: 72).

are prisoners in its spatial matrix. For cartography as much as other forms of knowledge, 'All social action flows through boundaries determined by classification schemes' (Darnton, 1984: 192–193). An analogy is to what happens to data in the cartographer's workshop and what happens to people in the disciplinary institutions – prisons, schools, armies, factories – described by Foucault (Rouse, 1987: 213–226): in both cases a process of normalization occurs. Or, similarly, just as in factories we standardize our manufactured goods, so in our cartographic workshops we standardize our images of the world. Just as in the laboratory we create formulaic understandings of the processes of the physical world, so, too, in the map, nature is reduced to a graphic formula.[32] The power of the map maker was not generally exercised over individuals but over the knowledge of the world made available to people in general. Yet this is not consciously done and it transcends the simple categories of 'intended' and 'unintended' altogether. I am not suggesting that power is deliberately or centrally exercised. It is a local knowledge which at the same time is universal. It usually passes unnoticed. The map is a silent arbiter of power.

What have been the effects of this 'logic of the map' upon human consciousness, if I may adapt Marshall McLuhan's (1962) phrase ('logic of print')? Like him I believe we have to consider for maps the effects of abstraction, uniformity, repeatability and visuality in shaping mental structures, and in imparting a sense of the places of the world. It is the disjunction between those senses of place, and many alternative visions of what the world is, or what it might be, that has raised questions about the effect of cartography in society. Thus, Theodore Roszak (1972: 410) writes:

> *The cartographers are talking about their maps and not landscapes. That is why what they say frequently becomes so paradoxical when translated into ordinary language. When they forget the difference between map and landscape – and when they permit or persuade us to forget that difference – all sorts of liabilities ensue.*[33]

One of these 'liabilities' is that maps, by articulating the world in mass-produced and stereotyped images, express an embedded social vision. Consider, for example, the fact that the ordinary road atlas is amongst the best selling paperback books in the United States (McNally, 1987: 389–392) and then try to gauge how this may have affected ordinary Americans' perception of their country. What sort of an image of America do these atlases promote? On the one hand, there is a patina of gross simplicity. Once off the interstate highways the landscape dissolves into a generic world of bare essentials that invites no exploration. Context is stripped away and place is no longer important. On the other hand, the maps reveal the ambivalence of all stereotypes. Their silences are also inscribed on the page: where, on the page, is the variety of nature, where is the

---

[32] Indeed, cartographers like to promote this metaphor of what they do: read, for example, Monmonier and Schnell (1988: 15), 'Geography thrives on cartographic generalization. The map is to the geographer what the microscope is to the microbiologist, for the ability to shrink the earth and generalize about it ... The microbiologist must choose a suitable objective lens, and the geographer must select a map scale appropriate to both the phenomenon in question and the 'regional laboratory' in which the geographer is studying it'.

[33] Roszak is using the map as a metaphor for scientific method in this argument, which again points to the widespread perception of how maps represent the world.

history of the landscape, and where is the space-time of human experience in such anonymised maps?[34]

The question has now become: do such empty images have their consequences in the way we think about the world? Because all the world is designed to look the same, is it easier to act upon it without realizing the social effects? It is in the posing of such questions that the strategies of Derrida and Foucault appear to clash. For Derrida, if meaning is undecidable so must be *pari passu*, the measurement of the force of the map as a discourse of symbolic action. In ending, I prefer to align myself with Foucault in seeing all knowledge (Rabinow, 1984: 6–7) – and hence cartography – as thoroughly enmeshed with the larger battles which constitute our world. Maps are not external to these struggles to alter power relations. The history of map use suggests that this may be so and that maps embody specific forms of power and authority. Since the Renaissance they have changed the way in which power was exercised. In colonial North America, for example, it was easy for Europeans to draw lines across the territories of Indian nations without sensing the reality of their political identity (Harley, 1988d). The map allowed them to say, 'This is mine; these are the boundaries'.[35] Similarly, in innumerable wars since the sixteenth century it has been equally easy for the generals to fight battles with coloured pins and dividers rather than sensing the slaughter of the battlefield.[36] Or again, in our own society, it is still easy for bureaucrats, developers and 'planners' to operate on the bodies of unique places without measuring the social dislocations of 'progress'. While the map is never the reality, in such ways it helps to create a different reality. Once embedded in the published text the lines on the map acquire an authority that may be hard to dislodge. Maps are authoritarian images. Without our being aware of it maps can reinforce and legitimate the status quo. Sometimes agents of change, they can equally become conservative documents. But in either case the map is never neutral. Where it seems to be neutral it is the sly 'rhetoric of neutrality'[37] that is trying to persuade us.

16.4 Conclusion

The interpretive act of deconstructing the map can serve three functions in a broad enquiry into the history of cartography. Firstly, it allows us to challenge the episte-

---

[34] This criticism is reminiscent of Barthes' (1973: 74–77) essay on 'The *Blue Guide*', where he writes of the *Guide* as 'reducing geography to the description of an uninhabited world of monuments' (we substitute 'roads'). More generally, this tendency is also the concern of Szegö (1987). See also Roszak (1972: 408), where he writes that 'We forfeit the whole value of a map if we forget that it is *not* the landscape itself or anything remotely like an exhaustive depiction of it. If we do forget, we grow rigid as a robot obeying a computer program; we lose the intelligent plasticity and intuitive judgement that every wayfarer must preserve. We may then know the map in fine detail, but our knowledge will be purely academic, inexperienced, shallow'.

[35] Boelhower (1984: 47), quoting François Wahl, 1980, 'Le désir d'espace', in *Cartes et Figures de la Terre* (Centre Georges Pompidou, Paris), 41.

[36] For a modern example relating to Vietnam see Muehrcke (1986: 394), where, however, such military examples are classified as 'abuse' rather than a normal aspect of actions with maps. The author retains 'maps mirror the world' as his central metaphor.

[37] There is a suggestive analogy to maps in the example of the railway timetable given by Kinross (1985).

mological myth (created by cartographers) of the cumulative progress of an objective science always producing better delineations of reality. Secondly, deconstructionist argument allows us to redefine the historical importance of maps. Rather than invalidating their study, it is enhanced by adding different nuances to our understanding of the power of cartographic representation as a way of building order into our world. If we can accept intertextuality then we can start to read our maps for alternative and sometimes competing discourses. Thirdly, a deconstructive turn of mind may allow map history to take a fuller place in the interdisciplinary study of text and knowledge. Intellectual strategies such as those of discourse in the Foucauldian sense, the Derridian notion of metaphor and rhetoric as inherent to scientific discourse, and the pervading concept of power-knowledge are shared by many subjects. As ways of looking at maps they are equally enriching. They are neither inimical to hermeneutic enquiry nor anti-historical in their thrust. By dismantling we build. The possibilities of discovering meaning in maps and of tracing the social mechanisms of cartographic change are enlarged. Postmodernism offers a challenge to read maps in ways that could reciprocally enrich the reading of other texts.

Acknowledgements

These arguments were presented in earlier versions at 'The Power of Places' Conference, Northwestern University, Chicago, in January 1989, and as a 'Brown Bag' lecture in the Department of Geography, University of Wisconsin at Milwaukee, in March 1989. I am grateful for the suggestions received on those occasions and for other helpful comments received from Sona Andrews, Catherine Delano Smith and Cordell Yee. I am also indebted to Howard Delier of the American Geographical Society Collection for a number of references and to Ellen Hanlon for editorial help in preparing the paper for press.

References

Andrews, S.K. (1989) Review of Cartography in the Media. *The American Cartographer*, **16**, 219–220.

Arnheim, R. (1986) The perception of maps, in *New Essays on the Psychology of Art* (ed. R. Arnheim), University of California Press, Berkeley, CA, pp. 194–202.

Bakhtin, M.M. (1981) *The Dialogic Imagination: Four Essays* (ed. M. Holquist), trans. C. Emerson and M. Holquist, University of Texas Press, Austin, TX.

Balchin, W.G.V. (1988) The media map watch in the United Kingdom, in *Cartographie dans les Médias* (ed. M. Gauthier), Presses de l'Université du Québec, Sillery, Québec, pp. 33–48.

Barthes, R. (1973) *Mythologies* (Selected and Translated from the French by Annette Lavers), Paladin, London.

Barthes, R. (1980) The Plates of the Encyclopedia, in *New Critical Essays*, Hill and Wang, New York.

Blocker, H.G. (1979) *Philosophy and Art*, Charles Scribner's Sons, New York.

Boelhower, W. (1984) *Through a Glass Darkly: Ethnic Semiosis in American Literature*, Edizioni Helvetia, Venezia.

Boelhower, W. (1988) Inventing America: A model of cartographic semiosis. *Word and Image*, **4**(2), 475–497.

Campbell, P.N. (1973) Scientific discourse. *Philosophy and Rhetoric*, **6**(1), 1–29.

Cartographic Journal (1985) The so-called Peters projection. *The Cartographic Journal*, **22**(2), 108–110.

Clarke, C.N.G. (1988) Taking possession: The cartouche as cultural text in eighteenth-century American maps. *Word and Image*, **4**(2), 455–474.

Crone, G.R. (1953) *Maps and Their Makers: An Introduction to the History of Cartography*, 1st edn, Dawson, Folkestone, Kent.

Crone, G.R. (1978) *Maps and Their Makers: An Introduction to the History of Cartography*, 5th edn, Archon Books, Hamden, CT.

Darnton, R. (1984) *The Great Cat Massacre and Other Episodes in French Cultural History*, Basic Books, New York.

Dear, M. (1988) The postmodern challenge: Reconstructing human geography. *Transactions of the Institute of British Geographers NS*, **13**, 262–274.

Derrida, J. (1976) *Of Grammatology* (trans. G.C. Spivak), The John Hopkins University Press, Baltimore, MD.

Eagleton, T. (1983) *Literary Theory: An Introduction*, University of Minnesota Press, Minneapolis, MN.

Eagleton, T. (1986) *Against the Grain*, Verso, London.

Eco, U. (1976) *A Theory of Semiotics*, Indiana University Press, Bloomington, IN.

Ferraris, M. (1988) Postmodernism and the deconstruction of modernism. *Design Issues*, **4** (1/2), 12–24.

Foucault, M. (1973) *The Order of Things: An Archaeology of the Human Sciences*, Vintage Books, New York.

Foucault, M. (1978) *The History of Sexuality: Volume I An Introduction* (trans. R. Hurley), Random House, New York.

Foucault, M. (1980) *Power/Knowledge: Selected Interviews and Other Writings, 1972–1977* (ed. C. Gordon) (trans. C. Gordon, L. Marshall, J. Mepham and K. Sopher), Pantheon Books, New York.

Gauthier, M. (1988) *Cartographie dans les Medias*, Presses de l'Université du Québec, Québec.

Geertz, G. (1983) Art as a cultural system, in *Local Knowledge: Further Essays in Interpretive Anthropology*, Basic Books, New York.

Goffart, W. (1988) The map of the barbarian invasions: A preliminary report. *Nottingham Medieval Studies*, **32**, 49–64.

Gombrich, E. (1975) Mirror and map: Theories of pictorial representation. *Philosophical Transactions of the Royal Society of London Series B Biological Sciences*, **270**, 119–149.

Goodman, N. (1968) *Languages of Art: An Approach to a Theory of Symbols*, Bobbs-Merrill, New York.

Gregory, D. (1987) Postmodernism and the politics of social theory. *Environment and Planning D: Society and Space*, **5**, 245–248.

Harley, J.B. (1984) Meaning and ambiguity in Tudor cartography, in *English Map-Making, 1500–1650: Historical Essays* (ed. S. Tyacke), The British Library Reference Division Publications, London, pp. 22–45.

Harley, J.B. (1988a) L'histoire de la cartographie comme discourse. *Préfaces* 5 December, 70–75.

Harley, J.B. (1988b) Maps, knowledge, and power, in *The Iconography of Landscape* (eds. D. Cosgrove and S. Daniels), Cambridge University Press, Cambridge.

Harley, J.B. (1988c) Silences and secrecy: The hidden agenda of cartography in early modern Europe. *Imago Mundi*, **40**, 57–76.

Harley, J.B. (1988d) Victims of a map: New England cartography and the native Americans. Paper read at the Land of Norumbega Conference, Portland, Maine.

Harley, J.B. (1997) Power and legitimation in the English geographical atlases of the eighteenth century, in *Images of the World: The Atlas Through History* (ed. J.A. Wolter), Library of Congress, Washington, DC.

Harley, J.B. and Woodward, D. (1987) *The History of Cartography, Volume 1: Cartography in Prehistoric, Ancient, and Medieval Europe and the Mediterranean*, University of Chicago Press, Chicago.

Harley, J.B. and Woodward, D. (1992) *The History of Cartography, Volume 2: Cartography in the Traditional Islamic and Asian Societies*, University of Chicago Press, Chicago.

Henrikson, A.K. (1975) The map as an 'idea': The role of cartographic imagery during the Second World War. *The American Cartographer*, **2**(1), 19–53.

Henrikson, A.K. (1987) Frameworks for the world, in *Scholars' Guide to Washington D.C. for Cartography and Remote Sensing Imagery* (ed. R.E. Ehrenberg), Smithsonian Institution Press, Washington, DC, pp. viii–xiii.

Hoy, D. (1985) Jacques Derrida, in *The Return of Grand Theory in the Human Sciences* (ed. Q. Skinner), Cambridge University Press, Cambridge, pp. 65–82.

Hooykaas, R. (1987) The rise of modern science: when and why? *The British Journal for the History of Science*, **20**(4), 453–473.

Kinross, R. (1985) The rhetoric of neutrality. *Design Issues*, **2**(2), 18–30.

Knox, P.L. (1988) *The Design Professions and the Built Environment*, Croom Helm, London.

Korzybski, A. (1948) *Science and Sanity: An Introduction to Non-Aristotelian Systems and General Semantics*, 3rd edn, The International Non-Aristotelian Library, Lakeville, CT.

Kuhn, T.S. (1962) *The Structure of Scientific Revolutions*, University of Chicago Press, Chicago.

Lakatos, I. and Musgrave, A. (1970) *Criticism and the Growth of Knowledge*, Cambridge University Press, Cambridge.

Laudan, L. (1977) *Progress and Its Problems: Toward a Theory of Scientific Growth*, University of California Press, Berkeley, CA.

Loxton, J. (1985) The Peters phenomenon. *The Cartographic Journal*, **22**(2), 106–108.

Lupton, E. (1986) Reading isotype. *Design Issues*, **3**(2), 47–58.

Marin, L. (1988) Portrait of the King, in *Theory and History of Literature 57* (trans. Marth M. Houle), University of Minnesota Press, Minneapolis, pp. 169–179.

Markham, B. (1983) *West With The Night*, North Point Press, New York.

McCloskey, D.N. (1985) *The Rhetoric of Economics*, University of Wisconsin Press, Madison, WI.

McKenzie, D.F. (1986) *Bibliography and the Sociology of Texts*, The British Library, London.

McLuhan, M. (1962) *The Gutenberg Galaxy: The Making of Typographic Man*, University of Toronto Press, Toronto.

McNally, A. (1987) You can't get there from here, with today's approach to geography. *The Professional Geographer*, **39**, 389–392.

Meynen, E. (1973) *Multilingual Dictionary of Technical Terms in Cartography*, International Cartographic Association, Franz Steiner Verlag, Wiesbaden.

Mitchell, W.J.T. (1986) *Iconology: Image, Text, Ideology*, University of Chicago Press, Chicago.

Monmonier, M. and Schnell, G.A. (1988) *Map Appreciation*, Prentice Hall, Englewood Cliffs, NJ.

Morris, J. (1982) *The Magic of Maps: The Art of Cartography*. Unpublished MA Dissertation, University of Hawaii.

Muehrcke, P.C. (1986) *Map Use: Reading, Analysis, and Interpretation*, 2nd edn. J.P. Publications, Madison, WI.

Nelson, J.S., Megill, A. and McCloskey, D.N. (1987) *The Rhetoric of the Human Sciences: Language and Argument in Scholarship and Public Affairs*, University of Wisconsin Press, Madison, WI.

Norris, C. (1982) *Deconstruction: Theory and Practice*, Methuen, London.

Norris, C. (1987) *Derrida*, Harvard University Press, Cambridge, MA.

Peters, A. (1983) *The New Cartography*, Friendship Press, New York.

Philp, M. (1985) Michel Foucault, in *The Return of Grand Theory in the Human Sciences* (ed. Q. Skinner), Cambridge University Press, Cambridge, pp. 41–64.

Porter, P. and Voxland, P. (1986) Distortion in maps: The Peters' projection and other devilments. *Focus*, **36**, 22–30.

Rabinow, P. (1984) *The Foucault Reader*, Pantheon Books, New York.

Robinson, A.H. (1985) Arno Peters and his new cartography. *American Cartographer*, **12**, 103–111.

Robinson, A.H. and Petchenik, B.B. (1984) *The Nature of Maps*, University of Chicago Press, Chicago.

Robinson, A.H., Sale, R.D., Morrison, J.L. and Muehrcke, P.C. (1984) *Elements of Cartography*, 5th edn, John Wiley & Sons, Inc., New York.

Rorty, R. (1979) *Philosophy and the Mirror of Nature*, Princeton University Press, Princeton, NJ.

Roszak, T. (1972) *Where the Wasteland Ends: Politics and Transcendence in Postindustrial Society*, Doubleday, New York.

Rouse, J. (1987) *Knowledge and Power: Toward a Political Philosophy of Science*, Cornell University Press, Ithaca, NY.

Said, E.W. (1978) The problem of textuality: Two exemplary positions. *Critical Inquiry*, **4** (4), 673–714.

Saarinen, T.F. (1987) *Centering of Mental Maps of the World*, Department of Geography and Regional Development, Tucson, AZ.

Soja, E.W. (1989) *Postmodern Geographies*, Verso, London.

Snyder, J.P. (1988) Social consciousness and world maps. *The Christian Century* 24 February, 190–192.

Szegö, J. (1987) *Human Cartography: Mapping the World of Man* (trans. T. Miller), Swedish Council for Building Research, Stockholm.

Todorov, T. (1984) *Mikhail Bakhtin: The Dialogical Principle* (trans. W. Godzich), University of Minnesota Press, Minneapolis, MN.

Tuan, Y.-F. (1974) *Topophilia: A Study of Environmental Perception, Attitudes, and Values*, Prentice-Hall, Englewood Cliffs, NJ.

Wallis, H.M. and Robinson, A.H. (1987) *Cartographical Innovations: An International Handbook of Mapping Terms to 1900*, Map Collector Publications and International Cartographic Association, Tring, Hertfordshire, UK.

Whitaker, A.P. (1954) *The Western Hemisphere Idea: Its Rise and Decline*, Cornell University Press, Ithaca, NY.

Whittemore Boggs, S. (1945) This hemisphere. *Department of State Bulletin*, **12** (306), 845–850.

Winner, L. (1980) Do artifacts have politics? *Daedalus*, **109**(1), 121–136.

Wolter, J.A. (1975a) *The Emerging Discipline of Cartography*. Unpublished PhD Dissertation, University of Minnesota.

Wolter, J.A. (1975b) Cartography – an emerging discipline. *The Canadian Cartographer*, **12** (2), 210–216.

Wood, D. (1987) Pleasure in the idea: The atlas as narrative form, in *Atlases for Schools: Design Principles and Curriculum Perspectives* (eds G.J.A. Carswell, N.M. de Leeuw and R.J.B. Waters), Cartographica, Monograph 36, pp. 24–45.

Wood, D. and Fels, J. (1986) Designs on signs/ myth and meaning in maps. *Cartographica*, **23** (3), 54–103.

Woolgar, S. (1988) *Science: The Very Idea*, Ellis Horwood, Chichester, Sussex.

17

Reflection Essay:
Deconstructing the Map

Jeremy W. Crampton
Georgia State University, USA

17.1 Introduction

When the British historical geographer Brian Harley died in December 1991, he was known for two seemingly divergent, if not opposite, areas of research. His most longstanding work includes some very empirical accounts of Ordnance Survey (OS) maps and cartographic production techniques (Harley, 1975). He also wrote the cartographic production notes for the facsimile reissue of nineteenth century OS one-inch series of maps and had begun a major project on maps of the Columbian encounter, which included a travelling map exhibition (Harley, 1990). This historical work reached its florescence in the still ongoing *History of Cartography* volumes (Harley and Woodward, 1987). In the last few years of his life, Harley also produced a number of influential essays that gave him a new audience. Perhaps the most well known of these articles (especially amongst his wider readership) is his 1989 article for *Cartographica* 'Deconstructing the Map'. These articles have given Harley a reputation as a theorist, a postmodernist, and he has been called 'the father of critical cartography' by the *New Yorker* (Lemann, 2001).

Yet this is not necessarily how he would have seen himself. The thing to bear in mind about him is that he was not a theoretician. Nor was he an ideologue, forcing people to agree with his point of view; the key to Harley is that he was a great one for trying things on [evidence]. He tried a variety of approaches including iconography, semiotics, art history, a taste of Foucault and a bit of Barthes – all of it could be picked up, used, half thought through and then dropped. There is no key theoretical position that I can offer you that would sum him up.

Classics in Cartography: Reflections on Influential Articles from Cartographica Edited by Martin Dodge
© 2011 John Wiley & Sons, Ltd

When Harley's friend and colleague Matthew Edney wrote his obituary it had the word 'questioning' in the title three times (Edney, 1992). This is a useful way to think of him. What Harley did was take prevailing assumptions apart and allow them to be opened up for question. In many ways this is similar to what Foucault calls 'problematizing' (if Harley had known Foucault's work a bit better he might have found this idea useful – though not the ugly word itself!). Neither Harley nor Foucault were the first to do this, of course, but what matters is that Harley's questions about cartography were innovatively different. He was not so much interested in a history of the maps *per se*, but rather a question of the production and effects of those mappings. To use another clunky word, he *historicized* his subject matter, putting maps into socio-political context as a production of knowledge.

Today, this concept is hardly radical, even if it is still contested by some. But it's not hard to recall that far-off day in 1989 when I read his article for the first time as a graduate student at Penn State. It turned my assumptions about cartography upside down. On top of that, the Associate Editor of *Cartographica*, Ed Dahl, did something clever. Anticipating that the article would generate comment he solicited the reactions of a range of people and published eleven of them in the next issue.[1] Most of these were laudatory, and recognized the impact that the article would have (along with his other work) on 'doing' the history of cartography. Some of them presciently warned that the article would be controversial. The art historian Robert Baldwin observed that 'Harley's work will be dismissed as a nihilistic or 'extreme leftist' attack on truth, reason, science and the presumption of a knowable, stable, mappable world' (Baldwin, 1989: 89). Others were more critical. Edney observed that Harley's use of Foucault and Derrida left open the question of how the map 'operates as a "text"' (Edney, 1989: 94). In an associated piece three years later, Barbara Belyea went further, observing that Harley largely drew on secondary sources for his understanding of Derrida and Foucault (a serious accusation for a historian) (Belyea, 1992).

Nevertheless, the special section underlined that not only was this a 'Serious Article', but also that its effects were wide-ranging and not confined to the subdiscipline of cartography (then as perhaps now, a minority interest).

From a personal point of view, the article also derailed my nascent graduate career, turning it radically away from what I thought I wanted to do. In 1989 I was in my second year as a doctoral candidate, having completed my masters degree in 1987 on the cognitive processes of wayfinding. Under the supervision of Roger Downs, I was trying to develop a do-able project that would complement my previous study of expertise and spatial problem solving by developing a model of how and why people get lost. As far as I remember, this would involve suggesting a model of how people interpret both the landscape around them and how they fit the map to that landscape, and vice versa. If my thesis had been on way*finding*, then I would do the opposite for my dissertation, losing your way! As I had with the Masters work, I turned here to studies of orienteers, who practice the rather unusual sport of competitive wayfinding. With Roger's help I had obtained a few classic books on orienteering that gave tips on how to read a map

---

[1] These can be accessed at *Cartographica* online: http://utpjournals.metapress.com/content/17x 5193757005125/

(*Finding Your Way With Map and Compass* was one I remember), and we decided that they would make excellent 'subjects' for my study. Others we considered and rejected included taxi cab drivers (the town of State College, where Penn State University (PSU) is located, has rather few of those, and they don't need to spend five years learning 'the knowledge' that London black cab drivers do), emergency vehicle drivers and aeroplane pilots. So I'd spent most of the spring and summer travelling to America's top-ranked orienteers, setting up a video camera and recording them doing wayfinding with the special orienteering maps in their kitchens. (No running around the woods for me!)

My interest in maps was therefore very practical, and in fact I think I knew very little about them apart from my youth hostelling trips in Britain with Ordnance Survey maps. I didn't have any special cartographic interests apart from the usual ones you might expect as a geographer. My interests were in human geography and subjectivity, that is, why people 'see' the world differently. During my teens and early twenties I had read quite a lot of science fiction, particularly people such as Philip K. Dick, Chris Priest and, of course, J.G. Ballard. The common feature of this writing was what you might call alternative realities, but ones which were not apart from normal perception, but a part of it, so to speak. You might be familiar with Dick's classic moment in his novel *Time Out Of Joint* when the protagonist finds the world he thought he was living in unravelling around him. As he approaches a soft-drink stand, it gradually dissolves, leaving behind only a small piece of paper labelled 'soft drink stand'. A similar idea was later used in the movie *The Truman Show* with Jim Carrey.

This perhaps quirky interest had led me to humanistic geography as an undergraduate at Liverpool University, although taking geography was a kind of default option rather than a deliberative choice. (This was true for many of my friends as well, although it was not necessarily something you admitted to your lecturers.) It was a subject I thought was broad enough to probably have some interesting stuff in it, while at the same time dealing with the 'real world', unlike history or chemistry. Even so, I still had a hard time since I didn't find it that inspiring. It was only in my third year that I was lucky enough to win that lottery which is the British university tutorial system and be assigned to Gerry Kearns.[2] He assigned us interesting and impossible books, such as Derek Gregory's *Ideology, Science and Human Geography* (Gregory, 1978, the first academic book I bought, for £3.50 in 1982), which he expected us to write an essay about by the following week; I cobbled together some bluffer's guide type stuff and hoped he wouldn't notice. I was also given the eye-opening book *Humanistic Geography* by David Ley and Marwyn Samuels (1978) and the incredible essay in it by Denis Wood called the 'Cartography of Reality'. Wood points out that even something as seemingly objective as scale is subject to human interpretation. He gives the example of looking for a carpet, which in the store seems to be just about the right size, but that when you get it home and put it in the room is actually too big. Until you drop your contact lens on it, and on your hands and knees as you search it seems to stretch away to the horizon. Even more important were Roger Downs' books on mental mapping, such as *Maps in Minds* (1977). Although Wood's essay was interesting, and fitted well with my sci-fi

---

[2] I saw Gerry again recently at the 2009 RGS-IBG Conference in Manchester and he gave a very inspiring talk on Mackinder and possibilities for peace.

proclivities, Downs' book provided a possible way to study my interests. It joined maps, cognition and human perception together, and with Gerry's help I applied for graduate school in the United States to study with Downs at Penn State or Reg Golledge at Santa Barbara. PSU was my first choice and luckily they accepted me on the strength of my British education and Gerry's letter (I still have their telegram letting me know, the only telegram I've received or am likely to receive). So I joined the department as a grad student in 1984, fully intending to do cognitive mapping, but without really knowing much about it or maps in general.

Shortly after I arrived, Penn State hired Alan MacEachren to its faculty as an associate professor, teaching cartography, and this certainly helped. As a product of the University of Kansas, however, we tended to think of Alan as a fairly mainstream cartographer.[3] After taking four years(!) to do my masters project with my orienteers, I wasn't really going anywhere with my proposed PhD research and, in fact, it was really boring me, but I was afraid to admit it. (Several people have since asked me rather sadly why I gave up my 'important research' on spatial cognition and wayfinding, but you can't do the same thing forever.)

So the timing was right for something to come along and set me on my ear. I don't remember actually reading the article, but sometime in the summer or autumn of 1989 I must have done so (as a poor student I didn't subscribe to the journal but maybe somebody leant me a copy or I saw it in the library). What I do remember, and for what the article has always had connotations for me, is going to a conference in October 1989 in Ann Arbor to deliver a presentation on maps that tried to update Arthur Robinson's arguments on map design, form and function (Robinson, 1952).[4] (Again, I'm sure this topic was due to Alan, as I'd never heard of 'Robbie' as he was known.) It was there that I met John Krygier (then at Wisconsin–Madison) and Matthew Edney (from SUNY-- Binghamton but previously also at Wisconsin–Madison) who has remained a close colleague.

Since Harley had joined the faculty at nearby Wisconsin–Milwaukee in 1986, they were already aware of his work (especially via David Woodward at Madison). Later, when John arrived at Penn State we were able to continue to talk about the paper. Somewhere I have a memory of stopping at a roadside café or fast food joint and chomping away at a plate of eggs where we bandied terms we barely understood, like postmodernism and power relations. (We must have driven from Pennsylvania to Michigan.)

In place of my planned PhD dissertation then, I changed my proposal to focus on a number of concepts that I understood Harley to be raising. I dropped the empirical component altogether. I divided the dissertation into chapters addressing my major topics; non-representational theory, politics in mapping, the ethics of GIS and so on.

---

[3] Alan agreed with us (sort of). In his monumental tome with the modest title *How Maps Work* he admits that he was one of 'those' who uncritically adopted the map communication model (a bête noir of Harley of course), which I like to think was the result of conversations we grad students had had with him since 1989 and the time the book was written (MacEachren, 1995). He also told us he wanted to call his book *How Maps Really Work,* in response to Denis Wood (1992). Now, of course, Alan has gone on to much grander things.
[4] This was the NACIS IX meeting. You can find the program in *Cartographic Perspectives* #3, www.nacis.org/ documents_upload/cp03fall1989.pdf

I called the whole thing 'Alternative Cartographies'; that is, alternatives (the plural was deliberate) to mainstream, progressivist scientific cartography. Stimulated by a footnote in one of Harley's papers, I also included a chapter on the Peters projection or world map. Harley had mentioned this map as part of his excoriation of cartographers for being blind to the socio-political components of mapping, and this aroused my curiosity. Rather than participating in the controversy itself however, I wrote a chapter about the actual controversy, about Arno Peters and about the Scottish clergyman and 'Preadamite' James Gall (Crampton, 1994). I found this much more interesting than the projection *per se*, because it revealed just how much a series of knowledges are related to their time and place. Harley's article is intimately tied to my early intellectual growth.

17.2 The Article's Influence

But what of its wider influence? It has been cited more than 500 times according to Google Scholar. It was chosen for an influential book on representation (Barnes and Duncan, 1992) and is one of the most cited articles in *Cartographica*.[5] Citations to it in other journal articles have remained remarkably constant according to citation indices such as Elsevier's Scopus (which does not collect references from books); about two-thirds of these coming from geography, but also engineering, computer science and the arts and humanities. It has reached a wider audience, I suspect, in its various reprints, first in Barnes and Duncan, then in Harley's posthumous collection (Harley, 2001), a collection on postmodernism (Taylor and Winquist, 1998) and readers (Agnew, Livingstone and Rogers, 1996; Dear and Flusty, 2002). On the other hand, you would have to say that a lot of the citations to it do not really engage with it, but mention it in passing. In other words it's a bit of celebrity piece, famous for being famous rather than for anything it affirmatively says.

Apart from my personal receptivity to its arguments at the time, if you reread the essay today (and it's still good value for money) Harley makes two significant claims that have been very influential. Firstly, he challenges the position of the map as an 'unquestionably 'scientific' or 'objective' form of knowledge creation' (Harley, 1989: 1). As the title of the piece promises, he 'deconstructs' this positionality, which he identifies as being delivered to us through an all too uncritical historical legacy. Secondly, Harley resituates maps within forms of power/knowledge, distinguishing as he does so between what he calls internal and external forms of power, that is, power arising from within mapping and cartography and power arising from outside. How successful are these arguments?

Your answer to this depends on whether you can separate Harley's overall thrust of his argument, which he makes not just here, but in several related pieces,[6] from the rather weak and contestable examples he provides. His identification of GIS with 'scientistic rhetoric' prefigures the GIS Wars of the 1990s, but is unable to foresee the

---

[5] The Barnes and Duncan version has some very minor differences, most notably Harley provides some background why he wrote the piece. The original version is reprinted in this volume.
[6] Available in his posthumous collection (Harley, 2001).

useful outcome of that debate, namely that most GIS is 'descriptive' (Pavlovskaya, 2006) and has tremendous potential as a qualitative method (Cope and Elwood, 2009). In other words, Harley rather overplays his antipathy to positivist science here. His observation that 'many cartographers are puzzled by the suggestion that political and sociological theory could throw light on their practices' (p. 4) is also not quite on the mark. Cartographers might be suspicious, or opposed, perhaps; but two decades of critical work in geography and allied disciplines have informed even the most ardent GIS practitioner that maps are not apolitical. (Whether this is to be regretted, as, for example, Peters projection denialists still seem to insist, is perhaps another matter.) As I suggested in my book on critical cartography (Crampton, 2010) what we see today are diverging interests within the field, with some vectors of research being identifiably descended from a more or less Harleian legacy, while others seek to formalize mapping and GIScience knowledge through the mechanism of certificates and 'bodies of knowledge'. However, it's not today the same landscape that Harley was writing about. Work on the so-called 'ontologies' in GIS, for example, acknowledges social and political effects on knowledge, while at the same time being resistant to philosophical critiques of objectivism (Crampton, 2009).

Harley's major objection to understanding mapping of science comes from what he sees as the masking effect that maps have. This relates to his second issue, that of power relations. In other words 'much of the power of the map. . .is that it operates behind a mask of a seemingly neutral science' (p. 7). His objection to the map's mask sprang not so much from theory, as from his careful historical work (he was trained as a historical geographer, not a cartographer). Instead of theory informing his empirical historical work then, it is clear, as Edney (2005) has argued, that his historical work informed his theory: the historical work remained empirically anchored, while the theoretical positions flexed around them. In this he resembled not only Foucault (who never developed 'a theory' either) but also and perhaps more provocatively J.K. Wright. As the former American Geographical Society (AGS) librarian and Director, Wright had written as early as 1942 'the trim, precise and clean-cut appearance that a well drawn map presents lends it an air of scientific authenticity that may or may not be deserved' (Wright, 1942: 527). The two men had much in common, not only deep historical interests (Wright's thesis was on the Crusades) but also a love of the 'human nature' aspects of maps, as Wright put it (Wright, 1966). In coining the term 'geosophy' Wright fore grounded the very modern idea that knowledges are what drive motivations and actions, and in splicing knowledges to power effects, Harley was merely updating this idea with a modern Foucaultian flavour. Indeed, wrote Wright, 'even those parts of it that deal with scientific geography must reckon with human desires, motives and prejudices' (1966: 83). Harley would surely agree.[7]

Furthermore, if there's a mask which is exercising power effects, then there must be something covered up underneath. This was Harley's (mis)understanding of

---

[7] The two men also shared an institutional affiliation with the AGS, Wright as Director, and Harley as director of the AGS Map Library, which was moved to Milwaukee along with his hire in 1986. Harley's PhD was on population and land use in Warwickshire between 1086–1300, awarded in 1960 by the University of Birmingham.

deconstruction. Not for him Foucault's satisfaction of identifying the politics of truth (Foucault, 1997) nor Derrida's endless chain of signifiers, but of stripping away the ideology to the truth of the matter. Here I would argue he is both radically modernist and severely cramped, for his focus in the 1989 paper is on cartography (and/or mapping/map making, he never clearly distinguishes them) rather than the effects of power relations. That is, with the 'internal' power effects, rather than the external. Harley does discuss, as you might expect him to, the productive nature of maps–productive of rationalities and actions, of truth, as well as its power to surveil and control; 'cartographers manufacture power' he concludes (p. 13). Drawing heavily on his understanding of Foucault (which was admittedly quite partial), he notes that mapping normalizes, standardizes and imparts a sense of place. He famously points out the silences of mapping, both in this essay and elsewhere (Harley, 1988), and the way that maps can justify territorial appropriation and renaming (Harley, 1992). But if Foucault's target is to use history to examine power/knowledge, Harley's is the contrary, and his focus remains resolutely on cartographic history, on the maps themselves once the power relations have been stripped away. Strip away history, says Foucault, and you'll have power/knowledge. No, says Harley, strip away power/knowledge, and you'll have truth.[8]

17.3 Legacies

I met Harley two times, both in 1991. As a graduate student at Penn State, we often benefited from a unique lecture series enabled by the late Peter Gould. Gould generously used an honorarium that came with his position to fund a number of speakers in a Distinguished Lecture series. In early 1991 we graduate students approached Gould with the suggestion that Brian Harley, then, as we thought, an emerging young and radical scholar (for so he seemed from his writing), come and visit. Gould agreed and in March Harley was able to visit and deliver four lectures that he planned to make the basis of a forthcoming book he was working on with John Pickles called *The Map as Ideology* (the title of his second lecture). The impact of these lectures, and those of other visitors to Penn State, such as Derek Gregory and Ed Soja, was quite profound for me as a graduate student. It's the same impact one can get from a good conference; that is, to meet people you are reading and see that they are ordinary people (in the best sense). As a student, one does not yet have a clear view of the discipline and meeting people brings them intellectually closer. Additionally, during the visit and the talk one can engage with the speaker in an active manner (asking questions, disagreeing), which again is important for clarifying and developing your own views. These encounters are highly contingent on time and place; if I had been at another department, or at Penn State at some other time, no doubt my interests would have developed differently. But they are also inspiring.

---

[8] Harley died at the age of 59, and we simply don't know what another decade or two of scholarship would have produced. It's also important to remember that even in this paper Harley is experimenting with new ideas, rather than settled positions.

The last and second time I saw him was coincidentally at the end of the AAG meetings that year. We were both in line to check in at the airport – he to Wisconsin, myself to Pennsylvania. We wished each other a safe flight as you do and got on our flights. In September of that year I had to return to the United Kingdom (basically my funding had run out and Penn State had rightly got tired of funding me). It was Christmas and I was sitting on my father's settee upstairs in his living room reading the newspaper. Suddenly a familiar (though younger) face appeared, that of Brian Harley. For a moment I didn't understand why I should see his face in the newspaper. I attended the memorial service, held at the Royal Geographical Society in March 1992, where I saw how many other scholars' lives had also intersected with that of Harley. Although this was a time of personal loss, it was clear that Harley had left an intellectual legacy, one in which we, his legatees, work on, sorting out our agreements and disagreements.

Harley was no philosopher, even less was he a 'postmodernist' as is sometimes claimed (Ormeling, 1992) (and as he sometimes claimed about himself) but really the most modernist of men. If, during the seventeenth and eighteenth centuries, the map was a mirror, and if for postmodernists (if they exist) truth is an illusion, then for the modernist of the twentieth century such as Harley truth is buried beneath illusion. Harley believed that truth in maps was essentially present, buried as it was beneath lies, propaganda and interests that had accreted around it. That was Harley's 'deconstruction' then; not that of Derrida, but the recovery and questioning of true meaning.

Further Reading

Edney, M.H. (2005) The origins and development of J.B. Harley's cartographic theories. *Cartographica*, **40**(1/2), 1–143. (This is the single most critically informed and sustained appreciation and analysis of Harley's entire work. Edney is now the Editor of the *History of Cartography* series, founded by Harley and David Woodward.)

Cosgrove, D. (2007) Epistemology, geography, and cartography: Matthew Edney on Brian Harley's cartographic theories. *Annals of the Association of American Geographers*, **97** (1), 202–209. (Denis Cosgrove's reply and elucidation of both Edney and Harley. Cosgrove was a historical and cultural geographer whose own work delved into some of the same issues as Harley, although by no means from the same perspective.)

Harley, J.B. (1987) The map as biography: Thoughts on Ordnance Survey map, six-inch sheet Devonshire CIX, SE, Newton Abbot. *The Map Collector*, (41), 18–20. (This rather hard to find piece succinctly summarizes Harley's view on how maps and personal lives are intertwined. A mere three pages, but almost poetically written.)

Harley, J.B. (1987) The map and the development of the history of cartography, in *Cartography in Prehistoric, Ancient, and Medieval Europe and the Mediterranean, Volume One of the History of Cartography* (eds. J.B. Harley and D. Woodward), University of Chicago Press, Chicago, pp. 1–42. (Harley's magisterial outline for a new history of cartography. Both manifesto and *tour-de-force*.)

The *Imago Mundi*, published by Routledge, www.tandf.co.uk/journals/titles/03085694.asp. (If you are interested in the history of cartography, this is the journal to read.)

References

Agnew, J.A., Livingstone, D.N. and Rogers, A. (1996) *Human Geography: An Essential Anthology*, Blackwell Publishers, Oxford.

Baldwin, R. (1989) Reader's response. *Cartographica*, **26**(3/4), 89–90.

Barnes, T.J. and Duncan, J.S. (1992) *Writing Worlds: Discourse, Text, and Metaphor in the Representation of Landscape*, Routledge, London.

Belyea, B. (1992) Images of power: Derrida/Foucault/Harley. *Cartographica*, **29**(2), 1–9.

Cope, M. and Elwood, S. (2009) *Qualitative GIS. A Mixed Methods Approach*, Sage, London.

Crampton, J.W. (1994) Cartography's defining moment: the Peters projection controversy, 1974–1990. *Cartographica*, **31**(4), 16–32.

Crampton, J.W. (2009) Being ontological. *Environment and Planning D: Society and Space*, **27** (4), 603–608.

Crampton, J.W. (2010) *Mapping: A Critical Introduction to Cartography and GIS*, Wiley-Blackwell, Oxford.

Dear, M.J. and Flusty, S. (2002) *The Spaces of Postmodernity: Readings in Human Geography*, Blackwell Publishers, Oxford.

Edney, M.H. (1989) Reader's Response. *Cartographica*, **26**(3/4), 93–96.

Edney, M.H. (1992) Harley, J.B. (1932–1991) – questioning maps, questioning cartography, questioning cartographers. *Cartography and Geographic Information Systems*, **19** (3), 175–178.

Foucault, M. (1997) *The Politics of Truth*, Semiotext(e), New York.

Gregory, D. (1978) *Ideology, Science and Human Geography*, Hutchinson, London.

Harley, J.B. (1975) *Ordnance Survey Maps: A Descriptive Manual*, Ordnance Survey, Southampton.

Harley, J.B. (1988) Silences and secrecy: the hidden agenda of cartography in early modern Europe. *Imago Mundi*, **40**, 57–76.

Harley, J.B. (1989) Deconstructing the map. *Cartographica*, **26**(2), 1–20 (Reproduced as Chapter 16 of this volume.)

Harley, J.B. (1990) *Maps and the Columbian Encounter: An Interpretive Guide to the Travelling Exhibition*, University of Wisconsin–Milwaukee, Golda Meir Library.

Harley, J.B. (1992) Rereading the maps of the Columbian encounter. *Annals of the Association of American Geographers*, **82**(3), 522–542.

Harley, J.B. (2001) *The New Nature of Maps: Essays in the History of Cartography*, Johns Hopkins University Press, Baltimore, MD.

Harley, J.B. and Woodward, D. (1987) *Cartography in Prehistoric, Ancient, and Medieval Europe and the Mediterranean*, University of Chicago Press, Chicago.

Lemann, N. (2001) Atlas shrugs: The new geography argues that maps have shaped the world, *The New Yorker*, 9 April 2001, pp. 131–134.

Ley, D. and Samuels, M.S. (1978) *Humanistic Geography: Prospects and Problems*, Maaroufa Press, Chicago.

MacEachren, A.M. (1995) *How Maps Work*, Guilford Press, New York.

Ormeling, F. (1992) The influence of Brian Harley on modern cartography. *Caert-Tresoor*, **11**(1), 2–6.

Pavlovskaya, M. (2006) Theorizing with GIS: A tool for critical geographies? *Environment and Planning A*, **38**(11), 2003–2020.

Robinson, A.H. (1952) *The Look of Maps: An Examination of Cartographic Design*, University of Wisconsin Press, Madison, WI.

Taylor, V.E. and Winquist, C.E. (1998) *Postmodernism: Critical Concepts*, Routledge, London.

Wood, D. (1992) *The Power of Maps*, Guilford Press, New York.

Wright, J.K. (1942) Map makers are human: Comments on the subjective in maps. *Geographical Review*, **32**, 527–544.

Wright, J.K. (1966) *Human Nature in Geography*, Harvard University Press, Cambridge, MA.

18

Cartography Without 'Progress': Reinterpreting the Nature and Historical Development of Map Making

Matthew H. Edney[1]

Abstract

This paper extends the current critique of cartography's empiricist presuppositions to the nature of cartography as a practice. After exploring the relevant aspects of empiricist cartography – the manner in which geographic data are treated as constituting a single, monolithic database and the reliance upon a linear and progressive view of cartographic history – a new interpretation of cartography's nature and of its history are presented. Cartography should be seen as a complex amalgam of cartographic *modes* rather than as a monolithic enterprise. Each mode comprises a set of cultural, social and technological relationships which determine cartographic practices. This concept is applied to modern European cartography in the period between 1500 and 1850, when map making appeared to progress from being an art to being a science (the 'cartographic reformation'). Approaching this period without prior assumptions of progress reveals that cartography's reformation is a myth created by our misunderstanding of the unified mode of Enlightenment cartography, 'mathematical cosmography'. Considering the history of cartography to be the history of the internal changes and external interactions of several modes would appear to be consistent with the complexity of both the historical record and the character of map making as an intellectual, technological, social and cultural process.

[1] Originally published: 1993, *Cartographica*, **30**(2/3), 54–68.
At the time of publication: Edney was Assistant Professor in the Department of Geography, State University of New York at Binghamton, Binghamton.

The recent literature of cartographic theory constitutes an extended critique of the supposition that modern cartography is an empiricist practice. Modern western culture has established a direct association between real-world phenomena and their cartographic representations, and has then privileged those representations with a correctness derived from the act of observation rather than from the social and cultural conditions within which the representations are grounded (Gregory, 1986). Harley (1988a, 1988b, 1989a, 1989b, 1990, 1991), Wood (1992a, 1992b; Wood and Fels, 1986) and others (Belyea, 1992; Pickles, 1992; Rundstrom, 1991; Woodward, 1992) have argued from a variety of philosophical positions that map making is inherently ideological and its 'facts' are not as value free as our culture has supposed. All maps – whatever their claims to the contrary – serve a larger purpose; map making is not a neutral activity divorced from the power relations of any human society, past or present; there is no single nor necessarily best way in which to represent either the social or physical worlds.

One issue that has yet to be addressed in detail by this critique is the constitution of cartography as a practice. We generally conceive of cartography as a singular and monolithic enterprise and we derive that conception from the history of cartography in a recursive manner. That is, the modern discipline of cartography justifies and legitimates its empiricist claims to objectivity and neutrality by pointing to its past progress (which it also extrapolates into the future); conversely, historians of cartography have defined their subject in terms of their a priori assumptions of map making's objectivity, neutrality and progressiveness (Harley, 1989c). This paper extends the ideological construction of cartography to redefine the nature of cartography as a practice without the empiricist emphasis upon observation and the sense of progress that it entails. In doing so, it also extends Rundstrom's (1991) idea of 'process cartography'. It first lays out the basic presuppositions of empiricist cartography and of its history. It then argues that cartography is composed of a number of *modes,* or sets of cultural, social and technological relations which define cartographic practices and which determine the character of cartographic information. The modes themselves are historically contingent, as demonstrated in the final section, which examines the broad outline of western European cartography between the sixteenth and nineteenth centuries, an outline constructed without the aid of the historiographic crutches which are required by an empiricist and progressive cartography.

I should stress at the outset that I conceive of 'cartography' in the most general sense possible: as the practice of constructing map artefacts, no matter how ephemeral (Wood, 1993; see Harley and Woodward, 1987: xvi, for a broad definition of 'map'). Having stated that, the following discussion is directed only towards the mainstream of modern cartography. More precisely, this paper considers only the 'formal' cartography, which is prosecuted within the commercial and governmental confines of the modern capitalist state. I recognize the significance of non-western cultural traditions (Harley and Woodward, 1992; Rundstrom, 1991) and also of 'alternative' or 'popular' cartographies (Crampton, 1993; Marsh, 1983, respectively), but my focus in this paper requires that I exclude them from discussion.

18.1 Cartography's Information Emphasis

The philosophical concept of most significance for empiricist cartography is the ontological assumption that the world possesses a quite unambiguous existence and can therefore be objectively known. There is a world of geographic facts 'out there' – separate and distant from the observer – which are to be 'discovered' by the explorer and surveyor, just as scientists 'discover' new facts about the way in which the world functions. Indeed, exploration and map making were important factors in the development of the modern western scientific tradition, in which respect maps serve as a paradigmatic document of empiricist science (Grafton, 1992; Livingstone, 1992: 32–101). Spatial location – the geographic fact – is deemed an independent attribute of discrete entities in the real world. The geometric expression of location might be inaccurate or imprecise, but it is never ambiguous; each place exists in only one location. The map, as the repository of unambiguous geographic facts, becomes itself an unambiguous, objective and factual document, subject to variation only with respect to the geometrical accuracy and comprehensiveness of those facts. Each map is thus defined by its content and is evaluated by comparing that content to an idealised, distortion-free replication of the world. 'We remain caught', Pickles (1992: 199) has observed, 'within the metaphysics of presence, which presupposes some foundational object against which the distortions and interpretations can be measured: that some interpretation-free image could be produced that does not distort the world'.

The mediation between the cartographic image and real world is through discrete geographic facts. Conceptually, the 'geographic quanta' constitute a single, monolithic corpus of data which is directly related to the world according to the world's spatial structure (quantified by their latitude and longitude). Each map is simply a selection of these autonomous data in graphic form. (More practically, the corpus can be considered as the informational sum of all the largest-scale maps existent for each region; smaller-scale maps are held as deriving from these.) While it is a logical absurdity for maps to be ideally correct, they are nonetheless held to replicate the world's essential order and structure. This is the crux of Lewis Carroll's *reductio ad absurdum* of a topographic map at 1:1.

"Have you used it much?" I enquired.

"It has never been spread out, yet," said Mein Herr: "the farmers objected: they said it would cover the whole country, and shut out the sunlight! So we now use the country itself, as its own map, and I assure you it does nearly as well". (Carroll, 1894: 169)

In this system, questions of scale influence the selection of the detailed information to be mapped, but they do not alter the sense that mere exists a pristine, scale-less database that is equivalent to the world itself.

It is in the emphasis upon geographic data – data which exist in their own world, like Platonic forms – that empiricist cartography is most fundamentally construed.

When academic cartographers sought in the 1960s and 1970s to define the essence of their discipline, they focused upon the flow of data from the world through the map to the user. It is significant that none of the proposed models of cartographic communication included scale as one of their parameters, a point which Keates (1982: 105) found rather paradoxical. Yet scale was unimportant for the communication modellers because geographic facts are infinitely precise when quantified, so that scale became an issue only in map design (i.e. the subjective and artistic side of cartography). If cartography was to be a science, then it needed to emphasize the objective and scientific aspects. Although the models of cartographic communication seem to have fallen by the wayside, they do still occasionally appear (King, 1990; Tyner, 1992) and 'information' and 'communication' remain at the core of the discipline's self-image. The rise of geographic information systems (GIS) has served only to reinforce the centrality of geographic data and its communication (Muller, 1991: 1–14; Visvalingham, 1989). Some theoretical discussions of GIS have claimed that digital databases are maps, an equivalency based only upon information content (Woodward, 1992). Furthermore, GIS has inherited from its parent subdiscipline, 'analytical cartography' (Goodchild, 1988; Tomlinson, 1988), a concept which originated in geography's quantitative revolution that 'premaps are a subset of maps and maps are a subset of mathematics' (Bunge, 1966: 71; Tobler, 1976). From this there is the assumption that given sufficiently sophisticated mathematics for modelling the data, any database can be properly sampled to give maps at any lesser scale. Indeed, current technological developments have led some cartographers to suggest that it is possible to create a single *actual* incarnation of cartography's monolithic database. A recent statement by federal US map makers identified 'a scale-independent database, with data from any number of sources, including maps of varying scales, capable of making any product' (quoted by the National Research Council, 1990: 36); although such a database might be beyond the timeframe of current US Geological Survey (USGS) research development, it is nonetheless implied as being a long-term goal. Finally, advocates of a more realistic approach to cartography nonetheless continue to stress geographic information but broaden the range of factors which determine the constitution of the database to include social and institutional as well as technological factors (Chrisman, 1991; Keates, 1982; Monmonier, 1982).

The monolithic corpus of geographic data is inherently accretive: it grows and develops in ways that maps themselves do not (Wood, 1992b). Each survey adds new data and it corrects existing data. The database becomes progressively more comprehensive, more precise and more accurate. Given that a map is defined by its content, no distinction is made between maps made of, and maps made of *and made in,* an area (criticized by Stone, 1988; Woodward, 1974). Any two maps of the same area can accordingly be compared with and judged against each other. Consider the example of a cartometric analysis of the South Carolina coastline on five maps from 1757 to 1865. This analysis started with the hypothesis that,

> *the accuracy with which the South Carolina coastline was represented on maps would improve through time, with each new map depicting it more accurately than earlier ones.* This may sound like a self-evident notion, *but other authors have noted that the eventual improvement in map accuracy over time is not a steady transition.*

Rather it is marked by examples of more recent, but less accurate, maps. (Lloyd and Gilmartin, 1987: 20, emphasis added)

The conclusion bore out the caveat: 'accuracy generally improved over time, but one map was found to be less accurate than expected, given its date' (Lloyd and Gilmartin, 1987: abstract). The study focused only upon information content. It disregarded several factors which usually indicate significant variation in the treatment of the cartographic line and in the degree of generalization: their variant scales (1:316 800 to 1:677 952) were glossed over as being 'reasonably similar'; their geographic origins (UK and USA) were set aside; and the different circumstances of their creation were ignored (one was a marine chart, one was derived from marine survey, and three were general land maps). The validity of the study's premise and conclusion is thrown into question when it is realized that it was the smallest-scale map of the series, which was also one of the terrestrial maps, which was found to be aberrant.

Lloyd and Gilmartin did not explicitly criticize the one map for its aberrancy, but that has usually been the next logical step in other (if less mathematically sophisticated) comparisons of this sort. By confusing maps of potentially wildly variant scale and source materials, historians and cartographers have created a linear trend defined by a canon of Great Maps which represent the great advances in geographic knowledge. Each map can be judged and ranked against other maps according to how well its content matches the overall corpus of data. Thus, Crone (1978[1953]: xi) stated that 'the history of cartography is largely that of the increase in the accuracy with which ... elements of distance and direction are determined and ... [in] the comprehensiveness of the map content', and Buczek (1982 [1964]: 7) hypothesized a 'normal' sequence of cartographic coverage for any European country, from small-scale general maps in the sixteenth century to medium-scale topographic mapping of particular provinces during the seventeenth and eighteenth centuries to the advent of large-scale national surveys after 1800. Most historians, however, pay closer attention to the historical record and agree with Skelton (1972: 5) that simplistic assumptions of *continuous* progress are perhaps unjustified and prefer a punctuated progress: 'the development of map making ... is marked by periods of rapidly accelerated advance, followed by periods of standstill or even retrogression'.

Historical vignettes are important features of the introductions to cartographic texts and atlases, where they serve to define the field as a progressive science for the benefit of the novitiates (Harley and Woodward, 1989). For example, Tyner (1992: 4–7) includes a graph of ever increasing 'cartographic activity'" over time; Hammond's new world atlas is wrapped in a rhetoric of progress which culminates in their own 'five-year effort to revolutionize how maps are made' (Hammond, 1993, 6–9 and back cover). Significantly, the key to these historical vignettes is the juxtaposition of maps from widely different eras. Hammond (1993: 6–7) set a 1570 map of The Netherlands next to a Landsat image of the same region; foregoing a lengthy sketch, Campbell (1991 [1984]: 1–3) nonetheless established a sense of historical progress by contrasting a 2500 BC Mesopotamian map against modern maps. As long as cartography defines its intellectual arena around issues of geometrically defined geographic data, cartography's progressive past will always be evident when we place, side by side, two maps of the same region but from such different eras.

Such contrasts rely for their effect upon the differences in *form* between the maps: clay tablet against modern draughting films and scribe-coat; decorative compass roses and ship-filled seas against austere and information-heavy pixels. These comparisons rely for their effect upon the idea that cartography has progressed from being an 'art' to being a 'science'. The occasion of this change is often called the 'cartographic reformation' and is broadly located in the century between 1670 and 1770 (Brannon, 1989; Rees, 1980). The decline in florid decoration and the rise of the factual neutrality of white space are used as surrogates for the decline in cartography's artistry (as the product of individual workmen) and the rise in cartography's science (as the product of large-scale, institutional surveys). Thus, Bagrow's (1985 [1964]: 22) oft-quoted statement – that he ended his general history in the late eighteenth century when 'maps ceased to be works of art, the products of individual minds, and [when] craftsmanship was finally superseded by specialized science and the machine' – has been reprised by Pelletier (1986: 26): 'from the end of the seventeenth century and in the eighteenth century the practice of cartography has moved out from the cabinet to the field'. Even David Harvey, in his theoretically sophisticated and highly acclaimed *The Condition of Postmodernity,* succumbed to the seductive lure of the cartographic reformation when he noted that by about 1700, maps had been 'stripped of all elements of fantasy and religious belief' and had become 'abstract and strictly functional systems for the factual ordering of phenomena in space' (Harvey, 1989: 249).

Once we decide, however, that there is more to geographic data than their geometrical definition, and that the practice of cartography is more than the collection and replication of these data, then the empiricist conception of cartography rapidly disintegrates. Hammond (1993: 7) might claim that its new atlas contains data for 'nearly every important geographic feature on earth', yet the layout reveals the usual biases so that 'important' continues to mean 'important for the Euro-American economies'. Geographical information is not scale-less (or 'scale independent') as has been supposed but is scale-defined; most fundamentally, a clear distinction can be drawn between the nature of small-scale and large-scale data (Carstensen, 1989; Robinson and Petchenik, 1976: 108–123; Woodward, 1992). There is no hard and fast distinction between the 'art' and the 'science' of cartography; nor is it that 'cartography is *both* an art and a science', as the usual vapid compromise would have it, but that each cartographic mode is 'socially constructed' in the same manner as art, science, technology and all their respective subdisciplines; as such, 'the boundary between' science, technology, art, cartography, and so on, 'is a matter for social negotiation and represents no underlying distinction' (Hughes, 1986: 284).

The linear view of cartographic history must also be rejected. Blakemore and Harley (1980: 14–32) present the most succinct critique of the manner in which presumptions of cartography's past progress have forcibly distorted the historical record to fit a simple and chronologically linear sequence. Perceived precocity is lauded whereas hysteresis is denigrated; spectral complexity is shunned in favour of monochromatic hindsight. More particularly, Stone (1988) argued that cartography's reformation was not reflected in the European mapping of Africa: neither map production techniques nor map use appreciably altered between 1650 and 1850; what did change in the eighteenth century was the *look* as opposed to the content of the maps. We might draw an analogy to the nineteenth century 'Whig' interpretation of history, an interpretation 'which celebrated [the English past] as revealing a continuous, on the whole uninterrupted, and generally glorious story of constitutional progress, all leading up to the triumph of liberty and representative institutions' (Clive, 1989: 129). Yet when Victorian historians turned from constitutional to social and economic history, the Whig interpretation collapsed under the weight of contrary evidence. Similar re-evaluations are currently occurring in the histories of science and technology. The 'quasi-linear model' of technological history – in which successful innovations form a single sequence from which stem failed offshoots – is being replaced by much more complex models which stress the interaction of society with technology (Bijker, Hughes and Pinch, 1987). Similarly, historians of science are now looking beyond the absolutes of scientific knowledge to the processes by which that knowledge was created. In doing so, they have realized that the once unambiguous 'scientific revolution', the defining event for modem science and the inspiration for cartography's own reformation, actually presents a mass of contradictions and frequently 'unscientific' behaviour (Lindberg and Westman, 1990; Toulmin, 1990). The scientific revolution, cartography's reformation and Whiggish history are all creations of modern historians and reflect neither continual progress nor any dramatic shifts in past practices.

18.2 Cartographic Modes

The view that cartography is a singular enterprise is true only in its most general sense as the construction of map artefacts. We habitually distinguish between specific types of map making. We identify conceptual differences in the information content and design of 'general' and 'thematic' maps (Castner, 1980; Petchenik, 1979). We use 'atlas' to define a very specific intermingling of written and cartographic texts whose whole is more man just the sum of its maps: the atlas is a symbol of both its maker's professional status and the social worth of its owner (because of the greater financial capital involved in atlas production) and is also a metaphor for the encyclopaedic sum of geographic knowledge (Akerman, 1991). We contrast the topographic map with marine and aeronautical charts, or terrestrial with celestial maps: even our normal, everyday words convey the differences. We also draw a boundary between the topographic map and other 'scientific illustration', such as engineering plans or blueprints, botanical or anatomical illustrations, and multivariate graphs. The list of different types of cartography is long, and is getting longer (Hall, 1992). These distinctions do not simply reflect variations in cartographic form which have been made to fulfil different functions. They also make manifest the requirements of different social organizations for graphic representations to aid their understanding of the human world. That is, they are the artefactual manifestations of different cartographic *modes*.

Each cartographic mode is a set of specific relations which determine a particular cartographic practice. There are three sorts of relations: cultural, social and technological. Each mode is thoroughly enmeshed in and is an integral part of these relations. In this respect, my classification of relations is a didactic device. On the other hand, when the interconnections between modes are considered, these relations form a hierarchy of definition.

Cultural relations are the most subtle, as they bind maps to spatial conceptions. Each human culture perceives space in its own way and those perceptions determine the culture's maps ... and those maps determine the culture's spatial perceptions: 'we see the earth's surface in terms of the cartographic convention we are familiar with: "[the discourse] constitutes its own object"' (Belyea, 1992: 6). Further relations exist between maps and other spatial representation (landscapes, poetry, dance, architecture, cosmography, travel diary, etc.). These relations govern the production of space: do we objectify and commodity space in order to exploit it and its contents, or do we personify and empathize with it? The cultural relations of cartography form the arena of the map's subliminal geometry (Harley, 1988b), of the map's potential as an iconic device, and of the map's very existence as a complex sign system (Wood and Fels, 1986) and as a discourse. They confer both cognitive and symbolic meaning on spatial configurations; they define the most fundamental constitution of 'geographic information'; they govern cartographic conventions.

Within the scope of a given culture, there are numerous social institutions which have a potential need for map use and so for map making, if only in a limited way. The relations between the institutions and cartographic practices are either enabling or constraining, but their precise nature will, of course, vary. The following statements concern the social formations only of modern Western culture. As a simple example, the concepts of property and the territorial state are rooted in a sense of commodified space, a sense which is often expressed in cartographic form. Social relations are the arena of maps as instruments of power wielded by the state, by educated and landed elites, by military officers, by city planners, by revolutionaries and by community activists. Social relations bind the individual cartographer not only into the larger society but also into the immediate hierarchy within which maps are made and used. Most fundamentally, it is the various social requirements for geographic information which define map scale. In sum, social relations constitute spheres of cartographic interest (Wood, 1992a).

Finally, there are the technological relations which govern the creation of the map artefact, or the practice of cartography *per se*. Those relations encompass intellectual techniques of manipulating geographical information as well as tool-based techniques for observing the world and producing the cartographic artefact. While these two elements might seem qualitatively different, they are usually so intertwined as to constitute the single activity of map compilation (Godlewska, 1988). The information embraced by each mode makes its own database, separated from the information of other modes by scale and the techniques of creation.

The mode is thus the combination of cartographic form and cartographic function, of the internal construction of the data and their representation on the one hand and the external *raison d'être* of the map on the other. No sense of priority is intended, lest we indulge in futile chicken-and-egg arguments. Rather, in cartography, as in any other process-orientated discipline, form follows function as much as function follows form.

Nor are cartographic modes independent. Within any given culture, there will probably be some overlap between the modes' cultural relations. Cartographic modes, the practices they determine and the geographic information they encompass do interact with each other. For example, when present-day cartographers set out to make small-scale maps of the United States, they do not start with the original USGS topographic quadrangles at 1:24 000 and generalize down to 1:5 000 000; instead, they use existing maps or databases already established for the small scale. Yet the present-day, small-scale database is the result of one or two centuries of refinement informed by large-scale surveys. The overall history of cartography is recast as the history of the creation, internal change, interaction, merging, bifurcation and over-lapping of cartographic modes. At certain periods in the history of cartography the various modes are easily distinguished; at other times they are so confused and interwoven as to *appear* to be one, as has been the case under the dominance of empiricist cartography during the past 200 years.

It is with respect to the issue of the interaction of concurrent modes that I find myself to be most in sympathy with Rundstrom's recent plea for a processual cartography. He seeks to broaden the study of maps from the current narrow focus on the 'end product' of a closed process to take into account the position that maps are contingent for their power upon continual, open-ended 'cultural, social, political and technical processes'. Following on, Rundstrom states that process cartography 'situates the map artefact within the map making process, *and* it places the entire map making process within the context of intracultural and intercultural dialogues . . .' (Rundstrom, 1991: 6). I would change these statements by subsuming the political under the social (a minor point) and by replacing 'the entire map making process' with 'all map making processes'. The difference is subtle yet significant. Also of interest is Rundstrom's argument that cultures cannot be strictly classified as either 'incorporating' or 'inscribing'. Even predominantly inscribing cultures (those, such as our own, which hold and fix meaningful information through material artefacts) possess incorporating traits (in which meaning is borne by oral communication and ephemeral human actions, e.g. dance and ritual). Much of the cultural relevance of the British mapping of India, for example, resided in the act (ceremony?) of surveying (Edney, 1993). That is, 'process cartography allows us to see that *acts* empower artifacts' (Rundstrom, 1991: 7). Cartographic practices not only lead to an end product which is wrapped in the various cultural, social and technological relations of its mode, but the practice itself (the process) is wrapped up in those relations.

Two other recent works present concepts similar to identified two contrasting intellectual 'archives' within eighteenth century cartography. Guillaume Delisle, Joseph-Nicolas Delisle and Philippe Buache – illustrative of the intellectual geographer – hypothesized about the geographical configuration of the interior of North America and gave 'tentative solutions' of rivers and an inland sea, which they deduced from known data. In contrast, James Cook and George Vancouver – illustrative of the empirical explorer or surveyor – proclaimed the limit of their directly observed knowledge with white space. In terms of the broad history outlined below, the first archive coincides with the established mode of 'mathematical cosmography' which dominated eighteenth century European cartography; the second archive coincides with the initial lurch towards systematic map making and charting which occurred towards the end of that century. The transition from the one to the other was not, however, as 'abrupt and unambiguous' as Belyea (1992: 6) states, for reasons discussed below. In a similar manner, Thongchai (1988) has produced a substantial account of the changing 'spatial discourses' (archives in Foucault's sense) which have underlain the creation of the self-image of the modern Thai nation state. In particular, he is intrigued by the reconciliation of indigenous Thai and Buddhist concepts of space with the European knowledge of space and territory as constructed by the *act* of science.

The historiographic identification of cartographic modes is a problematic task. As noted, the modes themselves are interrelated, sometimes very closely. The autonomy and distinctiveness of modes within a given society will also depend upon its degree of economic articulation and social complexity (Wood, 1993). It is accordingly impossible (and certainly undesirable) to create a formal list of attributes and properties whose permutations define any mode of map making. We cannot treat cartographic modes as being as distinct and as non-overlapping – as rigorous and as unambiguous – as we treat the categories of a choropleth map. Instead, modes must be defined radially and with the recognition that their interfaces are fuzzy and permeable and are not hard and discrete. There are prototypical instances of each mode about which we can place other instances in an abstract, conceptual space; we can then draw bounds about each set of instances, so defining the modes and their interrelations (Lakoff, 1987). As with mapping the world itself, the classification and delimitation of modes are the active responsibilities of the commentator, historical or otherwise, and are not inherent to the objects themselves.

18.3 The Modes of Formal European Cartography, 1500–1850

To illustrate the workings and interactions of cartographic modes, I turn to the broad sweep of cartographic history in Europe, between the sixteenth and the mid-nineteenth centuries, and in particular across the period of cartography's scientific reformation (Figure 18.1). As noted, my particular interest is in the formal cartographic modes. I do not wish to imply that the following sequence of cartographic history was teleologically determined; my intention is simply to summarize how the cartographic modes developed. I should also note that because I am not presenting a detailed exposition of several hundred years of cartographic history, I only provide citations for non-standard works or in support of potentially controversial statements.

18.3.1 Early Modern Europe

With the Renaissance, the introduction of the printing press, the rise of the modem territorial state, the establishment of mercantile classes and the creation of a commercial land market, there evolved three formal cartographic modes: chorography, charting and topography.[2] Europe's social and economic structures became increasingly

---

[2] I do not follow the usual distinction, first drawn by Ptolemy (1991 [1932], Section 1.1), between *geo*-graphy (description of the whole world) and *choro*-graphy (description of regions); the two were essentially the same with respect to cartographic practices.

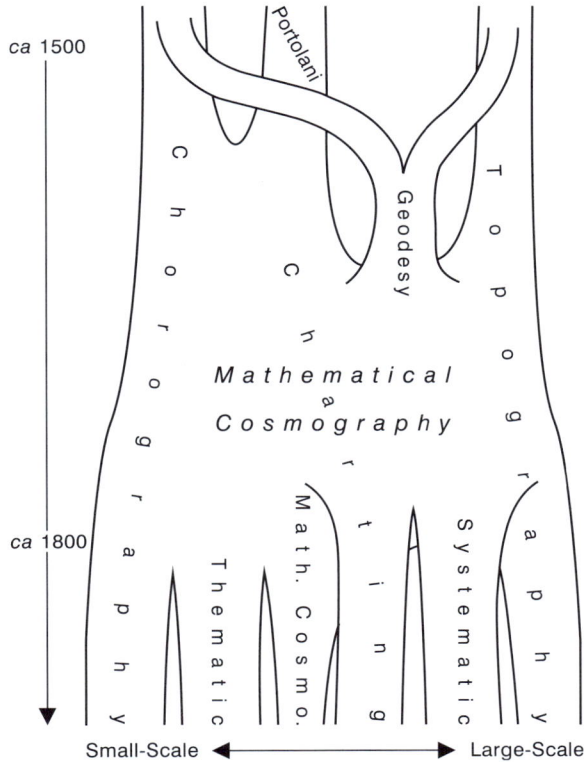

Figure 18.1 Schematic diagram of the eighteenth century convergence of the formal cartographic modes and their subsequent divergence after about 1800. Note that this figure is only schematic and should be taken not as an exact or precise image but as simply indicative of the general historical sweep of the interactions between cartographic modes. Both axes – time on the horizontal axis and cartographic scale on the vertical – are not uniform but only suggestive. Please refer to the text for more details.

complex and articulated. The burgeoning urban populations supported an increasingly specialized work force; the mediaeval artists' and scribes' guilds of the Italian cities spawned commercial print makers and cartographers. Commercial cartographers worked within complex networks of social and state patronage (Buisseret, 1992), a symbiosis that has continued in various forms to the present day. Early modern Europe was, however, not so sophisticated as to allow a great deal of specialization by cartographers within specific modes of map making, so that there was some sharing of personnel between modes and between cartography and non-cartographic activities. In the middle of the sixteenth century, for example, Giacomo Gastaldi worked as an engraver of regional maps and as the supervisor for Venice's hydrographic schemes; a half century later, John Norden was employed in England as an estate surveyor and as a regional geographer. Nonetheless, the three main modes are quite distinct in terms of their scales of enquiry, spatial conceptions, social institutions, technologies and corpora of information.

Chorography is the mode of small-scale mapping both of regions and the world (1:1 000 000 or less). While such map making occurred sporadically in mediaeval Europe, it began to expand in the fifteenth century. The translation of Ptolemy's *Geographia* into Latin (about 1406–1407) stimulated a number of applications of an abstract, projection-and-coordinate conception of cartographic space to regional maps and even to cosmographic *mappaemundi* (Woodward, 1987, 1990). The dramatic growth of Europe's geographic information through the explorations to Africa, Asia and the Americas, and the need to understand that information, led to a burgeoning map trade, which was only furthered by the application of printing to cartography in the 1470s. The information of chorography is concerned with places as points, with roads and rivers as lines, and with regions. Chorography itself is concerned as much with geographic generalization as with cartographic representation; its essence is the evaluation and reconciliation of geographic information of different types (textual, itinerary, other cartographic, pictorial) and of different dates (both ancient and contemporary) to produce a single map. In this respect, the small-scale mapping of the early modern era is the precursor of the modern discipline of geography. A map projection would first be constructed in a form that was aesthetically pleasing or easy to draw (whose mathematical rigour, if any, was a bonus), the key points whose coordinates were well known would be plotted, and then the rest of the map's data would be interpolated between. The fact that raw geographic data were now quantified (being ascribed latitudes and longitudes) did not mean that small-scale mapping itself became an objective practice; the imprecision of astronomical observations was well known, and the number of properly fixed places was very small, even into the middle of the eighteenth century. For example, Tobias Mayer had only 33 control points for his *Germaniae atque in ea locorum principaliorum Mappa Critica* of 1753, and of these none had an acceptable longitude value! (Forbes, 1980a: 65) Small-scale mapping was thus a consciously intellectual process – the 'Science of Princes' – which has been the preserve of geographers who have never visited the regions that they map yet who have nonetheless debated the relative merits and consistencies of different reports about those regions. Chorography is still the cartographic mode which defines how we perceive the overall structure of the world and its continents. Its institutions are those of the state and its educated elites, the scholar, and the commercial cartographer. The cartographer (chorographer?) takes existing data, selects those which are relevant, compiles and generalizes them into one image, and thereby creates a new map.

The early modern mode of charting was the lineal descendent of the mediaeval mode of constructing charts from *portolani,* or intricate lists of navigational instructions. The information base of each chart comprised sailing distances and directions, with very few positions located by their latitude and longitude. The social institutions of charting were, first and foremost, the maritime governments and the bureaucratic entities charged with regulating navigation and with preserving the secrecy of the new routes to Asia and the Americas. In contrast, the social and economic organization of navigators in guilds meant that much information, especially of local harbours and estuaries, was not written down, let alone charted, in order to maintain the power and independence of those guilds. There were two areas of interaction between charting and chorography. Just as some of the informational corpus of the mediaeval *portolani* had been used for depicting the Mediterranean on some later *mappaen undi,* so the generalized results of navigational data continued to be a major source of new information for chorography. In this respect, charting was a subset of maritime geography and oceanography. Conversely, charting slowly adopted Mercator's projection after 1569 and so acquired chorography's projective conception of space, at which point charting began to privilege the determination of latitude and longitude within its repertoire of techniques.

The third mode of early modern cartography – topography – is that of representing limited portions of land at a large scale (larger than 1:100 000). It is dominated by surveying, the process of directed enquiry for the purposes of the management and supervision of the place concerned. ('Survey' and 'surveillance' have the same etymological root.) Like marine navigation, surveying is not necessarily concerned with the graphic representation of land and the cartographic surveyor – the topographer *per se* – has often had other non-cartographic responsibilities, such as the qualitative assessment of land area and land value, estate management and engineering and architectural work. Thus, the creation of large-scale maps and plans is just one portion of a much larger agenda set by social concerns. The information base of topography is that of the boundaries (walls, edges and fences) which enclose and define fields, lakes, buildings, woodlands, estates, and so on, and that of the lie of the land for drainage and engineering purposes. It is the concrete and rarely abstract information of property ownership and of those human activities which are literally built into the landscape. The actual techniques employed depend upon the individual's level of education and experience, from simple pacing and sketching, through direct measurement with rod and chain, to complex techniques such as triangulation which require sophisticated trigonometrical and geometrical knowledge to be converted to a paper image. The representational strategies may ideally involve planimetric space but they merge into landscape images. Nor does the planimetric large-scale map necessarily share the projective space of chorography and charting. Topography's techniques of representation are also direct and involve little cartographic generalization: the decision of which features are important is made in the field so that they are measured and recorded; the map itself and the landscape painting can be sketched directly in the field before the final design and polish is added in the workroom. The social institutions of topography are those of land tenure, of the gentleman farmer and the farm labourer, of the engineer and architect, and of the power which accrues to the individual through the ownership of land and the control of its products. The establishment of commercial land markets in Renaissance Europe and the agricultural improvements which began the break with mediaeval subsistence farming (plus the transportation improvements which allowed crops to be moved to distant markets) dramatically increased the scope of topography, and continue to do so through to the present.

The projective space of chorography and the techniques of topography merged after 1500 in a new mode, that of geodesy. In terms of the amount of effort expended before the later seventeenth century, geodesy was quite a minor mode, yet it is intellectually significant. Geodetic surveys involve the comparison of the terrestrial and astronomical lengths of the same meridional arc. The initial system, established by the astronomers of caliph al-Ma'mūn in the early ninth century AD,[3] was to measure the arc's terrestrial length directly, a system used occasionally in Europe between 1525 and 1645. After Gemma Frisius formally described the process of triangulation in 1533, the terrestrial distance of each meridional arc was increasingly measured by that technique: first by Willebrod Snell in 1615 in the Netherlands, and then by Jean Picard (1669–1671) between Paris and Amiens. Thereafter, triangulation became the technique of choice for the eighteenth century push to determine the earth's figure (Airy, 1845; Bialas, 1982: especially 73–122). Geodesy was intimately tied to astronomy, mathematics and natural philosophy; its practitioners were scientists of international renown. While the geodetic surveys were concerned with the fundamental cartographic issue of the earth's figure, they rarely produced maps other than abstract triangulation diagrams. Geodesy as a cartographic practice was situated midway between the two modes of chorography and topography: it relied upon surveying techniques for the actual measurement, but its results were used in chorography, especially for the provision of scale to map projections.

18.3.2 The Enlightenment Convergence and Mathematical Cosmography

By the early eighteenth century, these four modes had all gradually converged into the single, if ambiguous, mode of 'mathematical cosmography' (Forbes, 1980b). The new mode's earliest manifestations are in the later seventeenth century (with, for example, the first Cassini survey) and reach a peak, as it were, in the period 1750–1800, but instances of the mode continue into the present. Again, while it is possible to identify 'pure' instances of cosmography, topography and charting throughout the 1700s, it is nonetheless impossible to draw hard and fast distinctions between each of the existing modes and the new, super-encompassing mathematical cosmography. The ambiguity derives from the varying degree of merger of each of the three components of the established modes. The culture of the Enlightenment promoted a unified and geometrical philosophy of map making. Socially, the eighteenth century expansion of the role of the state provided a more coherent social basis for cartography, although this was by no means total. Least homogeneously, the techniques and informational corpora of the existing modes did not fuse but instead each blurred and merged to create continua of information sets and of practices in parallel to the continuum of cartographic scale. These points are discussed in turn.

The differentiation of the early modern modes has been made according to the geographic scale at which they constructed their information and to their resultant techniques. Topographic mapping took data directly from the survey and represented them with minimal generalization; on the other hand, for surveyed data to be used in chorography, extensive generalization and reconstitution of those data would be

---

[3] Eratosthenes (second century BC) only calculated the earth's size, as indicated by his willingness to correct the initial value (itself suspiciously rounded) of 250 000 to 252 000 *stades* for the sake of computational convenience (Harley and Woodward, 1987: 154–155).

required before they could be meaningfully incorporated into the chorographic data sets. Thus the initially wide separation of these two modes in Figure 18.1. But they did not stay so separate for long. Regional surveys were initiated by European governments at scales smaller than those of the estate survey (larger scale surveys would have taken too long and would have cost too much). At the same time, the steady accretion (for a few regions) of chorographic and marine data allowed their representation at larger scales. The result was a convergence of the two modes in the confused realm of medium-scale cartography, between 1:100 000 and 1:1 000 000. Sufficiently large scale to allow the representation of surveyed data with only little generalization and simplification, medium-scale maps nonetheless *cover* substantially larger areas than are easily covered by a single survey and so still require the reconciliation of different and often conflicting sources. The process of reconciliation had developed into a high art by the early nineteenth century, as Godlewska (1988) has recently described for the Napoleonic *Carte topographique de l'Égypte*.

With respect to cartographic techniques, medium-scale mapping promoted the use of the traverse. A technique of the large-scale surveyor, it was quickly adapted to smaller-scale surveying as the measured itinerary, whether terrestrial along roads or maritime along coasts. For medium- and small-scale map making, the traverses were incorporated into the existing corpora of information by relating them to a few known places whose positions were determined by astronomical observation or by geodetic calculation. (This is the context of geodesy's contribution to mathematical cosmography.) The measured routes could be fitted graphically (directly on the map) or mathematically to these control points; other, unmeasured data could then be interpolated. Route information seems to have been standardized into a formal canon by the early centralized states to regulate the movements of peripatetic officials (judges and tax collectors) and military columns. Some sixteenth century states did not rely upon accumulated information, but set out to survey their entire territory in one geographic census; many of these state surveys have been subsequently praised by modern historians as having used triangulation, but it is more likely that they constituted elaborate 'sketches' set within the basic framework of known places and measured roads (Gasser 1975 [1907]; Ravenhill, 1983). Itineraries remained the favoured source of information for regional mapping even into the nineteenth century and claims by commercial cartographers that their maps were founded upon triangulation should be taken as rhetorical liberties without certain evidence to the contrary.

The third arena of convergence between topography, charting and chorography was their symbiotic relationship with the European states. Patronized map makers gained access to the state's otherwise secret or restricted information and increased their professional and intellectual status, which was in turn parlayed into financial and social gain. Chorographers were more socially mobile than topographers because they generally possessed a higher initial status and dealt with more overtly intellectual and socially acceptable matters. With the eighteenth and nineteenth century expansion of military map making, cartography became one way for officers to be noticed by their superiors and so gain promotion (Marshall, 1980). In return, the patronage system provided the state and its ruling elites with three items: power, through the acquisition and control of geographic information; legitimation, through the willing submission of loyal servants; and prestige, through the visible support of the arts and sciences. Map making was integral to the fiscal, political and cultural hegemony of Europe's ruling elites.

Finally, the cartographic modes of early modern Europe fused most completely in terms of the philosophical propositions which gave them a theoretical unity to match their practical convergence. The intellectual agenda of the Renaissance humanists had emphasized the particularity and contingency of both physical nature and of human existence and so recognized that there were many different avenues to knowledge. But this philosophical tolerance was destroyed by the stark divisions drawn between the religious dogmas constructed at the end of the sixteenth century. The horrors of the Thirty Years' War, and the search for a moral reconciliation between Protestants and Catholics, led Descartes and other natural philosophers to promote a logical, systematic and rational enquiry which did not depend on personal belief (Toulmin, 1990). The multiple perspectives of the 1500s gave way to a single perspective of rationality, which was explicitly sanctioned by the European states when they established scientific academies after 1660 (McClellan, 1985: 41–66). Ortelius could put two quite different maps of Denmark on one plate in his *Theatrum Orbis Terrarum* (1570) without fear of confusing a reader, but the new intellectual regime transformed this epistemological multiplicity into self contradiction (reflected in Hodgkiss, 1981: 14); the new epistemological singularity required that at any *one* time, there should be only *one* map of *one* territory.

From this perspective, maps and map making epitomized the Enlightenment practice of science. While the Enlightenment was a multifaceted and often contradictory intellectual period, we can still come to some broad understanding of this period as one which championed the application of rational thought in the form of experience and experiment, of observation and measurement (Porter, 1990: 1–11). Implicit within this mentality was the perspective of 'encyclopaedism', that properly conducted rational debate can reconcile conflicting points of view, so that all knowledge could ultimately be brought into a single whole and there be systematically described (Macintyre, 1990: 170–172). Combined with a mechanistic conception of the cosmos, this perspective led naturally to a formalization of the sciences – and indeed human society generally – as being inherently progressive.[4] Thus, in 1750, the naturalist Albrecht von Haller reaffirmed his faith in the universality and progressiveness of knowledge and in the methods of scientific observation and reasoning:

> *we perceive with precision from those which we do possess, those things which we lack … A theoretician of Nature acts like a land surveyor, who begins a map on which he has determined some locations, but lacks the positions of other places in between. [He] nevertheless makes an outline, and according to half-certain reports, indicates the remaining towns, of which he still has no mathematical knowledge. If he had made absolutely no sketch in which he combined the certain and uncertain [components] in*

---

[4] Bury (1987 [1932]) demonstrated that Progress is *not* a characteristic of the Judeo-Hellenic-Christian traditions but is a creation of the later sixteenth century. Belief in Progress 'took off' with the Enlightenment before becoming enshrined in positivism in the nineteenth century (Dale, 1989; Spadafora, 1990).

one composition, men his work of determining more exactly the locations and boundaries which still remained would be much more difficult and almost impossible. Indeed, it would not be possible, because [the work] would have no coherence, and would constitute no whole. (Quoted in Lyon and Sloan, 1981: 304)

The cosmographer's habitual reconciliation of information from varying sources and of varying quality to produce a single map of undeniable worth, comprehensiveness and utility has been recast as a metaphor for the fundamental belief of Enlightenment philosophy and the encyclopaedic mentality, that rational debate and enquiry will lead ultimately to unified knowledge.

Belyea's (1992) identification of contrasting archives of imaginative geographers and empirical explorers reflects a duality inherent to mathematical cosmography. Encyclopaedism encouraged the reconciliation of conflicting data through 'rational thought', which sometimes involved large conclusions being logically deduced from a small database, as when Guillaume Delisle 'discovered' the Mer de l'Ouest in the interior of North America. Ordered knowledge was achieved by reconstituting already existing knowledge within a single framework provided by the map. Cartographic truth is created by the cartographer. But at the same time, map making had incorporated astronomical and geodetic measurements, careful and meticulous observation, the newest instruments (Frängsmyr, Heilbron and Rider, 1990), mathematical principles and the Enlightenment's *ésprit géometrique,* all of which offered a second avenue to ordered knowledge via its systematic creation right from the start. This is the archive of Cook and Vancouver. By the end of the eighteenth century, this second approach to map making was widely accepted by European states as the proper way to create entirely new corpora of structured and ordered geographic information. The result was the birth of a new cartographic mode – systematic map making – which involves mapping an entire country at the same scale and with consistent survey techniques. More particularly, systematic surveys feature *new* detailed surveys which depend for their accuracy upon a prior-established control network of triangulation.

18.3.3 The Fragmentation of Mathematical Cosmography

Modern systematic map making was not the natural and inevitable development after cartography's supposed scientific reformation of the seventeenth and eighteenth centuries. It was instead the creation of Europe's burgeoning militaristic states (Brewer, 1988). The first systematic surveys were established by the French state to create a uniform and systematic corpus of knowledge about French territory and so allow an increasingly centralized government to support a huge military machine. From about 1730 to 1815, France was Europe's leader in creating and exporting the desideratum of a mathematically uniform space (Heilbron, 1990; Konvitz, 1987, 1990). The adoption of the French-style surveys throughout Europe after 1730 was a function of the existence of a centralized bureaucracy sufficiently willing and able to support an expensive procedure in order to make more efficient the ability to marshal

military resources and so wage war more effectively. The mode of systematic mapping –
with such instances as the Ordnance Survey, the US Geological Survey, and the Istituto
Geografico Militare – is the mode of centralized government mapping, of a uniform
'national space', of bureaucratic surveillance, of taxation, of the militaristic state and of
the state patronage of science (particularly geodesy, geography and geology). There
should be no fundamental distinction between systematic 'topographic' surveys for
overtly military and economic purposes and systematic cadastral surveys designed to
rationalize the tax burden and to make tax collection more efficient: they are simply
different aspects (or different submodes) of the one basic need to define and control
space for the purposes of the modem state.

When states lacked the bureaucratic and institutional ability to support complex
systematic surveys, those surveys simply did not succeed. Giuseppe Piazzi, for example,
attempted to establish a systematic survey of Sicily in 1808, but the state's bureaucracy
was *too* undeveloped to support the survey after the political demise of its patron in 1810
(Fodera'Serio and Nastasi, 1985). Those states perpetuated the chorographic aspects of
mathematical cartography and sought to reconstitute existing data according to a new
framework, thereby creating a single image. The provision of a new triangulation with
which to reconcile existing sources was popularized by the first Cassini survey in France
(intermittent between 1681 and 1744) and was a common occurrence throughout the
nineteenth century, in the United States (Edney, 1986), England and Wales (proposed;
Kain and Prince, 1985), Belgium (Danckaert, 1985), Russian Siberia (Postnikov,
Personal Communication), British India (Edney, 1991, 1993), and so on. These states
created standardized map *images* and surrounded them with a rhetoric claiming them
to have been based upon perfect triangulations and new surveys. In this sense,
mathematical cosmography has continued to be practiced to the present day, camou-
flaged by its rhetorical claims, and is still flourishing in those states which are peripheral
to the world economy (Stone, 1988; Jerie, 1972; Nittinger, 1975).

The needs of the centralized, militaristic state also spawned another new mode from
mathematical cosmography: thematic map making. While examples can be found
from before 1700 of several of the basic symbolization strategies commonly used on
thematic maps – for example, the isoline, the flow line, or the shading of socially
defined regions – the necessary institutional elements were not established until the
later 1700s. There were two salient factors. Firstly, the seventeenth century realization
that the physical world can best be studied through mathematics had merged early in
the eighteenth century with the desire to study society and its aspects in a rational,
systematic and encyclopaedic manner, a methodology which required the quantifi-
cation and measurement of the object of study (Frängsmyr, Heilbron and Rider, 1990;
Toulmin, 1990). Thus, there had developed the intellectual assumption that it is
meaningful to map data about the social world in the same manner as data about the
physical world. Secondly, the later Enlightenment featured a dramatic increase in the
collection of economic, demographic and social data by European states – or by
members of the educated elites working for the state's benefit – in order to better
understand and so control their societies and economies (Nadal and Urteaga 1990;
Porter, 1986: 18–39). In general, as Castner (1980) has described for Russia, the later
eighteenth century featured the increasing specialization of mapping and the steady

addition to the established, application-neutral reference maps of more application-specific 'special purpose' and 'thematic' maps. The connection between the development of systematic mapping and of thematic mapping is clear: while the first allowed the state to understand and control the physical territory of the European state or of European colonies, the second allowed the state to understand and control the social contents of those territories.

The development of the two new modes and the continuation of mathematical cosmography into the present does not mean that the older modes of chorography, charting and topography have ceased to exist. However, they underwent quite substantial changes. Topography was affected by increasing professional specialization during the nineteenth century, so that cartographic surveyors slowly surrendered their non-mapping functions to other groups. The modern surveyor who works outside the state dominated modes of systematic mapping and of inertia bound mathematical cosmography is once again concerned with very small areas for engineering purposes or for the affirmation and restoration of property boundaries. For its part, chorography is still tied closely to the state and to the established, educational elites. Increasing literacy in European and North American societies plus decreasing production costs (with lithography) have produced a submode of truly commercial cartographers unsubsidized by the state. Nonetheless, most commercial cartographers have remained close to the state, using government contracts to subsidize commercial ventures or repackaging and reselling government maps (today, digital geographic information) at a profit. Charting is very much a maritime version of the systematic surveys, although it too has a commercial component which repackages government information for the civilian mariner.

It should also be noted that none of the post-1800 formal cartographic modes enjoyed the same cultural relevancy and intellectual glitter that mathematical cosmography had at its peak. The Romantic movement and the early nineteenth century religious revival challenged most of the Enlightenment's assumptions and social philosophies. Certainty about being able to find the precise size and shape of the earth gave way to statistical representations under the weight of irreconcilable results (Stigler, 1986: 11–158), while the dramatic progress made in instrument manufacture during the eighteenth century made high accuracy and high precision available to many surveyors rather than the select few. Geographers and geodesists turned to new theoretical issues and left cartography behind as mere technique (Bialas, 1982; Godlewska, 1989).

18.4 Explaining the Rhetoric of Empiricist Cartography

The fragmentation of mathematical cosmography has left map making in a schizophrenic state. On the one hand, the systematic mapping activities of most of the world's wealthier countries have continued to advertise and promote the philosophical assumptions about maps and nature as established under mathematical cosmography (as Lewis Carroll realized). The official, systematic survey proclaims that there is one world which can be progressively described and known, and that the survey map is that description. It perpetuates the eighteenth century blurring of the boundary between the chorographic and topographic corpora of data by subsuming the generalized and abstracted data of the one beneath the concrete and hard edged data of the other. Geographic data are awarded an epistemological coherence to match the ontological coherence of the world itself. The other modes have adopted the same rhetoric of certainty, both in their representational strategies and in their social acceptance. Museums began in the 1800s to collect maps according to their content, rather than to their underlying processes (Harley, 1987: 15–16). All the formal modes of modern western cartography share the same cultural expectations that the map replicates the territory's structure precisely and accurately: *this* is the empiricist conception of cartography. On the other hand, none of those modes can entirely or properly fulfil those expectations.

The legacy of the Enlightenment's mathematical cosmography is cartography's empiricism. The key to the discipline's self-image and to its rhetoric is accordingly its perception of eighteenth century cartography. That is, cartography's supposed reformation is actually the historiographic misunderstanding of empiricism's ascendancy, a misunderstanding justified by the scientific revolution which serves both as the necessary precursor to the cartographic reformation and as a demonstration that such intellectual shifts can occur and that they have indeed occurred. Empiricist cartography presupposes that maps are graphic selections from a single corpus of geographic data which relate directly to the world via the global coordinate system. The history of such a cartography therefore stresses those cartographic modes which feature corpora of information which are organized by latitude and longitude. Before 1700, these were only chorography and charting; after 1800 the empiricist self-image postulates a single mode based on systematic mapping. Cartography's supposed reformation comprises the switch from one to the other, from small scale ('artistic') to apparently scale-less (and 'scientific') geography.

That these otherwise quite distinct cartographic modes are historiographically comparable is the result of mathematical cosmography's internal duality. Chorography, incorporated within mathematical cosmography via encyclopaedism, is made to be directly equivalent to the instrumentalist-driven systematic survey. The uncertain art of the regional map maker is held to be comparable to the exact science of the surveyor. Yet, in the formulation presented here, cartography comprises a number of modes which should be compared only with caution. Comparison between modes requires the consideration of all their constituent relations – cultural, social and technological – and of their respective corpora of data. It is insufficient to assume that commonality in one component indicates commonality in the rest. Although today's formal modes share a common cultural grounding, they do not constitute the monolithic enterprise that empiricism suggests. Because two temporally separated modes organize data in a similar manner, one is not the antecedent of the other.

The approach suggested here, of considering cartography to be constituted from many modes, is more satisfactory than the traditional means for envisioning the broad sweep of cartographic history. It is more consistent with the complexity of both the historical record and the intellectual character of map making as an intellectual, technological, social and cultural process. This approach is neither prescriptive nor proscriptive and seeks only to broaden our discussions of the nature and history of cartography to encompass the myriad forms in which maps have been – and in which they continue to be – constructed and used.

Acknowledgements

I am indebted to Jim Akerman for our initial discussions on this topic; John Krygier raised many interesting points which I have attempted to express here. I am most grateful for the helpful comments and critical reactions made to earlier drafts, especially by Mark Monmonier and Robert Rundstrom, and also by Barbara Belyea, Michael Blakemore, Catherine Delano Smith, Bill Gartner, Roger Kain, Pellervo Kokkonen, Richard Oliver, Scott Salmon and David Woodward. Some of the ideas were originally presented in 'The Conceptual Bases of Modem Cartography: The Encyclopaedic Mentality and the Idea of Progress', part of the special session on 'Alternative Cartographies' (organizer: Jeremy Crampton), Annual Convention, Association of American Geographers, Miami, April 1991; a summary was presented to the Department of Geography, Portsmouth University, UK, February 1993.

References

Airy, G.B. (1845) Figure of the Earth. *Encyclopaedia Metropolitana*, **5**, 165–240.

Akerman, J.R. (1991) *On the Shoulders of a Titan: Viewing the World of the Past in Atlas Structure.* Unpublished PhD Dissertation, Pennsylvania State University.

Bagrow, L. (1985) *The History Of Cartography*, 2nd edn, Precedent Publishing, Chicago.

Belyea, B. (1992) Images of power: Derrida/Foucault/Harley. *Cartographica*, **29**(2), 1–9.

Bialas, V. (1982) *Erdgestalt, Kosmologie und Weltanschauung: The Geschichte der Geodäsie Als Teil der Kulturgeschichte der Menschheit*, Konrad Winter, Stuttgart.

Bijker, W.E., Hughes, T.P. and Pinch, T.P. (1987) *The Social Construction of Technological Systems: New Directions in the Sociology and History of Technology*, MIT Press, Cambridge, MA.

Blakemore, M. and Harley, J.B. (1980) Concepts in the history of cartography: A review and perspective. *Cartographica*, **17**(4), 1–120.

Brannon, G. (1989) The artistry and science of map-making. *Geographical Magazine*, **61**(9), 37–40.

Brewer, J. (1988) *The Sinews of Power: War, Money and the English State*, Century Hutchinson, London.

Buczek, K. (1982) *The History of Polish Cartography from the 15th to the 18th Century* (Trans. A. Polocki), Meridian Publishing, Amsterdam.

Buisseret, D. (1992) *Monarchs, Ministers, and Maps: The Emergence of Cartography as a Tool of Government in Early Modern Europe*, University Of Chicago Press, Chicago.

Bunge, W. (1966) *Theoretical Geography*, Royal University of Lund, Department of Geography, Lund.

Bury, J.B. (1987) *The Idea of Progress: An Inquiry into its Origin and Growth*, Dover, New York.

Campbell, J. (1991) *Introductory Cartography*, 2nd edn, Prentice Hall, Englewood Cliffs, NJ.

Carroll, L. (1894) *Sylvie and Bruno Concluded*, Macmillan, London.

Carstensen, L.W. (1989) A fractal analysis of cartographic generalization. *The American Cartographer*, **16**, 181–189.

Castner, H.W. (1980) Special purpose mapping in 18th century Russia: A search for the beginnings of thematic mapping. *The American Cartographer*, **7**, 163–175.

Chrisman, N.R. (1991) Institutional and societal components of cartographic research, in *Advances in Cartography* (ed. J.C. Muller), Elsevier (for the International Cartographic Association), London, pp. 231–242.

Clive, J. (1989) *Not by Fact Alone: Essays on the Writing and Reading of History*, Houghton Mifflin, Boston, MA.

Crampton, J. (1993) *Alternative Cartographies*. Unpublished PhD Dissertation, Pennsylvania State University.

Crone, G.R. (1978) *Maps and Their Makers: An Introduction to the History of Cartography*, 5th edn, Dawson, Folkestone, UK.

Dale, P.A. (1989) *In Pursuit of a Scientific Culture: Science, Art, and Society in the Victorian Age*, University Of Wisconsin Press, Madison WI.

Danckaert, L. (1985) La Carte Topographique de la Belgique par Philippe Vandermaelen. Imago Et Mensura Mundi: Atti Del Ix Congresso Internazionale Di Storia Della Cartografia, Rome, vol. 1 (ed. C. Marzoli), pp. 171–176.

Edney, M.H. (1986) Politics, science, and government mapping policy in the United States, 1800–1925. *The American Cartographer*, **13**, 295–306.

Edney, M.H. (1991) The *Atlas of India, 1823–1947*: The natural history of a topographic map series. *Cartographica*, **28**(4), 59–91.

Edney, M.H. (1993) The patronage of science and the creation of imperial space: The British mapping of India, 1799–1843. *Cartographica*, **30**(1), 61–67.

Fodera'Serio, G. and Nastasi, P. (1985) Giuseppe Piazzi's survey of Sicily: The chronicle of a dream. *Vistas in Astronomy*, **28**, 269–276.

Forbes, E.G. (1980) *Tobias Mayer (1723–62): Pioneer of Enlightened Science in Germany*, Vandenhoeck and Ruprecht, Gottingen.

Forbes, E.G. (1980) Mathematical cosmography, in *The Ferment of Knowledge: Studies in the Historiography of Eighteenth-Century Science* (eds. G.S. Rousseau and R. Porter), Cambridge University Press, Cambridge, pp. 417–448.

Frängsmyr, T., Heilbron, J.L. and Rider, R. (1990) *The Quantifying Spirit in the Eighteenth Century*, University of California Press, Berkeley, CA.

Gasser, M. (1975) Zur Technik der Apianschen Karte von Bayern. *Acta Cartographica*, **20**, 279–300.

Godlewska, A. (1988) The Napoleonic survey of Egypt: A masterpiece of cartographic compilation and early nineteenth-century fieldwork. *Cartographica*, **25**(1/2).

Godlewska, A. (1989) Traditions, crisis, and new paradigms in the rise of the modern French discipline of geography, 1760-1850. *Annals of the Association Of American Geographers*, **79**, 192–213.

Goodchild, M. (1988) Stepping over the line: Technological constraints and the new cartography. *The American Cartographer*, **15**, 311–319.

Grafton, A. (1992) *New Worlds, Ancient Texts: The Power of Tradition and the Shock of Discovery*, Harvard University Press, Cambridge, MA.

Gregory, D. (1986) Epistemology, in *The Dictionary of Human Geography*, 2nd edn (eds. R.J. Johnston, D. Gregory and D.M. Smith), Blackwell, Oxford, p. 127.

Hall, S.S. (1992) *Mapping the Next Millennium: The Discovery of New Geographies*, Random House, New York.

Hammond Inc. (1993) *Atlas of the World, Concise Edition,* Paperback Edition, Hammond, Maplewood, NJ.

Harley, J.B. (1987) The map and the development of the history of cartography, in *Cartography in Prehistoric, Ancient, and Medieval Europe and the Mediterranean,* vol. 1 The History of Cartography (eds. J.B. Harley and D. Woodward), University of Chicago Press, Chicago, pp. 1–42.

Harley, J.B. (1988a) Silences and secrecy: The hidden agenda of cartography in early modern Europe. *Imago Mundi,* **40,** 57–76.

Harley, J.B. (1988b) Maps, knowledge and power, in *The Iconography Of Landscape* (eds. D. Cosgrove and S.J. Daniels), Cambridge University Press, Cambridge, pp. 277–312.

Harley, J.B. (1989a) Deconstructing the map. *Cartographica,* **26**(2), 1–20.

Harley, J.B. (1989b) Historical geography and the cartographic illusion. *Journal of Historical Geography,* **15,** 80–91.

Harley, J.B. (1989c) 'The myth of the great divide': Art, science, and text in the history of cartography. Paper presented at Thirteenth International Conference on the History of Cartography, Amsterdam, 26 June.

Harley, J.B. (1990) Cartography, ethics and social theory. *Cartographica,* **27**(2), 1–23.

Harley, J.B. (1991) Can there be a cartographic ethics? *Cartographic Perspectives,* **10,** 9–16.

Harley, J.B. and Woodward, D. (1987) *Cartography in Prehistoric, Ancient, and Medieval Europe and the Mediterranean,* vol. 1, The History of Cartography, University of Chicago Press, Chicago.

Harley, J.B. and Woodward, D. (1989) Why cartography needs its history. *The American Cartographer,* **16,** 5–15.

Harley, J.B. and Woodward, D. (1992) *Cartography in the Traditional Islamic and South Asian Societies,* vol. 2 (1), The History Of Cartography, University of Chicago Press, Chicago.

Harvey, D. (1989) *The Condition of Postmodernity: An Enquiry into the Origins of Cultural Change,* Blackwell, Oxford.

Heilbron, J.L. (1990) The measure of Enlightenment, in *The Quantifying Spirit in the Eighteenth Century* (eds. T. Frängsmyr, J.L. Heilbron and R. Rider), University of California Press, Berkeley, CA, pp. 207–242.

Hodgkiss, A.G. (1981) *Understanding Maps: A Systematic History of Their Use and Development,* Dawson, Folkestone, UK.

Hughes, T.P. (1986) The seamless web: Technology, science, etcetera, etcetera. *Social Studies of Science,* **16,** 281–292.

Jerie, H.G. (1972) New concepts of topographic mapping in developing countries. *World Cartography,* **12,** 3–20.

Kain, R.J.P. and Prince, H.C. (1985) *The Tithe Surveys of England and Wales,* Cambridge University Press, Cambridge.

Keates, J.S. (1982) *Understanding Maps,* Longman, London.

King, R. (1990) *Visions of the World and the Language of Maps, Trinity Papers in Geography,* **1,** Department of Geography, Trinity College, Dublin.

Konvitz, J.W. (1987) *Cartography in France, 1660-1848: Science, Engineering, and Statecraft,* University of Chicago Press, Chicago.

Konvitz, J.W. (1990) The nation-state, Paris, and cartography in eighteenth and nineteenth-century France. *Journal Of Historical Geography,* **16,** 3–16.

Lakoff, G. (1987) *Women, Fire, and Dangerous Things: What Categories Reveal About the Mind,* University Of Chicago Press, Chicago.

Lindberg, D.C. and Westman, R.S. (1990) *Reappraisals of the Scientific Revolution,* Cambridge University Press, Cambridge.

Livingstone, D.N. (1992) *The Geographical Tradition: Episodes in the History of a Contested Enterprise*, Blackwell, Oxford.

Lloyd, R. and Gilmartin, P. (1987) The South Carolina coastline on historical maps: A cartometric analysis. *The Cartographic Journal*, **24**(1), 19–26.

Lyon, J. and Sloan, P.R. (1981) *From Natural History to the History of Nature: Readings from Buffon and his Critics*, University of Notre Dame Press, Notre Dame, IN.

Macintyre, A. (1990) *Three Rival Versions of Moral Enquiry: Encyclopaedia, Genealogy, and Tradition*, University of Notre Dame Press, Notre Dame, IN.

Marsh, D.P. (1983) *A Cartography of Popular Maps*. Unpublished PhD Dissertation, Pennsylvania State University.

Marshall, D.W. (1980) Military maps of the eighteenth-century and the Tower of London drawing room. *Imago Mundi*, **32**, 21–44.

McClellan, J.E. (1985) *Science Reorganized: Scientific Societies in the Eighteenth Century*, Columbia University Press, New York.

Monmonier, M.S. (1982) Cartography, geographic information, and public policy. *Journal of Geography in Higher Education*, **6**(2), 99–107.

Muller, J.C. (1991) *Advances In Cartography*, Elsevier (for the International Cartographic Association), London.

Nadal, F. and Urteaga, L. (1990) *Cartography and State: National Topographic Maps and Territorial Statistics in the Nineteenth Century*, Cuadernos Criticos de Geografia Humana 88, English Parallel Series 2, Catedra de Geografia Humana, Facultad de Geografia e Historia, Universitat de Barcelona, pp. 9–67.

National Research Council, Mapping Science Committee (1990) *Spatial Data Needs: The Future of the National Mapping Program*, National Academy Press, Washington, DC.

Nittinger, J. (1975) Cadastral surveying as an instrument of political, economic, and social development. *World Cartography*, **13**, 21–25.

Pelletier, M. (1986) La France Mesuree. *Mappemonde*, **1986**(3), 26–32.

Petchenik, B.B. (1979) From place to space: The psychological achievement of thematic mapping. *The American Cartographer*, **6**, 5–12.

Pickles, J. (1992) Texts, hermeneutics and propaganda maps, in *Writing Worlds: Discourse, Text and Metaphor in the Representation of Landscape* (eds. T.J. Barnes and J.S. Duncan), Routledge, London, pp. 193–230.

Porter, T.M. (1986) *The Rise of Statistical Thinking, 1820-1900*, Princeton University Press, Princeton, NJ.

Porter, R. (1990) *The Enlightenment. Studies in European History*, Humanities Press International, Atlantic Highlands, NJ.

Ptolemy, C. (1991) *The Geography* (Trans. E.L. Stevenson), Dover, New York.

Ravenhill, W. (1983) Christopher Saxton's surveying: An enigma, in *English Map-Making, 1500–1650: Historical Essays* (ed. S. Tyacke), British Library, London, pp. 112–119.

Rees, R. (1980) Historical links between cartography and art. *The Geographical Review*, **70**, 60–78.

Robinson, A.H. and Petchenik, B.B. (1976) *The Nature of Maps: Essays Toward Understanding Maps and Mapping*, University of Chicago Press, Chicago.

Rundstrom, R.A. (1991) Mapping, postmodernism, indigenous people and the changing direction of North American cartography. *Cartographica*, **28**(2), 1–12.

Skelton, R.A. (1972) *Maps: A Historical Survey of their Study and Collecting*, University of Chicago Press, Chicago.

Spadafora, D. (1990) *The Idea of Progress in Eighteenth-Century Britain*, Yale University Press, New Haven, CT.

Stigler, S.M. (1986) *The History of Statistics: The Measurement of Uncertainty Before 1900*, Harvard University Press, Cambridge, MA.

Stone, J.C. (1988) Imperialism, colonialism and cartography. *Transactions of the Institute of British Geographers NS*, **13**, 57–64.

Thongchai, W. (1988) *Siam Mapped: A History of the Geo-Body of Siam*. Unpublished PhD Dissertation, University of Sydney.

Tobler, W. (1976) Analytical cartography. *The American Cartographer*, **3**, 21–32.

Tomlinson, R.F. (1988) The impact of the transition from analogue to digital cartographic representation. *The American Cartographer*, **15**, 249–261.

Toulmin, S. (1990) *Cosmopolis: The Hidden Agenda of Modernity*, The Free Press, New York.

Tyner, J. (1992) *Introduction to Thematic Cartography*, Prentice-Hall, Englewood Cliffs, NJ.

Visvalingham, M. (1989) Cartography, GIS and maps in perspective. *The Cartographic Journal*, **26**, 26–32.

Wood, D. (1992a) *The Power of Maps*, Guilford Press, New York.

Wood, D. (1992b) How maps work. *Cartographica*, **29**(3/4), 66–74.

Wood, D. (1993) Maps and mapmaking. *Cartographica*, **30**(1), 1–9.

Wood, D. and Fels, J. (1986) Designs on signs: Myth and meaning in maps. *Cartographica*, **23**(3), 54–103.

Woodward, D. (1974) The study of the history of cartography: A suggested framework. *The American Cartographer*, **1**, 101–115.

Woodward, D. (1987) Medieval *mappaemundi*, in *Cartography in Prehistoric, Ancient, and Medieval Europe and the Mediterranean*, vol. 1, The History of Cartography (eds. J.B. Harley and D. Woodward), University of Chicago Press, Chicago, pp. 286–370.

Woodward, D. (1990) Roger Bacon's terrestrial coordinate system. *Annals of the Association of American Geographers*, **80**, 109–122.

Woodward, D. (1992) Representations of the world, in *Geography's Inner Worlds: Pervasive Themes in Contemporary American Geography* (eds. R.F. Abler, M.G. Marcus and J.M. Olson), Rutgers University Press, New Brunswick, NJ, pp. 50–73.

19

Reflection Essay: Progress and the Nature of 'Cartography'

Matthew H. Edney

Osher Professor in the History of Cartography, University of Southern Maine, USA;

Director, History of Cartography Project, University of Wisconsin–Madison, USA

My 1993 essay 'Cartography without "progress"' (reproduced as Chapter 18 in this volume) originated in the disconnect between what my own studies unveiled about past cartographies and the history of cartography as presented in the literature. Indeed, given how the modern idealization of cartography – as well as its further abstractions by academic cartographers – depended upon the established view of map history, I was struggling to understand the very nature of this thing we call 'cartography'.

What the literature told me was that cartography was a uniform practice that had steadily progressed throughout the history of human (i.e. Western) civilisation, except for some abnormal periods when cartographic practice stagnated or even regressed. It enshrined cartography's innately progressive character in a canon of key maps. The automatic first step in the analysis of any old map was its comparison against other maps of the same region, including those of the present day, so as to situate it along cartography's supposed progressive trend line. The few maps marking major advances in knowledge of a region were canonized; the many simple, derivative maps were dismissed from consideration as inadequate or abnormal; the remainder, incorporating minor advances, were listed in cartobibliographies but otherwise left unstudied. The logic was cyclical, unacknowledged and immensely powerful. The result was an a priori meta-narrative that constrained scholars to narrate only the progressive development of map making.

I think I was opposed to the idealized meta-narrative from the start. Thinking I would become a professional surveyor, I had focused about a third of my undergraduate degree on land surveying. This training was fundamentally pragmatic: surveying comprises a wide variety of techniques that must be selectively deployed to meet the various needs (and budgets) of clients. When I began my graduate studies with David Woodward in 1983, I inevitably brought this perspective to bear on cartography and its history. For the MS thesis, I explored a particular constellation of practice, need and community in the federal mapping of US states in the later 1800s (Edney, 1986). For the doctoral dissertation, I turned to the constellation involved in the systematic mapping of India by the British before 1843 (Edney, 1990, 1997). Not that I actually used the term 'constellation' at that time, and the concept remained inchoate. But as I worked on the dissertation, and encouraged by the then emergent critical map scholarship, I increasingly realized the need for a more formal explication of the concept.

It seemed to me that a formal approach to understanding mapping practices and communities would form the basis for *explaining* cartographic history without resorting to any necessary presumptions of progress. Such a conceptual framework would serve to counter the apparent eternal verities of the modern cartographic ideal. Other critical scholars had yet to dispel the ideal's meta-narrative of progress, the cartographic canon that legitimated the meta-narrative, and the very idea of 'cartography' that the meta-narrative in turn sustained. A concern for practices and communities was – and remains – essential for the successful development of critical map studies.

This is not to say that mapping has *never* progressed. Rather, if scholars employ as underdetermined a word as 'progress', then they must be absolutely clear about how they define it, how they use it, how they measure it (quantitatively or qualitatively) and how it is limited to specific aspects of certain mapping practices. Cartographic progress must be demonstrated, not presumed.

19.1 The Modern Cartographic Ideal and its Critics

Since the 1870s, historians have blithely talked about the cartographies of ancient Greece, Renaissance Europe and modern North America as if they were all one and the same thing. The underlying presumption is that what we perceive today as cartography existed in essentially the same form in each of these widely differentiated societies. Such historical writing is a key manifestation of modern Western culture's pervasive and persistent idealization of 'cartography' as a coherent and moral body of practice and knowledge that is properly pursued by trained and disciplined individuals for the betterment of their own societies and indeed for human civilisation, that is applied uniformly across all geographic scales of social organization, and that exemplifies the strictly experiential creation of knowledge (and is thus 'empiricist' in character). The problem is that this idealized cartography bears little, if any, resemblance to the multiple ways in which people have actually produced and consumed maps.

A great deal of intellectual energy was expended in the 1980s and early 1990s to point out the many inadequacies of the modern cartographic ideal. Its claims for the inherent objectivity of all maps – excluding, of course, inadequate maps and mere 'map-like objects' – were now roundly trounced, so that it became possible to proclaim loudly that all maps are 'mental' maps, constitute propaganda, and are just plain 'subjective' (Axelsen and Jones, 1987). The personal, cognitive processes of mapping were clearly distinguished from the socio-linguistic practices of making maps, while the ethnocentrism (some might say racism) underlying the ideal's confusion of the cognitive with the social was exposed (Wood, 1993a, 1993b). The faith in cartography's moral purity, already strained (Monmonier, 1991), was exploded by analyses of the social functions filled by maps (especially Harley, 1989, reproduced as Chapter 16 in this volume; Harley, 1988, 1990). New approaches were developed to read maps as culturally symbolic texts, drawing especially upon iconography (Blakemore and Harley, 1980: 76–86; Harley, 1983; Harley, 1985) and semiotics (especially Wood and Fels, 1986, reproduced as Chapter 14 in this volume). The fundamental definition of 'map' was recast along open, flexible and culturally sensitive lines (Harley and Woodward, 1987: xvi).

Yet this incredible intellectual frenzy remained limited in extent and implication. Having successfully exposed the mythic nature of modern cartography and demonstrated what cartography is *not*, there was little interest in developing any sense of the larger picture of how map making and map use *have* functioned and developed. This is perhaps the real point of contention for those map scholars who have resisted critical approaches. It is not that they have been unable to appreciate the critique; rather, they see focused textual interpretations and precise contextual analyses as having no relevance for their own understanding of progressive map history. As John Andrews (2001: 31–32) complained, for example, an assessment of the power relations manifested in a given map or survey says little about the relationship of that work to other cartographic works, before and since.

Critical map historians had hitherto lacked interest in the large sweep of map history largely because of their intellectual trajectory (Edney, 2005a, 2011b). For most of the twentieth century, map historians had worked entirely within the traditional paradigm to make the content of old maps accessible to a variety of other scholars, ranging from historians to diplomats. In doing so, they relied upon the overwhelming presumption of cartography's inherently progressive nature to situate each map into a broader context and so give their studies broader intellectual significance; the self-evidently progressive nature of cartography meant that map historians did not actually have to think about or examine that context. As they adopted critical approaches and, as their work has been augmented by that of literary and art historical scholars, the main focus of the post-1980 socio-cultural paradigm of map history has remained on textual analysis, albeit now of the cultural meanings of old maps, supported by some analyses of social context. Close study of cartographic context was discouraged by the manner in which many critical commentators, led in part by Harley (1988, 1989), were content to perpetuate an older, transcendentalist criticism of modern maps as sterile and culturally impoverished (Roszak, 1972: 407–411; Tuan, 1977), especially in comparison to the evident cultural sophistication and environmental awareness of non-modern cartographies (e.g. Harvey, 1989: 241–259; Duncan and Ley, 1993; but see Edwards, 2006: 1–15). More pragmatically, contextual analyses have been discouraged by the difficulty and time required for the necessary archival research.

Meanwhile, some academic cartographers, especially after 1950, had sought to justify their academic status by studying cartography's professional history; while the resultant internal paradigm of map history tended to be self-serving, it did produce many studies of past map practices and it prompted many of the questions that would in the 1980s displace the traditional paradigm (Edney, 2005b, 2011b). But the internal paradigm declined rapidly after 1985 as academic cartographers increasingly focused on digital technologies; the historical study of mapping processes and cartographic contexts lost momentum. At the same time, academic cartography's fundamental concern to improve map design so as to perfect the delivery of each (small-scale) map's message served only to perpetuate the modern conviction that the cartographer is an empowered individual; but now, while once it was argued that cartographers need to be properly disciplined to maintain the integrity of maps (Wright, 1942), the choices made by the individual cartographer became the source of each map's inevitable 'subjectivity' (Black, 1997). Only Robert Rundstrom (1991) had tried to argue for a consideration of mapping practices as the core to a complex understanding of cartography that would not depend upon a naïve duality of objectivity versus subjectivity. His arguments about the performative qualities of mapping have been broadly accepted by other scholars of indigenous cartographies (Woodward and Malcolm Lewis, 1998: 1–10). But, because he did not extend the crucially important implications of a processual analysis from individuals functioning within small communities to the mapping activities of broader impersonal societies, scholars of formal cartographies have failed to appreciate the significance of his work.

19.2 'Cartography without "Progress"'

If there was a precise start to my formal conceptualization of cartography's constituent modes, it was in my discussions with James Akerman in 1988 and 1989. At the same time, I saw a number of scholars struggling with the divisions evident in the empirical record between cartography's genres (Hodgkiss, 1981), archives (Belyea, 1992) and spatial discourses (Thongchai, 1988). This was also when Penny Richards, then also in the geography graduate programme at UW–Madison, produced an illuminating visualization of the interactions of multiple streams and threads of spatial information and perception concerning northern Pennsylvania in the eighteenth century, which she likened to a loosely braided rope (Figure 19.1). Her complex diagram was the model for my own, much cruder interpretation of the intersections, convergences and divergences between cartographic modes (Figure 18.1). Finally, Rundstrom's essay on a processual approach prompted me to clarify and write up my ideas about cartography's discrete constellations – which I now formally called modes – of communities and practices. The result was 'Cartography without "progress"'.

The essay's basic argument holds up very well. There is not one universal endeavour of 'cartography'. That is a fiction created in the 1800s to distinguish and privilege Western knowledge practices from those of other peoples as a key element of the structures of sentiment that underpinned modern imperialism (Edney, 2009: 41–43). Rather, there are many ways to conceptualize and graphically organize the spatial complexity of the world between and within societies (Mukerji, 2002 came independently to much the same conclusion). Each mode depends upon a functional

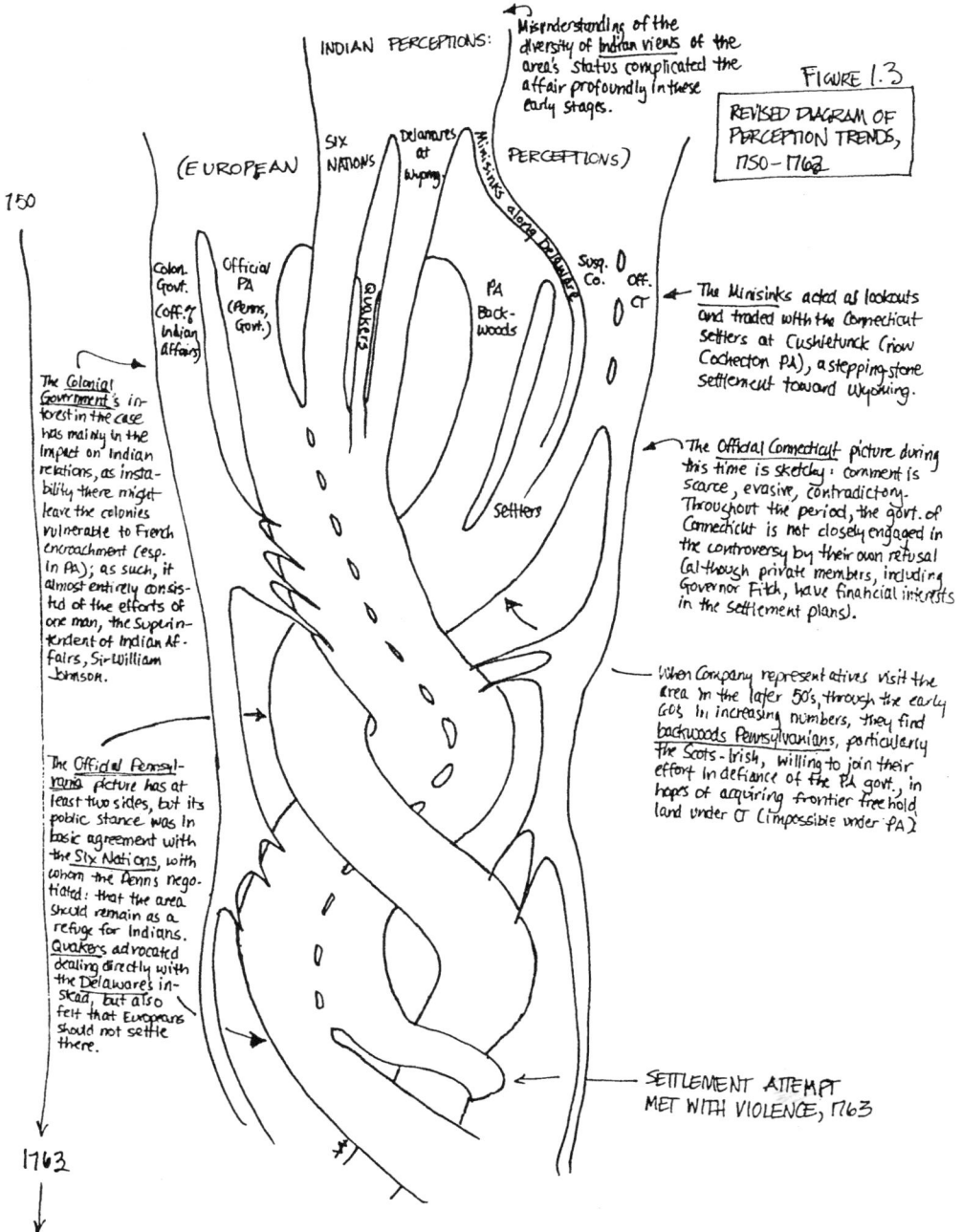

Figure 19.1 Penny Richards' 'Revised Diagram of Perception Trends, 1750–1762' (Source: Figure 1.3 in Richards, 1990: 12–13; Reproduced by permission of Ms Richards.).

conception of the world in which a set of institutions produces and consumes maps for specific purposes and at specific scales. Key implications include:

- mapping practices, from surveying technologies to symbolization strategies to the formation of archives of knowledge, are *all* scale dependent;

- no single set of practices or epistemologies has ever been common to all mapping, even in the present day;

- mapping practices are socially determined, or rather, to follow Bruno Latour's (2005) reconfiguration of sociology, they are constitutive of social relations;

- each mode might exhibit a certain internal logic or dynamic, but the intersections between modes are contingent;

- intersections can occur in the production of maps, their circulation, or their consumption; and

- the history of cartography becomes the history both of individual modes and of the intersections between them.

I have found these principles to be crucial when synthesizing cartographic history (Edney, 2007a, 2011a). Indeed, my colleagues and I have designed the last three volumes of *The History of Cartography* (Harley and Woodward, 1987*et seq.*), which will cover the 250 years of cartography since 1650, in an encyclopaedic manner, around the cartographic modes in each era. No other conceptual framework is able to manage the complex narratives of such a huge enterprise in a historically meaningful manner.

In this respect, 'Cartography without "progress"' is a classic study. It still has great relevance and significance today. Indeed, comments by graduate students pursuing studies in map history suggest that it has something of a cult following, even if it is not cited as frequently as essays by Harley or Wood. However, fifteen years of further research and reflection mean that I inevitably now see problems with how I implemented the argument.

In explaining how the modern cartographic ideal had evolved, I drew heavily on Eric Forbes' (1980) elucidation of the eighteenth century concept of 'mathematical cosmography'. I presented mathematical cosmography as a mode in and of itself, one that encompassed all other modes with the intent of creating a scale-less archive (Edney, 1994). I would now argue that mathematical cosmography is best considered as a cultural ideal that coloured several modes, thereby giving the appearance of a unified archive. Moreover, further analysis of the formation of the modern cartographic ideal indicates that it was created as a complex ideological palimpsest over the course of not the eighteenth but rather the nineteenth century. In privileging mathematical cosmography I attempted to adhere to the common assertion (which unfortunately persists) that modern cartography acquired its idealized character as part of the 'Enlightenment Project' (Horkheimer and Adorno, 1972). That assertion correlates to cartography's supposed 'scientific reformation' in the eighteenth century. Yet that reformation was construed only in 1900 as the final rhetorical element of the modern cartographic ideal and might be said to bind the whole together (Edney, 2011b).

A second issue is that some of the labels I applied to modes have conflicted with existing usages. I used 'topography' in its original meaning as the inscription of place (τοπος [place] + γραφειν [to write, describe]), but it has proved impossible to supplant its modern meaning as both actual relief and the representation of relief. Early modern commentators made much of the scale-based distinction between 'chorography' (the description of a region [χορος]) and 'geography' (the description of the entire earth [γε]). In 1993, I sought to get around the problems of distinguishing between them by using 'chorography' for the mode of mapping supra-personal space; I subsequently decided the mode should be called geographical mapping. But maybe regional and world mappings are different modes, or were so in some societies at certain times. I have recently, therefore, begun to avoid specific terms and instead to write more usefully about mappings of place, mappings of property, mappings of regions and mappings of territory, which is to say mappings of regions as if they were places (Edney, 2007a).

The key problem with 'Cartography without "progress"' is that I then thought in terms of predefined structures (the modes) into which I sought to fit each map or survey. This approach might have promoted coherency, but it has not dealt well with complexity. A poststructural flexibility is far more appropriate: the variety and variability of symbolic conventions and practices found in the empirical record need to be respected; each mode needs to be able to be reconfigured through internal change and external influence; there needs to be mechanisms to permit the formation of new modes. That is, I should have thought in terms of discourse as well as practice. (Harley had talked about the idea of discourse but did not properly develop it – Edney, 2005a: 85–111 – and it was not really until the later 1990s that I came to an appreciation of Foucauldian analysis that satisfied me).

In working to integrate discourse and practice, I have been greatly influenced by several studies in the history of the book that have focused on the circulation of materials as a means to produce a 'social history of culture' (Hall, 1996: 1; Darnton, 1982, McKenzie, 1999). I have also found three works whose introductions stand as scholarly exemplars for any processual consideration of map history and the nature of maps: simply replace the key term in each with 'mapping' or 'cartography' to produce compelling arguments for properly poststructural map studies: Tagg (1988: 1–33) exploded the modern belief in photography as a 'natural' system of representation; Warner (1990: 1–33) demonstrated that the technologies of printing are culturally constituted and do not determine modern culture (as McLuhan, 1962 had influentially claimed); Sigel (2002: 1–14) explored the legal and private processes by which pornography has been constituted, which is to say how meanings for representations are discursively created; Sigel also outlined effective strategies for analysing the consumption of texts in poorly documented circumstances.

By tracing the circulation of maps through private and official channels and through the market place, we can delineate precise spatial discourses through which particular communities of producers and consumers constitute themselves

(Edney, 2008). More important, we can see how those maps were consumed in conjunction with other kinds of texts (constituted from the written word, images, physical monuments, rituals etc.) so that 'the map' loses its unfounded and entirely unwarranted privilege as *the* means to represent spatial relations (consider, in this respect, Meece, 2006). These precise spatial discourses are constituents of larger discourses that are grounded both in general practices of map production and consumption (i.e. modes) and in the contingent coincidences that permit specific practices to be transferred between discourses. Such analyses of the precise processes of production, circulation and consumption permit us to say something meaningful about the development, spread and merging of conventions and practices. They permit us to consider fully the role of the consumer in defining a map's meaning without having to get all bent out of shape about degrading agency of the map's producer, even as they permit room for personal idiosyncrasy.

19.3 Perspective

These issues remain woefully unappreciated in map studies, whether historical or contemporary. The critique (Petchenik, 1977; Guelke, 1976, reproduced as Chapter 10 of this volume) and subsequent abandonment of communication models did not prevent map scholars from continuing to privilege the map maker as the primary determinant of a map's meaning. Only recently have some sought to come to terms with the manner in which map users imbue maps with meaning and the implications for understanding mapping processes (Del Casino and Hanna, 2006; Kitchin and Dodge, 2007; Edney, 2009). There remains little understanding of the regulated networks of representation (i.e. discourses) within which map makers and users function together. Admittedly it does take time and a certain creativity to study map circulation and map use in an effective manner, but it is feasible. As a result, map scholars still rely instinctively on concepts underpinned by the meta-narrative of cartographic progress.

Some map scholars have thought through some of the implications of 'Cartography without "progress"' (Monmonier and Puhl, 2000; Crampton, 2003: 33–36) and I am especially heartened that my understanding of modes forms the starting point of Dodge, Kitchin and Perkins' (2009) manifesto for future map studies. This is because effective map studies must be founded on the critical investigation of the complex nature of maps and cartography. All too often, map scholars resort to idealized verities. Maps continue to be portrayed as individual interpretations based on idiosyncratic values and concerns. 'Cartography' is still depicted as a coherent and singular practice, especially within the modern West. Indeed, most scholars insist on referring to 'the map' as a meaningful concept, as if one unqualified concept can embrace the multiplicity of textual forms – performative and material, ephemeral or durable – through which humans have sought to organize and simplify the world's spatial complexity in order to comprehend and manipulate it. This seems to be increasingly evident with the burgeoning creativity in online mapping, the emergence of maps as promiscuous digital artefacts, and the apparent collapse of distinctions between modes (Dodge, Kitchin and Perkins, 2009: 221). The now ubiquitous filmic convention of zooming in on the earth from some distant point in space all the way down to a street of field, and even further down to cellular levels, replicates the high modernity of *Powers of Ten* (Morrison *et al.*, 1982) to celebrate the apparently scale-less character of spatial data and so re-establish the modern cartographic ideal.

I have come to realize that the socio-cultural study of maps as texts is ultimately inadequate, despite the amount of excellent scholarship that has generated true insights into certain cultures and societies and their spatial concerns (Edney, 2007b). There is a need for a new agenda, to explore the practices by which maps are produced, circulated and consumed, and the spatial discourses that require and promote those practices in conjunction with other representational strategies. This is no small task and defies easy methodological statement. We might trace how map artefacts physically circulate to reveal precise spatial discourses; within those discourses, we can explore how maps are produced and consumed in conjunction with other texts, and we can model the social and spatial variability in the discourse's participants. We can group discourses together to understand common and differential cartographic practices, and so trace the intersections between modes. Rather than wall off maps and mapping behind professional and disciplinary barriers and privilege 'the map' as *the* means to represent spatial relationships, we must explore the ways in which maps are produced and consumed as one element of human spatial practice. It is an approach that seeks to build understanding from the ground up, within a broad and flexible conceptual framework, rather than impose a priori presumptions and meta-narratives. There is thus room for major reinterpretations of map history (for example, what *was* the relationship of regional and world mapping in eighteenth century Europe?) and for precise studies of individual conventions (such as of the changing signification of the compass rose as deployed over time and across modes). This is where I see my future work heading. I very much hope that many other people will be joining me.

Further Reading

Del Casino, V.J. and Hanna, S.P. (2006) Beyond the 'binaries': A methodological intervention for interrogating maps as representational practices. *ACME: An International E-Journal for Critical Geographies*, **4**(1), 34–56 http://www.acme-journal.org/vol4/VDCSPH.pdf. (Explodes the modern cartographic ideal by showing, *inter alia*, that once we accept that the meaning of a map is re-established each time it is read, it becomes impossible to draw a hard and fast line between map production and consumption, author and reader, and, most importantly, representation and practice.)

Edney, M.H. (2007) Mapping parts of the world, in *Maps: Finding Our Place in the World* (eds. J.R. Akerman and R.W. Karrow), University of Chicago Press, Chicago, 117–157 (A worked example, with many details, of the primary modes of cartographic activity, as they have been pursued across multiple societies and several millennia, to provide an effective, concise history of cartography.)

Edney, M.H. (2009) The irony of imperial mapping, in *The Imperial Map: Cartography and the Mastery of Empire* (ed. J.R. Akerman), University of Chicago Press, Chicago, IL, 11–45. (An examination of the discursive construction of both 'empire' and 'cartography' in modern Western culture, which begins by arguing that mapping practices vary more between modes than they do between political contexts.)

Kitchin, R. and Dodge, M. (2007) Rethinking Maps. *Progress in Human Geography*, **31**(3), 331–344. (An effective argument for the need to consider the role of the reader/consumer in creating map meaning, and therefore for a discursive approach to cartographic analysis.)

Rundstrom, R.A. (1991) Mapping, postmodernism, indigenous people and the changing direction of North American cartography. *Cartographica*, **28**(2), 1–12. (The crucial essay for understanding the variety of strategies for creating map texts, whether inscriptive ('written') or incorporative ('performed'), and advocating a processual approach to studying cartography.)

References

Andrews, J.H. (2001) Introduction: Meaning, knowledge, and power in the map philosophy of J B Harley, in *The New Nature of Maps: Essays in the History of Cartography* (ed. P. Laxton), The Johns Hopkins University Press, Baltimore, MD, pp. 1–32.

Axelsen, B. and Jones, M. (1987) Are all maps mental maps? *GeoJournal*, **14**, 447–464.

Belyea, B. (1992) Images of power: Derrida/Foucault/Harley. *Cartographica*, **29**(2), 1–9.

Black, J. (1997) *Maps and Politics*, Reaktion Books, London.

Blakemore, M.J. and Harley, J.B. (1980) *Concepts in the History of Cartography: A Review and Perspective*, Cartographica Monograph 26, University of Toronto Press, Toronto.

Crampton, J.W. (2003) *The Political Mapping of Cyberspace*, University of Chicago Press, Chicago, IL.

Darnton, R. (1982) What is the history of books? *Daedalus: Proceedings of the American Academy of Arts and Sciences*, **111**(3), 65–83.

Del Casino, V.J. and Hanna, S.P. (2006) Beyond the 'binaries': A methodological intervention for interrogating maps as representational practices. *ACME: An International E-Journal for Critical Geographies*, **4**(1), 34–56, http://www.acme-journal.org/vol4/VDCSPH.pdf.

Dodge, M., Kitchin, R. and Perkins, C. (2009) Mapping modes, methods and moments: A manifesto for map studies, in *Rethinking Maps: New Frontiers in Cartographic Theory* (eds. M. Dodge, R. Kitchin and C. Perkins), Routledge, London, pp. 220–243.

Duncan, J.S. and Ley, D. (1993) Introduction, in *Place, Culture, Representation* (eds. J.S. Duncan and D. Ley), Routledge, London, pp. 1–21.

Edney, M.H. (1986) Politics, science, and government mapping policy in the United States, 1800–1925. *The American Cartographer*, **13**, 295–306.

Edney, M.H. (1990) *Mapping and Empire: British Trigonometrical Surveys in India and the European Concept of Systematic Survey, 1799–1843*. Unpublished PhD Thesis, Department of Geography, University of Wisconsin–Madison.

Edney, M.H. (1994) Mathematical cosmography and the social ideology of British cartography, 1780–1820. *Imago Mundi*, **46**, 101–116.

Edney, M.H. (1997) *Mapping an Empire: The Geographical Construction of British India, 1765–1843*, University of Chicago Press, Chicago.

Edney, M.H. (2005a) *The Origins and Development of J.B. Harley's Cartographic Theories*, Cartographica Monograph 54; *Cartographica*, **40**(1/2), University of Toronto Press, Toronto.

Edney, M.H. (2005b) Putting 'cartography' into the history of cartography: Arthur H. Robinson, David Woodward, and the creation of a discipline. *Cartographic Perspectives*, **51**, pp. 14–29 (Reprinted with corrections in eds. S. Engel-Di Mauro and H. Bauder, eds (2008), *A Reader in Critical Geographies*, Praxis ePress, www.praxis-epress.org, pp. 711–728).

Edney, M.H. (2007a) Mapping parts of the world, in *Maps: Finding Our Place in the World* (eds. J.R. Akerman and R.W. Karrow), University of Chicago Press, Chicago, pp. 117–157.

Edney, M.H. (2007b) Recent trends in the history of cartography: A selective, annotated bibliography to the English-language literature [Version 2.1]. *Coordinates: Online Journal of the Map and Geography Round Table, American Library Association, Series B*, **6**, http://purl.oclc.org/coordinates/b6.pdf.

Edney, M.H. (2008) John Mitchell's map of North America (1755): A study of the use and publication of official maps in eighteenth-century Britain. *Imago Mundi*, **60**(1), 63–85.

Edney, M.H. (2009) The irony of imperial mapping, in *The Imperial Map: Cartography and the Mastery of Empire* (ed. J.R. Akerman), University of Chicago Press, Chicago, pp. 11–45.

Edney, M.H. (2011a) Knowledge and cartography in the early Atlantic, in *Oxford Handbook on the Atlantic World, c. 1450–1820* (eds N. Canny and P.D. Morgan), Oxford University Press, Oxford.

Edney, M.H. (2011b) Field/Map: An historiographic review and reconsideration, in *Ways of Knowing the Field: Studies in the History and Sociology of Scientific Field Work and Expeditions* (eds. C.J. Ries, K.H. Nielsen and M. Harbsmeier), Aarhus University Press, Aarhus.

Edwards, J. (2006) *Writing, Geometry and Space in Seventeenth-Century England and America: Circles in the Sand*, Routledge, London.

Forbes, E.G. (1980) Mathematical cosmography, in *The Ferment of Knowledge: Studies in the Historiography of Eighteenth-Century Science* (eds. G.S. Rousseau and R. Porter), Cambridge University Press, Cambridge, pp. 417–448.

Guelke, L. (1976) Cartographic communication and geographic understanding. *The Canadian Cartographer*, **13**(2), 107–122 (Reproduced as Chapter 10 of this volume).

Hall, D.D. (1996) *Cultures of Print: Essays in the History of the Book*, University of Massachusetts Press, Amherst, MA.

Harley, J.B. (1983) Meaning and ambiguity in Tudor cartography, in *English Map-Making, 1500–1650: Historical Essays* (ed. S. Tyacke), The British Library, London, pp. 22–45.

Harley, J.B. (1985) The Iconology of early maps, in *Imago et Mensura Mundi: Atti del IX Congresso Internazionale di Storia della Cartografia*, vol. 1 (ed. C.C. Marzoli), Istituto della Enciclopedia Italiana, Rome, pp. 29–38.

Harley, J.B. (1988) Maps, knowledge, and power, in *The Iconography of Landscape: Essays on the Symbolic Representation, Design and Use of Past Environments* (eds. D. Cosgrove and S. Daniels), Cambridge University Press, Cambridge, pp. 277–312.

Harley, J.B. (1989) Deconstructing the map. *Cartographica*, **26**(2), 1–20 (Reproduced as Chapter 16 of this volume.)

Harley, J.B. (1990) Cartography, ethics and social theory. *Cartographica*, **27**(2), 1–23.

Harley, J.B. and Woodward, D. (1987) *Cartography in Prehistoric, Ancient, and Medieval Europe and the Mediterranean*, vol. 1, The History of Cartography, University of Chicago Press, Chicago.

Harvey, D. (1989) *The Condition of Postmodernity: An Enquiry into the Origins of Cultural Change*, Blackwell, Oxford.

Hodgkiss, A.G. (1981) *Understanding Maps: A Systematic Enquiry of their Use and Development*, Dawson, Folkestone, UK.

Horkheimer, M. and Adorno, T.W. (1972) *The Dialectic of Enlightenment* (Trans. J. Cumming), Herder and Herder, New York.

Kitchin, R. and Dodge, M. (2007) Rethinking maps. *Progress in Human Geography*, **31**(3), 331–344.

Latour, B. (2005) *Reassembling the Social: An Introduction to Actor-Network-Theory*, Oxford University Press, Oxford.

McKenzie, D.F. (1999) *Bibliography and the Sociology of Texts*, 2nd edn, Cambridge University Press, Cambridge.

McLuhan, M. (1962) *The Gutenberg Galaxy*, University of Toronto Press, Toronto.

Meece, S. (2006) A bird's eye view – of a Leopard's Spots: The Çatalhöyük 'map' and the development of cartographic representation in prehistory. *Anatolian Studies*, **56**, 1–16.

Monmonier, M. (1991) *How to Lie with Maps*, University of Chicago Press, Chicago.

Monmonier, M. and Puhl, E. (2000) The way cartography was: A snapshot of mapping and map use in 1900. *Historical Geography*, **28**, 157–178.

Morrison, P., Morrison, P.and the Office of Charles and Ray Eames (1982) *Powers of Ten: About the Relative Size of Things in the Universe*, W.H. Freeman, San Francisco.

Mukerji, C. (2002) Cartography, entrepreneurialism, and power in the reign of Louis XIV: The case of the Canal du Midi, in *Merchants & Marvels: Commerce, Science, and Art in Early Modern Europe* (eds. P.H. Smith and P. Findlen), Routledge, New York, pp. 248–276.

Petchenik, B.B. (1977) Cognition in cartography, in *The Nature of Cartographic Communication*, Cartographica Monograph 19 (ed. L. Guelke), B.V. Gutsell, Toronto, pp. 117–128.

Richards, P.L. (1990) *Perception and Cartographic Depiction in the Eighteenth Century: Pennsylvania's Northern Frontier, 1750–1782*. Unpublished MA Thesis, Department of Geography, University of Wisconsin–Madison.

Roszak, T. (1972) *Where the Wasteland Ends: Politics and Transcendence in Postindustrial Society*, Doubleday, Garden City, NY.

Rundstrom, R.A. (1991) Mapping, postmodernism, indigenous people and the changing direction of North American cartography. *Cartographica*, **28**(2), 1–12.

Sigel, L.Z. (2002) *Governing Pleasures: Pornography and Social Change in England, 1815–1914*, Rutgers University Press, New Brunswick, NJ.

Tagg, J. (1988) *The Burden of Representation: Essays on Photographies and Histories*, University of Minnesota Press, Minneapolis, MN.

Thongchai, W. (1988) *Siam Mapped: A History of the Geo-Body of Siam*. Unpublished PhD Thesis, University of Sydney.

Tuan, Y.-F. (1977) *Space and Place: The Perspective of Experience*, University of Minnesota Press, Minneapolis, MN.

Warner, M. (1990) *The Letters of the Republic: Publication and the Public Sphere in Eighteenth-Century America*, Harvard University Press, Cambridge, MA.

Wood, D. (1993) The fine line between mapping and mapmaking. *Cartographica*, **30**(4), 50–60.

Wood, D. (1993b) Maps and mapmaking. *Cartographica*, **30**(1), 1–9.

Wood, D. and Fels, J. (1986) Designs on signs: Myth and meaning in maps. *Cartographica*, **23**(3), 54–103 (Reproduced as Chapter 14 of this volume.)

Woodward, D. and Malcolm Lewis, G. (1998) *Cartography in the Traditional African, American, Arctic, Australian, and Pacific Societies*, vol. 2, Book 3, The History of Cartography, University of Chicago Press, Chicago.

Wright, J.K. (1942) Map makers are human. *The Geographical Review*, **32**, 527–544.

20

Between Demythologizing and Deconstructing the Map: Shawnadithit's New-found-land and the Alienation of Canada

Matthew Sparke[1]

Abstract

Understood as the product of an agent of knowledge, the cartographic work of Shawnadithit (the last known Beothuk survivor in Newfoundland) questions a whole set of essential and Eurocentric notions of identity, space and history. Geographically, it displaces the 'new'-ness and emptiness of the Europeans' Newfoundland. Historically, it disrupts a traditional treatment of native people as at once sacrificial victims and heroic proxies in Canadian national history. And epistemologically, it serves to put into question some of the dominant modes of classifying 'indigenous cartography' within cartographic history. In order for such disruptive effects to be realized, it is necessary to shuttle between demythologizing and deconstructing the map – two modes of analysis that need to be better distinguished in scholarship that addresses maps as technologies of power/knowledge.

[Colonial boundaries drawn on maps] provide perhaps the most spectacular illustrations of how an anticipatory geography served to frame colonial territories in the minds of statesmen and territorial speculators back in Europe. Maps were the first step in the appropriation of territory. Such visualizations from a distance became critical in choreographing the colonial expansion of early modern Europe.

[1] Originally published: 1995, *Cartographica*, **32**(1), 1–218.
At the time of publication: Sparke was Assistant Professor in Geography and International Studies at the University of Washington.

[However the map as] an instrument of colonial power could be re-appropriated by colonized people.

Brian Harley (1992: 532, 528)

[T]he look of surveillance returns as the displacing gaze of the disciplined, [a process in which] the observer becomes the observed and the partial representation rearticulates the whole notion of identity and alienates it from essence.

Homi Bhabha (1984: 129)

20.1 Introduction

Between the sweeping spatial arrogance of imperialism and the localized struggles of native resistance, the map was and remains a highly ambivalent part and product of colonial encounters. The above epigraphs – the first from one of Brian Harley's last articles, and the second from Homi Bhabha's description of colonial subversion – capture some of the politics of this fraught ambivalence structuring colonial cartography. To be sure, as Harley argued himself in many other essays, cartographic craft seems to have been far more effective in the service of imagining and implementing imperial rule. 'As much as guns and warships', he underlined (1988a: 282),

> maps have been the weapons of imperialism.... Surveyors marched alongside soldiers, initially mapping for reconnaissance, then for general information, and eventually as a tool of pacification, civilization, and exploitation in the defined colonies.

The ambivalence, then, seems to have been violently skewed against the interests of those colonized by Europeans. Yet reading European maps against the grain – and paying attention to different, sometimes autonomous, traditions of non-European cartography – has also led scholars such as Harley to an increasing awareness of how people living in the lands mapped as Empire could and can 'map back'. Whether they 'reappropriated' and, in Bhabha's terminology, 'mimicked' European cartographic techniques, or whether, as was often the more complicated case, they adapted already-existing native genres, there can be no doubt that those negotiating with imperial rule sometimes used maps to present to the colonialists a pre-European understanding of the land. In so doing, they demonstrated indigenous cartographic skills, and also reaffirmed – in a way that remains vital for contemporary struggles to decolonize – the land's deep inscription through millennial pre-contact historical geographies.

   In this paper, I discuss such questions surrounding colonial cartography in relation to the mapping of Newfoundland. The island's name itself bears witness to the erasure of the pre-colonial native presence. Whether it was as Newe Found Lande, Terra Nova,

Tierra Nueva or Terre-Neufsve, the naming of the place on sixteenth century English, Portuguese, Spanish and French charts announced the common theme of novelty and discovery.

The way in which this toponymy dissembled the fact that the land was already well known by other people – in the words of Gayatri Chakravorty Spivak, the toponymy's 'successful cognitive failure'[2] – represented a form of interested forgetfulness that went on, as Walter Mignolo (1993) has indicated, to be generalized across the whole continent, through European cartography. Names and images on maps were instrumental to the generalization of the New World as 'New'. Similarly, they played a part in disseminating the Colombian cartographic error that led to the misnaming of the native inhabitants as 'Indian'. In addition, the 'new-ness' invented by European cartography had a far more hegemonic effect insofar as it served, in Harley's (1992a: 531) words, 'to dispossess the [so-called] Indians by engulfing them with blank spaces'. In this sense at least, the 'New', as it is commemorated in Newfoundland's name, can also serve to remind us that the island, as it began to emerge on the horizon of the European geographical imagination, was something radically discontinuous from the place understood and experienced by those who had lived there previously.

Right from the European arrival, cartography was central to the wider processes of colonization. When John Cabot claimed that he had mapped the place for England, the Spanish Ambassador complained to Ferdinand and Isabella that this was mere cartographic robbery of an island already on Spanish charts. 'I have it here', he said, examining Cabot's map, 'and to me it seems very false, to give out that they are not the same islands'.[3]

Whatever the politicians and diplomats claimed, it was the terms of their discussion that are most interesting today.[4] From its outline on the famously detailed Desceliers map of 1550, and onwards – as Fabian O'Dea (1971) has documented, into and through the seventeenth century – the emerging European depiction of Newfoundland was as a basically empty space, a space seen from sea, delineated by coastline, and as a people-less void within. Even coves and bays were not fully explored and, as O'Dea (1971: v) notes in studiously Eurocentric terms, it was not until the 1760s that Captain James Cook came 'to establish a scientific basis for an understanding of the true shape of Newfoundland'. Whatever the merits of Cook's theodolitic and telescopic quadrant, this 'true shape' was even then largely a matter of outline (Figure 20.1). Joseph Banks, who was travelling with the famed surveyor, noted that

---

[2] Spivak (1988: 199) connects this critically loaded formula most immediately with what she calls the 'sanctioned ignorance' of colonial and neo-colonial knowledge about India.

[3] Quoted in Howley (1915). This book has since been republished, and it is to the later, 1974, Coles Publishing Company, Toronto edition to which all subsequent citations refer.

[4] For a learned discussion of the evolution of such terms in renaissance cosmography, a discussion which directly examines André Thevet's 'fictions of new-found lands', see Lestringant, 1994, especially 116–121.

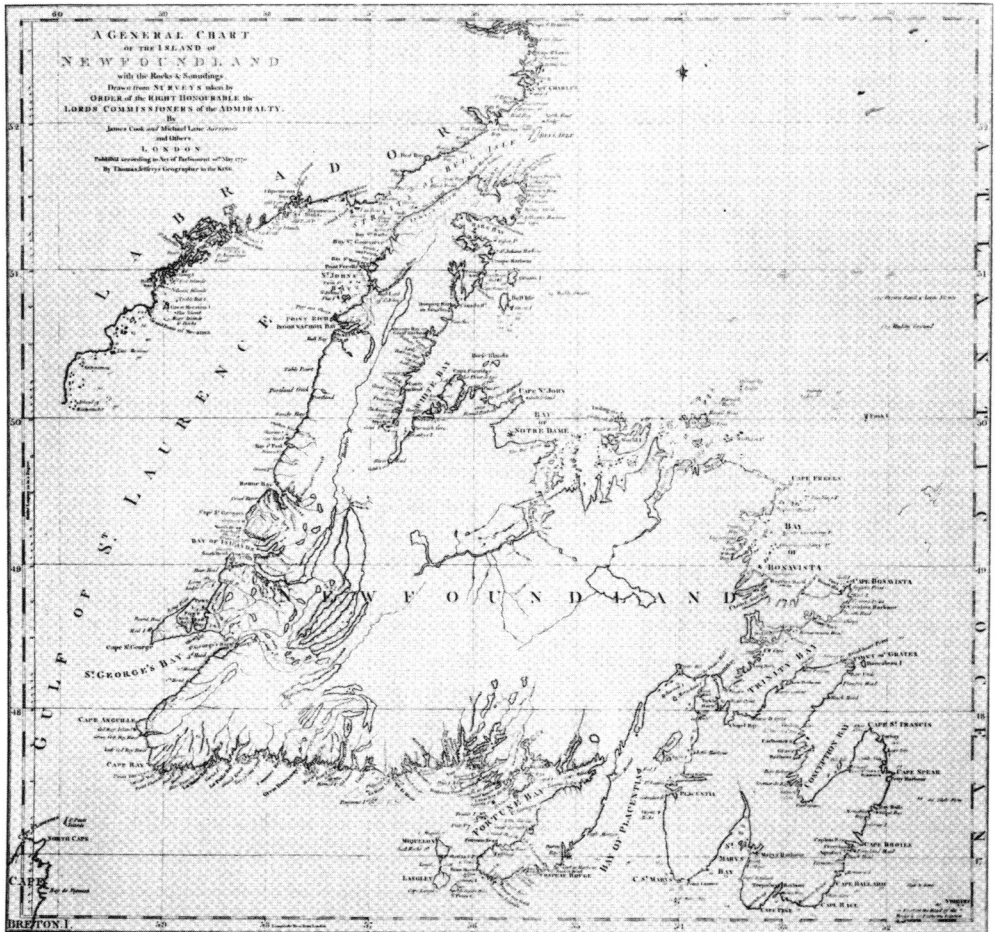

Figure 20.1   Cook's general chart of the island of Newfoundland.

'we know nothing at all of the Interior Parts of the Island' (quoted Pastore, 1989: 55). The ongoing evolution of the new 'true shape' was thus slow and uneven, and, for the same reason, is worthy of study in and of itself.[5] However, in the following pages, I seek to highlight the culturally-specific quality of such original European 'truths' by examining another sort of map drawn from within a very different and aboriginal

---

[5] For a study aimed to do precisely this *vis-à-vis* the emergence of Australia – also a significant product of Cook's explorations – within the non-aboriginal geographical imagination, see Carter (1988). In reading 'new'-ness as a sign of Newfoundland's construction within a new geographical imagination, I follow Carter on Cook's toponymy. 'Implicit in Cook's names', he states (1988: 331), 'is the irony implicit in geography itself – that it is a travelling discourse, a historical discipline, which at the same time, aspires to the transcendent placelessness and timelessness of the map.'

geographical imagination; a map in which, to use Bhabha's more lyrical terms, 'the look of surveillance returns as the displacing gaze of the disciplined'.

The contestatory cartographic gaze that I want to highlight is that of Shawnadithit, the native woman described by the monument in St John's Harbour, Newfoundland, as 'very probably the last of the beothics who died on June 6th, 1829' (a photograph of the monument is included in Such, 1978: 86). The Beothuk[6] were the native population of the island, who painted their skin with red ochre, and whom the Europeans labelled 'red Indians'. That this 'red-ness', like the 'new-ness', became indiscriminately extended to describe the native population of the whole continent, points to the spatial reach of European misunderstanding and arrogance. That the Beothuk, a whole human population and culture, were all dead by 1829, points in turn to the violence through which such misunderstanding and arrogance became history. Shawnadithit's death from tuberculosis, while in custody, is significant as the historical marker of this oppressed end of a people. In addition, it is of particular note because it has been transformed into what revisionist history characterizes as the last part of a gruesome 'tragedy' in the pageant of Canadian national development. As such, Shawnadithit, whose grave was dug up to make way for a new road, has also been dug up and moved cursively. She has been transported into the pantheon of Canada's famous figures, a pantheon that otherwise includes explorers, fur traders, politicians and railwaymen whose more general discursive duty in death remains as heroes and heroines in the romanticization of the very processes that caused the genocidal demise of native people such as the Beothuk in the first place.

One of the specific side effects of this exercise has been to have turned the few statements and drawings Shawnadithit made for her custodians in St John's into anthropologically nationalized relics.[7] As I will explain more thoroughly in later sections, this process is especially significant for historians of cartography because it affects the way in which the story-maps Shawnadithit drew are interpreted. As relics of a national tragedy, they tend to be entombed in books and articles as mawkish mementos, the last gasps of a native informant whose people have gone forever. I want to question this process of entombment, problematizing the convention by examining just one of her maps in order to draw attention to its agency, as opposed to its tragedy. My aim is to point to the cartography as the product of an *agent* of knowledge. It documents the view of a native woman who returned the surveiller's gaze – who, by observing the observers,

---

[6] William Kirwin (1992) describes three historical phases in the English rendering of the name 'Beothuk': the first (1828–1885) as 'Beothick'; the second (1908–1952) as 'Beothuck'; and the third as 'Beothic': see 'A note on Beothuk names in Newfoundland', *Onomastica Canadiana* (1992), **74**(1), 39–45. The fact that 'Beothic' was used in 1992 in the naming of the research vessel *Beothic Endeavour* indicates the radical, and, as in this case, ironic, rupture severing the word from any possible pre-European meaning. Moreover, the fact that contemporary spellings like mine necessarily take part in this history of European transformation – I use 'Beothuk' following Kirwin and the *Historical Atlas of Canada* – can further be read as a sign of the broader rupture separating our representations from Beothuk life 'as it was'. I should also note that, following the more general contemporary English usage, I employ the equally discontinuous spelling 'Shawnadithit' as opposed to the local 'Shananditti' that has evolved in Newfoundland as a consequence of Anglo-Irish pronunciation.

[7] Her maps and other drawings were first published as part of the mini-archive produced as a book by the geological cartographer of Newfoundland, James P. Howley (1915).

can be understood in Bhabha's abstract terms as having rearticulated the whole notion of identity, alienating it from essence. In short, this paper asks a cartographic question: what happens to essentialized notions of national identity, to natives and to Newfoundland, if we take a Beothuk woman's maps seriously as the work of an agent of knowledge?

One of the essentialized notions of identity displaced by Shawnadithit's gaze is that of Canada as a mappable national whole. I argue that the critical dynamic exemplified by her surveying of the surveyors also displaces the whole historical-cartographic tradition of identifying maps – especially native American maps – as artefacts to be collected and codified. Against such antiquarian interests, I suggest that her work underlines the need to come to terms with what Robert Rundstrom (1991: 2) has described as 'the crucial importance of *process* in cartography'. Rundstrom's own writing, as well as that of a host of other cartographic critics, clearly provides the background to this argument. For that reason, the next section is intended to clarify the paper's theoretical charge by situating it within the broader scholarly debate over the social and cultural contexts of cartography.

20.2 The Palimpsest in Process

We must understand the map ... as simultaneously constituting a stock of information for a collective memory and instituting a signaling tool for scrambling previous territorializations.

José Rabasa (1985: 3)

The growing academic attention to the social and cultural contexts of cartography provides us with some useful ways of addressing questions surrounding maps, colonialism and the possibilities for de-colonization. Harley's so-called 'deconstruction' (1992b) of the 'hidden agendas' of cartography (1988b) has, of course, been critical amongst these scholarly projects. As a recent debate in this journal served to underline (see 'Commentary', *Cartographica* 1989, 26(3/4): 89–121), his examinations of the social and cultural power relations concealed behind the touted neutrality of the cartographic enterprise have inspired a considerable range of new work now being conducted by anthropologists (Orlove, 1991, 1993), geographers (Pickles, 1995; Katz and Smith, 1993; Gregory, 1994), literary critics (Helgerson, 1992 Chapter 2, 1993) and other scholars of cartography (Wood and Fels, 1992; Edney, 1993; Akerman, 1993). With this increasing volume of scholarly interest has also come further theoretical nuance.

As Barbara Belyea (1992a) has pointed out, Harley's own presentation of his investigations as 'deconstruction' was somewhat misleading. Certainly, he brought to bear a critical form of historicization and a Foucauldian sensibility toward the power relations underpinning, and expressed in, maps. It is, after all, the appeal of this approach that has enabled so many other projects – including, at one level, this present one – to begin the critical work of putting cartography in its cultural and social context. But such analysis is probably better thought of as *demythologization*. It takes as its central target the *myth* of cartographic objectivity, critiquing the notion of the map as a transparent window on the world by revealing the web of power dynamics that under-gird it, that make it an interpretative as opposed to a reflective document, and that thus, finally, enable it to show one thing while concealing another.[8] The repeated gesture of such demythologizing, then, is one of critique through revelation. The critic points to the cover-up effected by the map's paper-thin authority, and, as Harley (1988: 69) sought to do in his demythologization of early modern European maps of North America, reveals the hidden truth: '[these maps] remain silent about the *true* America', he complained.

Following Derrida down the path cleared by Belyea leads in a rather different direction, or – to put the case another way – it at least enables us to demythologize while being more wary, and less romantic, about uncovering 'true' places and constructing 'true' maps. This form of deconstruction is, therefore, not the gospel of reckless relativism it is often said to be. To be as clear as possible, I would argue that, in contrast to demythologization, a deconstructive approach to maps is better understood as a responsibility towards the context of cartography.[9] In a sense, this is precisely what cartographic critics as different as Rundstrom, Denis Wood and Nicholas Chrisman have already suggested, when – despite their differences over 'postmodernism' – they have highlighted the need to come to terms with cartography as a dynamic component of wider social dynamics.[10] Harley, too, made similar arguments.[11] However, this same reflexive attention to process is also given a more philosophical and, to this extent, critically open-ended twist in the very arguments of Derrida's early work, which Harley, like many others, dismissed as disconnected textualism. When he protested, 'I do not accept Derrida's view that nothing lies outside the text', he was misinterpreting arguments that, as Derek Gregory (1994) has pointed out, can be better read as philosophical guidelines for exactly the social history of cartography that Harley himself was demanding.

Gayatri Chakravorty Spivak, the translator, makes clear that the line 'there is nothing outside the text' (*'il n'y a pas de hors-texte'*) can also be translated as, 'there is no outside text' (Derrida, 1976: 158). In this vein, I read Derrida as arguing that any reference to a so-called 'real' life behind the work of a writer like Rousseau would itself have to be predicated on an interpretation of that life – a reading that, in turn, would inevitably

---

[8] Following Roland Barthes (1972), we might say that the role of the myth of cartographic objectivity is not to deny things, but rather 'to talk about them; simply, it purifies them, it makes them innocent, it gives them a natural and eternal justification, it gives them a clarity which is not that of an explanation but that of a statement of fact'. It is this claim to facticity and eternal justification that the majority of demythologizations attack.

[9] I have elaborated on the connections between deconstruction and a critical ethic of persistent responsibility elsewhere (Sparke, 1994).

[10] On process and social responsibility, see Rundstrom (1991); on process and the ethics of mandates and custodians, see Chrisman (1987); and, on process and a critique of what he calls 'the alibi' (pages 18 and 20) intended to absolve cartography of its responsibility to context, see Wood (1992) especially Chapter 2, 'Maps are Embedded in a History They Help Construct'.

[11] 'Any self-respecting history', he said, 'must systematically embrace the structures or contexts within which individuals acted to produce their maps.' (Harley, 1990).

exclude aspects of that life, and, as a result, be open to the same charges levelled at a narrow reading of his books – namely, that it was 'leaving something out'. (For further discussion of the interiority/exteriority question see Nealon, 1992). In such arguments, there is clearly some of the same scepticism towards full – or value-free – representation that Harley brought to the question of the ways in which maps represent the world. The difference is that while his cartographic critiques focused on what was concealed and how, Derrida's attention is drawn to the problem that there will always be something left out, and the critic must therefore remain alert to any complacency that might follow in the wake of demythologization. Put this way, a deconstructive approach offers less of a critique targeted at flawed representation, and more of a reflexive ethic and critical attitude towards the possibility of full representation in general. Rather than exposing errors and myths, Spivak (1989: 214) argues, 'deconstruction notices how we produce truths'. As such, I suggest, it is best seen not as a stand-alone methodology, but rather as a disruptive supplement to the project of demythologization. It prevents the cartographic demythologizers from treating the problem as solved, the power relations as revealed, and the map as finally fully understood. Instead, it urges us to go back and look for other ways in which the map, and what is supposed to lie outside of it – power relations, interpretation, the 'real' world etc. – might actually be still more complexly interrelated.

I am suggesting that a project of shuttling between demythologization and deconstruction can add to, rather than take away from, the sorts of cartographic critique that Harley and others have done so much to elaborate. It moves the debate beyond a narrowly representationalist concern with what maps conceal and what individual map makers 'intend', and opens up instead a series of questions about the exclusions that make cartography appear coherent. This amounts to a concern with the process of cartography; not only with the way maps are palimpsests, layered over-writings that represent an always already-inscribed world in particular ways, but also with how such representation, having been given authoritative clout by the authenticating seal of seeming cartographic coherence, can take part in the 'worlding' of the world. Certainly, Harley himself indicated the role of cartography in the shaping of worldly affairs. However, because of his inclination, in Belyea's (1992a: 2) terms, to maintain a fairly orthodox view of 'maps as graphic representations of the world' – he tended to treat the world so represented as something outside of the cartographic process, as something somehow fixed, and thus as not always previously inscribed or 'cartographed' before. It is particularly towards this pre-inscribed or generalized textual quality of what the cartographer portrays and, from this, towards the whole layered process of cartographic erasure and overwriting, that deconstruction draws our attention.

In the context of colonial cartography, a deconstructive attentiveness to the way what is mapped might already be inscribed has political import. It is not without significance that two of the more canny essays deconstructing cartography have both chosen the scene of colonialism as the place to illustrate the critical ethic's liberating potential. In Graham Huggan's (1989: 117) work, where the coherence of the map is depicted as something riven by 'the discrepancy between its authoritative status and its approximative function', the map becomes legible 'as a palimpsest covering over alternative spatial configurations which, once brought to light, indicate both the plurality of possible perspectives on, and the inadequacy of any single model of, the world' (p. 120).

Prior to this, Rabasa (1985: 1) had already marked out the ensuing political implications:

> *The transposition of the image of the palimpsest becomes an illuminative metaphor for understanding geography as a series of erasures and overwritings that have transformed the world. The imperfect erasures are, in turn, a source of hope for the reconstitution or reinvention of the world from native points of view.*

The question of 'reconstitution' brings me back now to the case of Shawnadithit's work as an inscription of the land from a Beothuk point of view. My approach involves moving between both demythologization and deconstruction as I have outlined them – re-situating the map as a moment of contestation within the colonizing process of cartographic overwriting that has comprised the mapping of Newfoundland. This means that – rather than focusing on how European maps such as Captain Cook's simply erased the native presence through an arrogant optics of empire (a claim that would have to take close account of Cook's sea-bound blindness to the Beothuk), and rather than saying that the 'true' land of the Beothuk was thereby occluded by the empty spaces of modern maps – I highlight, instead, how the place was already far from uninscribed, because Shawnadithit's people also surveyed the land, and she herself continued this vital process with an innovative cartography that retraced Beothuk travelling routes, all the while taking cognisance of the colonisers' presence.

20.3 Something Inexpressibly Forlorn: Artefacts, Artifice and Appropriation

> *The story of the Beothuk Indians of Newfoundland is indescribably tragic – and on several levels. It is tragic, of course, because they were wiped out by white furriers and fishermen and Micmac Indians – hunted down like wild geese during a two-hundred-year cycle of haphazard genocide. But it is also tragic in other, subtle ways. There is, for instance, convincing evidence that the bloodbath began as a result of an accident; that had it not been for a chance encounter, these Indians might easily have survived as partners in the fur trade. There is also the tragedy of good, but failed, intentions. There were, in Newfoundland, perfectly sincere and humane people who tried, after their fashion, to save the Beothuks from extinction; they simply had no sensible idea of how to do it. Finally, there is the tragedy of a culture lost. Our knowledge of the Beothuks is abysmal, because scarcely anybody bothered to find out anything about them until they were all gone. There is something inexpressibly forlorn in the final picture of the lovely Shawnadithit, the last surviving member of her race (her features already sallowed by the ravages of consumption), trying as best she could, through a series of story-maps, to explain how her people lived, hunted and worshipped, and how, one by one, they died.*

Pierre Berton (1976: 165)

So begins the eloquent lament for what he calls 'The Last of the Red Indians' by one of Canada's more popular and populist historians. Pierre Berton's humanistic account of the genocide in Newfoundland – boldly inserted in *My Country* between colourful stories about the cult figure Brother XII of British Columbia and the Catholic-turned-Protestant Charles Chiniquy of Quebec – manages with all the skill of a best-selling national raconteur to piece together a story-book strength narrative from rather patchy evidence and, in driving that tragic narrative home, to repeat all the basic gestures of benevolent, but Eurocentric, appropriation. Just as with the earlier painful accounts of Harold Horwood, Ernest Kelly and Keith Winter, there can be no doubting the sincerity or anguish evident in Berton's (1976: 168) criticisms of what he describes as 'the slaughter' of the Beothuk. Like Horwood, who in a 1959 *Maclean's* 'Flashback', made an angry appeal to Canadians to remember '[t]he people who were murdered for fun', Berton (1976: 168) describes for another generation how fishermen 'set out on shooting parties, as they would for deer or wolves, to bag themselves a few head'. Like Winter (1975: 4), who romanticized the Beothuk as stereotypical noble savages – 'They had a mellifluous language, loved to sing and dance, and made a habit of welcoming all strangers with feasts and friendship' – Berton, too, portrays them in terms that say more about the narcissism of the Euro-Christian imagination: 'They were tall by earlier European standards, handsome, fairer than most Indians, with large expressive eyes . . .; a shy, unwarlike race, who treated their women with respect, believed in the importance of marriage and were said to reject both adultery and polygamy'. Finally, they made for what Berton goes on to describe, like Kelly,[12] as 'perfect victims'. They were, he says, 'ripe for killing' (Berton, 1976: 166). However, as he accomplishes all this, Berton, like the other three authors, seems oblivious to the way in which he himself is, in turn, appropriating and manipulating the story of the Beothuk as 'perfect victims' in the staging of a national tragedy.[13]

Clearly, the general re-examination of the brutality of colonialism in Newfoundland has not been without value. It has sensitized some of the more scholarly anthropological and historical writing about the island to the politics of colonial knowledge (Leslie and Upton, 1977; Raynauld, 1984), and it has also led to some more detailed research – most especially the rigorous work of Ralph Pastore (1989, 1990) – aimed less to dismiss than to complexify the mono-causal, intentionalist accounts of extermination offered by the popular historians. (Even Pastore, though, never elaborates in these pieces on the cultural politics of why the question of extermination is so contentious.) Moreover, it has upset the justifications offered by apologists for empire who would, following Diamond Jenness, the most imperial of Canada's anthropologists,[14] blame the bloodshed on the Beothuk themselves: 'But very soon trouble arose for the Indians stole . . . [and t]he fishermen retaliated by shooting every native that dared to show his face' (Jenness, 1934: 27). Even after Berton's book – indeed, it seems, as a partial response to what are

---

[12] 'Any self-respecting history', he said, 'must systematically embrace the structures or contexts within which individuals acted to produce their maps.' (Harley, 1990)

[13] As Watson (1994: 98) notes in relation to the Group of Seven artists, lamenting the 'dead Indians' who 'inhabit the wilderness as ghosts' is a classic gesture of Canadian nationalism.

[14] For a valuable contemporary critique of Jenness's construction of 'native barbarism', see Kulchyski (1992).

dismissed as its 'intemperate allegations' – Frederick Rowe (1977: 2, 25) like Jenness, repeated this same tired alibi, claiming: 'if there was any chance of a permanent friendly relationship . . . the Beothuks themselves probably destroyed it through their persistent habit of stealing'. Rowe, a senator from Newfoundland, still saw fit in 1977 to defend the island's settlers by rehashing the rhetoric of empire and blaming the Beothuk.[15] A singularized script has emerged for the Beothuk, their experience has become homogenized, and their place in the process of colonization has been diminished to that of either a cute, or a criminal, bit-part in the drama of their own destruction. In the more romanticized dramas of Berton and his three predecessors, the script so written at least makes for a picture of the Beothuk with which the modern Canadian reader can partially sympathize. Yet, it is this very same appeal to the contemporary citizen that holds within it the problematic gesture of misappropriation. Presenting Shawnadithit as an attractive tragic Figure, as '[h]andsome, with beautiful teeth and a swarthy complexion,' (Berton, 1976: 182) the historians simultaneously lock up and reify her cartography as a hallowed artefact within the national pantheon. It is such complex cultural negotiations that cartographic critics need to unpack if they are to address the process of cartographic inscription, rather than stand by as so many antiquarian spectators.

Central to the appeal that Berton's story affords his contemporary Canadians, is the gesture of epistemic colonization that sets up the Beothuk as a special form of what Spivak, glossing Edward Said, has dubbed the West's 'self consolidating Other'. (Spivak uses this phrase repeatedly, 1988b.) As Barbara Godard (1990: 190) argues in relation to the Canadian context, 'it is through this encounter with the Other who is Native to this land, that a 'totem transfer' occurs and the stranger in North America 'goes native' to possess the land, to be Native'. In this role as Other, the Beothuk do indeed make for 'perfect victims'. They are, after all, tall and 'fairer than most Indians', the almost-white-but-not-quite natives who have nuclear family values and who, more fortunately still, are not around any more to be asking troubling questions of the Canadian state about returning their ancient lands. Moreover, the documented horrors of the genocide create such an overpowering picture of violence that they seem almost to give an alibi to other, supposedly more 'civilized', forms of colonialism in the rest of Canada [16] It is as if by repeating the litany of tragedy and by angrily berating Newfoundland's fishermen that the modern historians can somehow find general redemption. In the moment of passionately narrating the tragedy, they can, as Godard put it, go native. At the same time, Berton himself is able to spread some of the blame to the Micmac, suggesting too that the tragedy originated in an accident, and indicating, most significantly, that, had

---

[15] Reinventing the 'Red Indians' in such a way makes it possible for the otherwise tendentious criticisms of a writer like James Clifton to have some descriptive merit. In his polemical introduction to the 'Invented Indian', Clifton (1990: 40) outlines the problem with a story like Berton's when he notes: 'The standard Indian narrative is factious because in the minds of narrators and audiences it divides the whole population into adversarial groups and explains and justifies their opposition to one another.'

[16] This was certainly Howley's prefatory contention in 1915. He describes the 'blotting out' of the Beothuk, concluding with a comparison: 'It is a dark part in the history of British colonization in America, and contrasts very unfavourably with that of the French nation in Canada and the Acadian provinces, where the equally barbarous savages were treated with so much consideration, that they are still met with in no inconsiderable numbers, and in very appreciable condition of civilization and advancement'.

the Beothuk been incorporated into the fur trade, they might easily have survived as partners. In short, great good could possibly have been done.[17]

At the heart, perhaps one might say the bleeding heart, of this whole process remains Shawnadithit herself, and, most especially, the tension between her position as a native informant and what Berton describes as the fourth and final part of the tragedy, 'the tragedy of a culture lost'. There is indeed something inexpressibly forlorn here, but it is not what was undoubtedly Shawnadithit's very real pain. Instead, it is the tragic lengths to which the modern historians will go to try to recapture and retell her story for the modern Canadian audience. Berton (1976: 184) criticizes what he calls the 'unctuous meddling' of the so-called 'Boeothuck Institution' founded in 1827 by settlers who, having failed in their imperial mission of finding and civilizing the 'Red Indians', began instead to try to educate Shawnadithit for the purposes of transforming her into a native informant. However, we must surely ask if there is anything less unctuous about the way Berton (1976: 184) himself describes what he calls the 'one valuable by-product' of the meddling, namely, the 'remarkable series of drawings'. Compare his description of 'the lovely Shawnadithit', 'trying as best she could through a series of story maps, to explain how her people lived, hunted and worshipped . . .', with the account of William Epps Cormack, Shawnadithit's custodian in her last few months in St John's. Exchanging notes with Bishop John of Nova Scotia, Cormack commented:

> *Shawnandithit is now becoming very interesting as she improves in the English language, and gains confidence in people around. I keep her pretty busily employed in drawing historical representations of everything that suggests itself relating to her tribe, which I find is the best and readiest way of gathering information from her. (Quoted in Howley, 1915: 210.)*

In both cases, the appropriative assumptions of what the anthropological critics Marcus and Fischer (1986) call 'salvage culture' are in full operative force. (How far these authors themselves avoid salvaging anthropology as cultural critique on this same basis is open to question.) The notion that Shawnadithit is 'interesting' insofar as she begins to learn English highlights the more general imperial angst played out in nearly all the historical treatments of her people since. It is this angst, one that is ultimately preoccupied more with the lack of ethnological information than with the lack of people, that repeatedly transforms Shawnadithit's maps into artefacts. Their connections to a process of cartographic inscription are thus neglected, and instead they are turned into a quaintly visual source of data for narratives detailing the ill-fated journeys

---

[17] By employing here the vocabulary of 'great good done', I mean to invoke the liberal 'colonize to trade' sentiments of a much earlier description of the Beothuks by one Thomas Rowley, writing to inform Sir Percival Willoughby, a major investor in the London and Bristol company for the plantation of Newfoundland (Gilbert, 1992). On the question of incorporation, see the telling assessment of Upton (1977: 153), who asks: 'Could it be that the Beothucks died because they did not have enough contact with whites? There was no missionary to plead for their souls, no trader anxious to barter for their furs, no soldier to arm and use them as auxiliaries in his wars, no government to restrain the settlers.'

of Captain David Buchan (who was sent out by the Governor to communicate with the Beothuk) and John Peyton (who became Shawnadithit's master). Like the Chipewyan (Dene) woman Thanadelthur, whose position in the nation-building deployment of the historical archive has been critiqued by Julia Emberly (1993: 101), Shawnadithit thus becomes transformed into another Canadian native informant, another slave woman in a national narrative for which she serves simultaneously as 'heroic proxy and sacrificial victim'. It is, I think, the responsibility of the cartographic critic to refuse such nationalizing artifactualization by returning to study the cartography as the work of an agent of knowledge.

20.4 Clever with a Pencil: Shawnadithit's New-found-land

All savage Nations, whose language is necessarily defective, are accustomed to symbols; ingenious in the use of them and quick in ascertaining their meaning …. Any that more particularly belong to the Boeothuk may probably be painted out and explained with Mr. Peyton's help by Shawnawdithit. She may also assist in depicting her own tribe and their dress and habits as she is clever with a pencil – Bishop John of Nova Scotia.[18]

> *Who would have thought death was warm*
> *plump with meat and men who smile too much,*
> *who ask questions with pencils, wanting you to draw the canoes, the tents, the chasms*
> *dug for winter houses. They ask you to speak your language so they can study its*
> *sound.*
> *How full of holes it is, subterranean tunnels*
> *echo around your failing lungs.*
> *Can they hear?*
>
> *Joan Crate (1990: 17).*

Joan Crate's evocative poem, itself – like this paper – imposed on the past, nevertheless asks the critical question, 'Can they hear?' Bishop John's comments, along with the endeavours of the information hunters described above, would suggest a chronic and interested deafness. In what follows, I want to challenge this failing, and to contest in particular the way in which the assumption put forward by the Bishop – of a 'necessarily defective' language – has also, in various ways, been extended to treatments of Shawnadithit's cartography. Introducing here her map depicting the River Exploits and the journeys of the Beothuk and Captain Buchan's party, I will outline four ways in

---

[18] Bishop John of Nova Scotia to William Cormack, quoted in Howley (1915: 207).

which the map's treatment as an artefact has suppressed attention to its lively and contested context. Such suppression, I will argue, effectively silences Shawnadithit's surveyor's voice. I realize, following Spivak (1991: 105), that such an attempt to reconsider a native informant as someone capable of producing definitive descriptions demands a certain '(im)possible perspective'. However, I am also persuaded by Spivak's parallel argument that such attempts must continue, albeit self-critically, in order to counter the tendency of simultaneously incorporating and silencing the marginalized in evolutionary narratives.

These are the same narratives that characterize the very skills of those like Shawnadithit as the final proof of skillessness, savagery and general historical backwardness (Spivak, 1991: 105–106). At one level, therefore, they self-deconstruct. Yet, at the same time, they also dominate the historical record. They prescribe a disciplinary structure with which critics inevitably have to negotiate if they are to disclose some of the processes of cartographic inscription. The oxymoron of the title 'Shawnadithit's new-found-land' is meant to reflect this (im)possible negotiation. On the one side, it refers to the modern island's name, and thereby speaks to the constraints that the historicist narrative of the 'New' puts around any attempt to interpret Shawnadithit as something other than a native informant of Newfoundland's 'Prehistory'. On the other side, it recalls what must have been the terrible newness for the Beothuk, of seeing white men come upriver into the interior. The trails of such colonial search parties probably crisscrossed older Beothuk routes, thereby constituting a newly found spatiality when surveyed by an observer like Shawnadithit. Looking at her map of the river and those who travelled along it in 1820, what new-found-land do we see? (Figure 20.2)

20.4.1 The Context of Drawing the Map: Or, Unpacking the Artefact as Curio

Shawnadithit drew the map[19] (along with at least three others) in the winter of 1829. Afterwards, it was only a matter of months before she died on the 6th of June, her burial being recorded in the register of the Anglican Cathedral. By that time, her presence in St John's, and her work, had attracted a great deal of curiosity, so much so that *The Times* in faraway London posted an obituary on 14 September.

> *Died – At St John's Newfoundland on the 6th of June in the 29th year of her age, Shawnadithit, supposed to be the last of the Red Indians or Beothuks. This interesting female lived six years a captive amongst the English, and when taken notice of latterly exhibited extraordinary mental talents. (Quoted in Howley, 1915:231.)*

---

[19] The originals are in the Newfoundland museum in St John's: reference numbers NF 3304, NF 3308, NF 3307 and NF 3300.

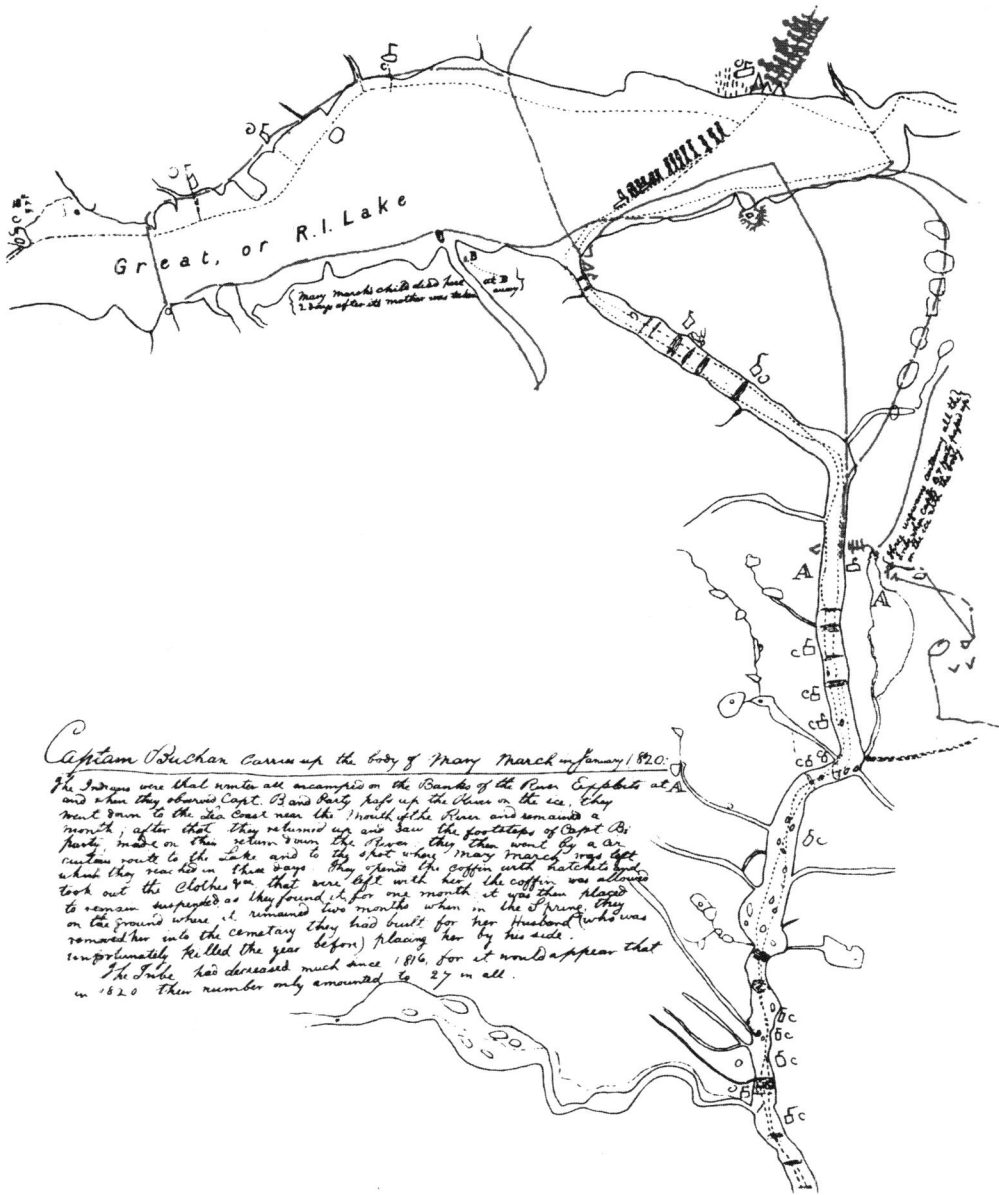

Figure 20.2 Shawnadithit's map with Cormack's annotations. (Courtesy of the Newfoundland Museum.)

The primary way in which her cartography was first received and reconstituted as an artefact was as testimony to such 'extraordinary mental talents' reported by *The Times*. Of course, *The Times* did not explicitly say 'extraordinary, for a savage, a "Red Indian",' that was an understanding carried by the context. It was also the guiding (mis) understanding of the Englishmen in charge in Newfoundland. Captured Beothuk women were often referred to in ethnological terms as 'interesting females',[20] and the Reverend Wilson, much like Bishop John, had already testified to Shawnadithit's 'surprisingly' clever skills with a pencil.

> *She made a few marks on the paper apparently to try the pencil; then in one flourish she drew a deer perfectly, and what is most surprising, she began at the tip of the tail. (Quoted Howley, 1915: 171.)*

Cormack, an anthropologist trained at the University of Edinburgh, whose written annotations cover Shawnadithit's cartography, thought in turn that he was getting an 'interesting' information source when he took charge of her as president and founder of the salvage-orientated 'Boeothuk Institution'. Even contemporary writers still speak of his good fortune with an anthropological gusto: 'Luckily he arranged to have Shawnadithit sent to him in St John's where he questioned her extensively about her people' (Such, 1978: 83). However, what gets lost in all this talk of luck, extraordinary talents, and surprising interest are the interests of Shawnadithit herself. Like the ethnocentric fascination with Australian Aboriginal relics described by Paul Carter (1988: 345), such 'activities divorce the project of study from the context of [its] production, that living space in which places have histories and implements are put to use'. I do not think it is useful, indeed it is just as arrogant, to diagnose Shawnadithit's interests long-distance in the sensationalist style of Winter's *Shananditti*.[21] Nevertheless, it is possible to at least take more careful cognisance of the circumstances, the living spaces, in which she was obliged to put pencil to paper.

Up to the time that Cormack 'sent for her', Shawnadithit had lived almost six years as an unpaid servant in the house of John Peyton, Jr, the magistrate of the fishing community of Twillingate. This was the same house from which John Peyton, Sr, had set out on many expeditions to kill the Beothuk, and also from which Peyton, Jr, had began his own infamously brutal assault in which the apparent Beothuk chief, Nonosabasut, was shot and bayonetted to death while pleading with Peyton not to take his partner, Demasduit, captive (a story mapped out in another of Shawnadithit's maps). Her baby left behind to die as well, Demasduit was nonetheless captured and taken to the house, where she was named Mary March. Shawnadithit's own arrival at this house was no less forced. By that time, Demasduit had died of consumption, her body having been carried back by Captain Buchan to the site of her capture. Shawnadithit, as her map shows (and as I shall discuss in the next section) observed the funeral cortege moving up and down the river in 1820. Three years later, she, her mother and her sister were taken captive, presented as reward material to Peyton, and transported to the courthouse in St John's. A failed attempt to let them go back to the interior only led to the more speedy deaths of

---

[20] The touted care and consideration of Captain Buchan was, for instance, couched in these terms: 'I am much pleased to find that these interesting females are under the care of Mr Peyton.' (Quoted in Howley, 1915: 169.).

[21] Such personalist story-telling seems only to lead to assumptions, and, with it, in Winter's (1975: 100) case, sexist generalization. 'Like women the world over', he avers, 'Shananditti was interested in clothes'.

her mother and sister from consumption. Shawnadithit, alone, returned to Twillingate to work for nothing but her keep for one of the families centrally involved in the final demise of her people.

Five years later, she was taken to St John's a second time to be made the object of further governmental, missionary and philanthropic-turned-anthropological interest. This was the context in which she drew the map. Clearly, these were circumstances very different from those enjoyed by a cartographer like Cook, and, for that matter, different from the conditions experienced by native map makers like the Inuit described by Rundstrom (1990) Shawnadithit's cartography was instead drawn under duress, when she was already ill, her world and family destroyed. Considering this context at least makes it possible to move beyond treating the map as a curio. The situation was alive with colonial power relations. Urging her on was Cormack, eager to salvage information; reporting how through his 'persevering attention' and constant tending of 'paper and pencils of various colours', Shawnadithit 'was enabled to communicate what would otherwise have been lost' (quoted by Hewson, 1978: 7). Far from enabled, however, Shawnadithit may very well have felt disabled by such circumstances, surrounded by people who, while soliciting and gathering information, treated her language as 'gibberish' (quoted by Hewson, 1978: 7). From this perspective, other topics than simply camping sites, numbers of people and colonial expeditions can be read in her map. There is a record of pain and misery, perhaps; a will to mark the truth; and, most likely of all, a need to communicate, and communicate through, a Beothuk representation of space.[22]

20.4.2 The Context of the Mapped Movements: Or, Unpacking the Artefact as History

Shawnadithit's cartographic expression of a Beothuk geographical imagination has also been hidden because of the way the maps have been treated as artefacts of History. I use the capitalized form of the word here, following Robert Young (1990), in order to represent the sort of western history that disavows its geographical provenance and claims the space of universal truth as the whole world's History. Converted into an artefact in the service of such History, Shawnadithit's work has been treated as a source of raw information about the various expeditions of Peyton and Buchan into the interior. Howley set the pattern for instrumentalizing the maps in this way, and others, such as Berton, Winter, Such and Rowe have followed. The History of these ill-fated missions is, as a result, often recounted in great detail by the historians. The directions taken by the different settler sorties are listed with exactitude. From Shawnadithit's map

---

[22] I have not addressed here the problematic possibility of deception and co-optation as it has elsewhere been broached by Arthur Miller (1991) in his discussion of the maps drawn by the Zapotec for the Spanish in New Spain. These, too, were maps created by people in the clutches of colonialism, but unlike Shawnadithit's, their experience was one of more systemic and economically oriented cartographic appropriation. The Spanish wanted the details of tracts of land and their resources, and it was in this context that deception in the maps became a strategy of native resistance. Miller argues that such resistance was limited because it actually worked within – and thereby reinforced – the proprietary terms of abstraction of the Spanish. Shawnadithit's map, while concerned to mark the colonial presence, did so in far less European terms.

of the River Exploits, for example, the information gatherers follow her depiction of the British party led by Buchan – the group she depicts in black. They retrieve data about the numbers in the group, the number of stops that were made, the camp sites, the place where Buchan left the coffin of Demasduit, the vain search for the Beothuk around the edge of the frozen lake, and the trip back. Occasionally, as in Berton's account, it is noted that all of this adventuring seems pathetic and ill considered. The irony of Buchan's mission, a mission to return a dead body of a Beothuk woman who had been captured to be used as a go-between, becomes used in turn as a tragic metaphor of the wider tragedy of miscommunication.

Only very occasionally, and then only briefly, are the actual positions of the Beothuk mentioned. It seems that, because the historians find it impossible to corroborate these map-depicted positions and journeys with any written records, their acceptance as History is denied. Because, by contrast, Buchan's journey is also recorded in conventional archives, Shawnadithit's mapping of it is used as a supplement. In this way, the treatment of her cartography parallels the more general way in which British imperialists have looked upon indigenous geographical knowledge elsewhere. As Matthew Edney (1993: 63) notes in the case of India:

> *The British had a very low opinion of Indian knowledge as a whole, and of their geography in particular, thinking it too indefinite, too imprecise, or completely false. They had no qualms about using geographical knowledge from native sources, but they discarded it as soon as even the sparsest survey had been completed.*[23]

The same seems to be true in Newfoundland. There is evidence that explorers used maps drawn for them by the Beothuk and the Micmac for moving around.[24] However, when it comes to the writing of History, such maps can only ever serve, it seems, as artefacts supplementing the colonial archive. Nonetheless, as Derrida (1982) has famously pointed out, there is a disruptive potential lying within such moments of supplementation, and, in the case of Shawnadithit's cartography, such disruption comes in the shape of a geography that History hides. Clearly marked on the map are the travels and camping sites of the Beothuk, a historical geography inscribed in red. Shawnadithit's drawing shows where they camped, the places from where they observed Buchan's party and the tracks along which they followed him. It documents the overland routes the Beothuk made to move back and forth from the lake, the way in which they arrived to take down Demasduit's coffin and the place they took her body to bury it next to that of Nonosabasut. Moreover, apart from just these movements – the meaning of which we cannot pretend to understand in Beothuk terms – Shawnadithit's

---

[23] As Barbara Belyea also notes in Inland Journeys, Native Maps: Hudson's Bay Company Exploration 1754–1802 (unpublished manuscript, p. 17): 'Native maps were useful until a survey could be made which would anchor their geographical features to a spatial grid.'

[24] There are four copies of anonymous undated map parts kept in the Canadian cartographic archives in Ottawa. The catalogue entry reads: 'thought to be of the Exploits River, ... drawn on birch bark by a Newfoundland Indian'. A further annotation suggests that, according to Ed Tompkins of the Newfoundland Provincial archives, this 'Indian' might have been the guide to Cormack, whose name might have been 'Sylvester Joe'. Ottawa: H3/110/ R. Exploits/n.d.

cartography documents the fact that the Beothuk were also observers and agents of geographic interpretation.

Despite numerous references to her maps, no effort is ever made by commentators to draw attention to the processes of observation and territorial inscription that they reflect. There is a point in his book at which Howley (1915: 96) notes in passing how:

> *One tall birch tree on the summit of [Canoe Hill] was pointed out by Shawnadithit as the lookout from whence the Indians observed Peyton's movements.*

However, he makes little of this as an indication of Beothuk knowledge-making, and makes no connection at all with the more systematic observation and inscription of the land represented in Shawnadithit's work. It seems to me, by contrast, that we can see in her map a whole geographical imagination, a coherent assemblage of conceptions of space and time that relate intimately to peoples, lives, deaths and travel.

Again, I do not think it is useful to read too much into the map and 'anthropologize' these conceptions. Rather, the important point is to simply acknowledge their existence. From this perspective, we can at least begin to take direct account of how the island, whose New 'true shape' Cook inscribed with such diligence from the sea, was already far from uninscribed by the Beothuk on the land.

Shawnadithit's cartography documents how their inscriptions had another shape, one that was constituted and understood in the terms of another, less sea-bound, geographic discourse. Her mapping of the route of Buchan's party also documents how the arrival and movements of the colonists presented new spatial developments that were nonetheless interpretable within these older and indigenous geographic terms. Clearly, the Beothuk observed the men in Buchan's party – men who were to become the new and official observers of Newfoundland's interior – and they saw them not just, as Cormack notes, from point 'A' on the river, but also, as the map itself underscores, from a specifically Beothuk point of view. The map does not tell us much about how the party and its route were interpreted in such terms as fear or friendship. However, it is possible to note that, despite the coffin-carrying cortege, Buchan's group is still depicted by Shawnadithit as made up of people, albeit smaller people, and that the geography of their journey is marked with great care. In the context of a cartography that charts so close a connection between knowledge of the land and travel through it, these new colonial tracks surveyed by Shawnadithit constitute new found land: travelling that inscribes a new colonial geography. Such new found land, as seen from the perspective of those who were colonized, is so rarely documented that historians concerned with the 'New' of History seem to have forgotten that it can exist.

20.4.3 The Context of Cartographic Re-Cognition: Or, Unpacking the Anthropological Artefact

It could be argued that Cormack himself, unlike the latter-day historians, was more respectful of the autonomy of the native knowledge presented to him. Certainly, he took more careful note of Shawnadithit's observations, annotating the work to mark observations she must have mentioned. He also seemed to take her knowledge as a form of reliable knowledge about the Beothuk, from which he could estimate such exact records as the numbers of those surviving in 1820. However, this same attention to exactness also betrays how Cormack in turn misappropriated the spatial specificity of Shawnadithit's maps by treating them as anthropological artefacts. The care, the concern, the attention to detail, were all directed toward the higher Historical purpose of anthropological salvage. Cormack's regard for Shawnadithit's drawing became thus his way of satisfying the need that he had outlined in *The Edinburgh New Philosophical Journal:* a way for him to procure what he called 'an authentic history of th[is] unhappy race of people, in order that their language, customs and pursuits might be contrasted with those of other Indians' (Howley, 1915: 189). Comparative anthropology, was in this sense, another formula for sublating Shawnadithit's geography into History.

More recent treatments have continued this transfiguration of the cartography as anthropological artefact. In such treatments there generally appear to be two basic rhetorical gestures, in fact, the two same basic gestures that were part of Bishop John's compliment about Shawnadithit's cleverness with a pencil. One of these elevates the maps, praising their topographical accuracy, their artistic skill and their generally high quality as specimens of the supposed genre. The other gesture radically diminishes the work, pouring scorn on its 'Indian' character, intimating its apparent sameness *vis-à-vis* some homogenized 'Indian' norm, and firmly marking its difference as inferior *vis-à-vis* the strict scale and grid-orientated rigour of properly Cartesian cartography. Howley (1915: 238), who was himself a celebrated cartographer of Newfoundland's geology (Tompkins, undated: 11), begins and ends his own description of Shawnadithit's maps with the second of these two gestures.

> *Although rude and truly Indian in character, they nevertheless display no small amount of artistic skill, and there is an extraordinary minuteness of topographical detail in those having reference to the Exploits river and adjacent country. These latter bear a striking resemblance to Micmac sketches of a similar character.*

By contrast, Winter's (1975: 128–129) description puts the negative comparison with Western cartography in the middle.

> *The drawings are accurate in topographical details, but they lack regular scale: rivers and lakes appear larger than they really are. Nevertheless, the details of shoreline, islands, bends in the river, falls, rapids, and junctions of rivers are accurate and the relation of each of these to the other is correct.*

It seems unfortunate that Winter's avowedly pro- Beothuk account should end up sounding more like the patronizing praise of a paternalistic school report. Indeed, perhaps Howley's racist view of the maps' 'rude and truly Indian' character is more honest. Winter does not say the details are 'correct' by Western standards, but the implication is there all the same.

Another more scholastic turn to Shawnadithit's maps by the cartographic historian G. Malcolm Lewis also suffers the same ethnocentrism.[25] Lewis (1979) includes the map depicting the Beothuk observations of Buchan's party as one amongst many examples of mapping grouped together as 'North American Indian'. He, too, praises the 'remarkable and readily interpretable detail' (Lewis, 1979: 28) in the work. However, he also begins his essay with a sweeping assessment of the deficiencies in 'Indian' mapping:

> *Indian representations of actual networks (whether drainage, routes or boundaries) were not drawn to scale and were characterized by gross distortions of direction (Lewis, 1979: 25).*

Overall, the general approach of such commentaries shares the same imperial will-to-knowledge evident in Cormack's search for an 'authentic history'. It also operates like the White Australian accounts of Aboriginal spatial history critiqued by Carter. In these accounts, argues Carter (1988: 344), '[p]ools, pastures and tracks were taken out of context and used like quotations'. The comparisons that treat Shawnadithit's work as a decontextualized anthropological artefact are no less co-optively appropriative. Theirs is an approach that delights in the artefact's quotable accuracies, all the while placing it low down in a supposedly larger comparative hierarchy.[26] Such a hierarchy inevitably elevates the 'true' cartography of those like Cook to the top, setting it up implicitly or explicitly as a standard against which all below is judged deficient. Lewis's (1993) own more recent multicultural epiphany may at first blush seem to afford away of avoiding such imperial hierarchization. But his overarching desire to compare, contrast and, most of all, codify maps as artefacts representing 'different cultures . . . at different stages of development' remains the same (Lewis, 1993: 98). If this is the end point – a reconfirmation of an epistemic empire that places European maps at the end point of some developmental timeline – there seems little point in even beginning such a multicultural project. Nevertheless, an attention to the same differences noted by Lewis may lead somewhere else, once the project is freed from the temptations of timing and taxonomy.

I find a more useful way forward in Brody's (1988) account of what he entitles *Maps and Dreams*. His work relates to a late-1970s land-rights struggle in British Columbia, in which the Beaver people represented their lands cartographically. Although he was there as a researcher and assistant in the production of contemporary maps concerning the people's land-use patterns, Brody (1988: 45–46) reports Atsin – a Beaver hunter – telling him the following:

> *Oh yes, Indians made maps. You would not take any notice of them. You might say such maps are crazy. But maybe the Indians would say that is what your maps are: the same thing. Different maps from different people – different ways.*

---

[25] For more on what she calls Lewis's 'schizophrenic approach', along with its tendency to operate with a 'working sense of "true" and "false"' founded in a 'European cartographic convention', see Belyea (1992: especially 269–270).

[26] In Belyea's (1992b: 270) words, such approaches 'still adhere, unconsciously if not deliberately, to a notion of progress towards increasing cartographic accuracy, and they persist in measuring that accuracy in European terms'.

It is by maintaining Atsin's unhierarchical attention to different maps and different ways that Brody goes in a direction different from the anthropologizing and antiquarian treatments of Shawnadithit's maps I have cited above. In this way, comparative questions like those broached by Lewis might still be pursued, but without the invidious desire to establish a developmental chronology.

In relation to Shawnadithit's work, Brody's attention to what he underlines are the different 'dreams' behind different maps opens a whole series of questions about the guiding assumptions of recent cartographic and cultural history. Some of this schol-arship has examined and critiqued the connections between the imperial gaze, cartography and what Timothy Mitchell calls the 'enframing' of territory.[27] Mitchell's work itself functions primarily as a critique of modernity. However, as Sami Zubai-da (1990) has noted, there are some unfortunate assumptions about 'pre-modern' culture – including the historicist notion of pre-modernity itself – implied by the critical argument. One of these is the suggestion that 'pre-modern' cultures do not have a capacity, however differently organized, to picture people and places from a distance. What Brody might label 'the geographical dream of Shawnadithit's map' clearly problematizes any such simple generalization. It demands of critics a more nuanced reading of the differences distinguishing the disciplinary picturings of Europeans from the picturings of the people they colonized. In this vein, we can go back to Cook's map and compare its exact outline and empty centre with Shawnadithit's detailed depiction of pathways through the interior.

Cook's map can clearly be argued to represent an enframing moment in which we see, in Mitchell's (1998: 44) terms, a 'conjuring up [of] a neutral surface or volume called "space."'[28] Such colonial conjuring of abstract space provided a seemingly blank frame of reference through which important parts (ports and other information vital for fishing and navigation) could be distinguished from the unimportant (the empty interior).[29] Moreover, the emptiness so conjured can also be read as a representation of the lack of British interest in the interior – a gap in governmental knowledge – that, combined with the related lack of disciplinary power in this arena, ultimately made possible the violent processes of its actual emptying.[30] Shawnadithit's map, by contrast,

---

[27] See Mitchell (1988. For an example of such scholarship in geography, see Derek Gregory (1995); and Edney (1993).

[28] Back in Europe, there was another, more direct exhibition of power involved here. Cook was the first European navigator to have a chronometer to measure longitude. His work, therefore, was also a form of technological exhibition, adding, to a would-be scientific-cum-colonial mission, a military demonstration. I owe this point to Nick Chrisman.

[29] This is a view that the geographer Staveley (1987: 247) suggests has persisted – a view that, as such, was lyrically summarized by J.D. Rogers: 'Newfoundland from within reveals only a fraction of its nature. Its heart is on the outside; there its pulse beats, and whatever is alive inside its exoskeleton is alive by accident. The sea clothes the island as with a garment, and that garment contains the vital principle and soul of the national life of Newfoundland.'

[30] The Chief Justice of Newfoundland, Mr John Reeves, reported this non-space of disciplinary order to the Parliamentary Committee in Westminster: 'This is a lawless part of the island, where there are no Magistrates resident for many miles, nor any control, as in other parts, from the short visits of a man-of-war during a few days in the summer; so that people do as they like, and there is hardly any time of account for their actions.' (Quoted in Howley, 1915: 55.).

pictures a peopled interior, a space of action, movement, life and death. Certainly, this is not the modern 'world-as-exhibition' approach to picturing examined by Mitchell – although, as I have shown, this is precisely what the historians have attempted to make it. However, it does nonetheless embody a particular world view that divides the important from the irrelevant. It is in this vein that I think we can return to the assessments of the taxonomists about the lack of scale. Rather than a deficiency, the uneven scale in Shawnadithit's map can be read as a rigorous and reliable picturing of the uneven possibilities for travel by foot across so uneven a landscape. Rather than set up an imaginary abstract grid of space on which all else falls into synchronic place, the complex time–space the map depicts traces the spatiality of people and movement. It is still a representation of space, still a picturing, and still an embodiment of a geographical imagination. Yet it is also a spatiality, rooted specifically in the practical pathways of Beothuk journeying and knowing.[31]

20.4.4 The Context of National Mapping; Or, Unpacking the Artefact's Nationalization

I have already outlined the ways in which national historians have transformed Shawnadithit's maps into national relics. In this last section on the context of her cartography, I want to continue this process of problematization by examining the more substantive connections between mapping and national identity. As Harley, Benjamin Orlove and Richard Helgerson have all pointed out in the pages of this journal, and as Thongchai Winichakul (1994) has examined at length in his book *Siam Mapped,* there is a long and globally varied history connecting mapping and nation-building (Harley, 1992b; Orlove, 1991; Helgerson, 1993). Clearly, the connections go far beyond the practical business of charting the length and breadth of national territory. They also extend to the complex power relations underpinning the imagination and organization of the nation as a spatially coherent community. Helgerson (1992: 114) argues that in Elizabethan England, for example, cartography was central to the developing break with the regime of dynastic loyalty. 'Maps', he says, 'opened a conceptual gap between the land and its ruler, a gap that would eventually span battlefields'. Even today, national maps and atlases continue to serve as a major vehicle for teaching citizens the spatial reach of their nationality, allowing them – as Benedict Anderson (1991) emphasizes in the revised edition of his famous book – to dream the secular, national dream of continuous, horizontal community. Beneath such unbroken horizons, many other

---

[31] Belyea makes a similar point in relation to the native maps recorded by agents of the Hudson's Bay Company. 'We need', she says, 'to look again at the defining characteristics of Native cartography, and instead of reading them as "distorted" spatial indicators, see them as consistent and reliable signs of a non-spatial principle of linear coherence.' (Belyea, Inland Journeys, 7.) Clearly, I would want to replace her notion of a 'non-spatial principle' with a diversified attention to *different* spatial principles. I would not, for example, argue that Shawnadithit's map is unframed. Obviously, it is not framed by a Cartesian grid, but decisions were clearly made about what to show and what not to show. To use a more Derridean formula, I would say that while her map is not logo-centric, it nevertheless remains a graphing of the geo.

dreams of community may, of course, exist and yet the flatness of the map, what Anderson calls its 'logoization of political space', would seem to hide them away. Returning to the differential spatiality of a map like Shawnadithit's allows such dreams of alienation (and, indeed, of alien nations) to break the unbroken surface. Comparing her cartography with attempts to map Canada as a complex and regionalized, yet mappable, whole illustrates a form of negotiation with nationalizing discipline that Brody (1988: xx) evokes in his account of the Beaver people's struggle. 'Dreams collide: new kinds of maps are made'.[32]

In Canada, in particular, as the chair of the Bank of Montreal noted in Volume Three of the *Historical Atlas of Canada*, '[m]ap making and nation-building are inextricably Linked' (Kerr and Holdsworth, 1990: donor's page). As the banker proceeded to suggest, and as numerous reviewers have agreed, the *Historical Atlas* itself has ably continued this tradition.[33] At the same time, its pursuit of national cartographic representation (and, through this, integration) in the contemporary period of first nations' struggle, has taken an uneven, but sometimes new, approach to the question of first nations within the nation of Canada. Unlike most previous national atlases of Canada, the *Historical Atlas* departs from a geographically exclusionary understanding of the country as just a 'white settler colony' and, for the first time, goes a considerable way towards putting native peoples on the map.[34] Indeed, despite the biblical boast of mapping Canada 'From the Beginning to 1800' – the subtitle given to the English edition of Volume One by the publishers – there is a sense in which the thorough attempt made to document a native presence on the land in the first volume fundamentally displaces the white settler colony thesis, and, with it, the very notion that the land had always been Canadian (Harris, 1987). Unhappily, the teleology of the three volume series as a whole concludes in Volume Three by addressing the twentieth century with only one obvious attempt – a Gitksan and Wet'suwet'en map – to mark the continuation of the same native presence into the places and politics of the present. But this, I think, creates all the more reason to re-examine the remarkable effort in Volume One to put native people, and amongst them the Beothuk, on the map. The question marks of Shawnadithit's cartography, I will suggest, ask us to consider how even this effort itself is not without dangers.

Putting first nations on maps that chart a story of national development may well be inclusionary, but it may also be incorporative, co-optive and controlling. Put another way, it may historically nationalize people who now, seeking de-colonization, refuse this very nationalizing principle. In the miserable context of trap lines, reserves and rejected land claims, being on a Canadian (as opposed to a first nations) map has most often coincided for native people with the violent disciplinary force of

---

[32] I think this notion of negotiation with a disciplinary structure is missed by Huggan (1991: 58) in an otherwise interesting discussion of the book's 'manipulation of time–space metaphors'.

[33] For a valuable institutional analysis of the *Atlas* as a federally funded national project see, Anne B. Piternick (1993).

[34] On the limited geographical imagination of the white settler colony thesis, see Frances Abele and Stasiulis (1989).

colonialism. It was with this same disciplinary structure of violence that the scholars working on the *Historical Atlas* inevitably had to negotiate. Committed as they were to using only the most reliable and academically respected data, they turned chiefly to the disciplined information sources of Canadian archaeology and anthropology, rather than to the oral histories of first nations. According to Conrad Heidenreich (1981: 6), one of the *Historical Atlas's* most careful and conscientious researchers of native movements, 'memory ethnography ... was ruled out as being virtually worthless.' In its stead, then, the Historical Atlas sought to put native people on the map by turning firstly to archaeologists' assessments of so-called 'Prehistory' and, secondly, to colonial records. Such assessments and records are, of course, interested and as Cole Harris, the editor, warns in his preface to Volume One: '[m]ore than goodwill is required to penetrate an Indian realm glimpsed through white eyes'. Certainly, there was plenty of good will in Volume One – 'we have tried to accord full place to native peoples' notes Harris – but in negotiating how 'to penetrate an Indian realm', the scholars negotiating with the penetrative power/knowledge apparatus of the colonial archive still continued to gaze through white eyes – their own and those of the colonial record keepers before them. Thus, albeit marginally and supplementally, Shawnadithit's gaze provides a different and less penetrative perspective on a small corner of the terrain mapped by the *Historical Atlas* as Newfoundland. Apart from the inclusion of a Beothuk pendant, and an archaeological generalization of Beothuk space in the 'Prehistory' section (Plate 9), the first major treatment of the Beothuk in the *Historical Atlas* comes in the second section on 'The Atlantic Realm' (Plate 20).

The most detailed part of the plate in question illustrates the findings of a comprehensive and rigorous attempt by Ralph Pastore to map Beothuk habitation and burial sites (Figure 20.3). This, I think, is a fine effort 'to accord full place to native peoples', and, placed as it is between a plate showing the routes of various European explorers (*Historical Atlas,* The Atlantic Realm, Plate 19) and a plate describing the migratory fisheries of Newfoundland (*Historical Atlas,* The Atlantic Realm, Plate 21), it has a certain displacing force on the *Historical Atlas's* narrative of nation. Nevertheless, this displacement, I think, is brought back into Historical line by the scene settings that enframe Pastore's map. The scene is seen, or at least glimpsed, after all, through 'white eyes'. In the introduction to the so-called 'Atlantic Realm', it is noted, for example, that the 'advantages and disadvantages of settlement in Newfoundland were argued throughout the seventeenth century in both England and France'. Nothing is said here about the possibility of a considerable discourse amongst the Beothuk about these same 'advantages and disadvantages' of European settlement. The way that Shawnadithit's surveys represent a continuation of such a Beothuk discourse on into the nineteenth century is neglected, and this despite the fact that, as maps, they employ the same cartographic medium as the *Historical Atlas* itself. The narrative of national development in the Atlantic Realm concludes instead that 'ravaged by malnutrition and disease and lacking guns, their summer fishing stations pre-empted by fishermen and their winter villages subject to the depredations of furriers, the Beothuk of Newfoundland became extinct'. To be sure, the words above Pasture's map also re-emphasize the Beothuk suffering: 'Harassed by trappers and fishermen, weakened by European disease, and excluded from coastal resources, they died out early in the nineteenth

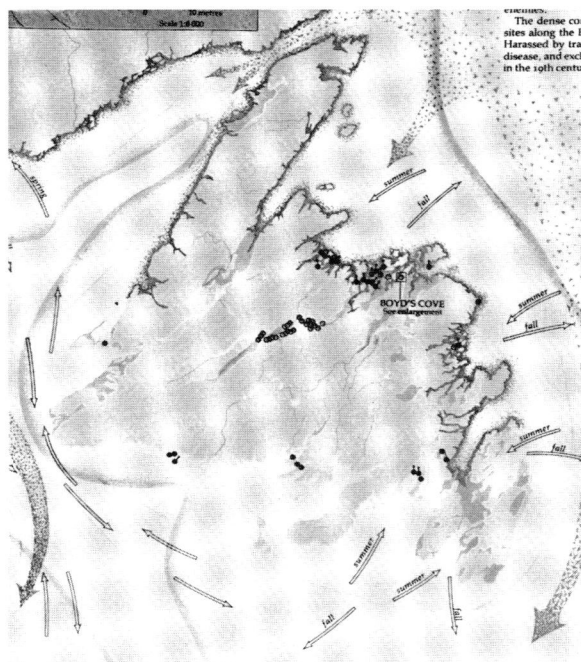

Figure 20.3 A black and white copy of the Newfoundland portion of Plate 20 in the *Historical Atlas of Canada, Volume One*. (Plain circles represent habitation sites. Circles with protruding additions represent burial sites. The arrows in the sea represent the directions of seasonal cod migrations.).

century'. In other words, the colonial violence is definitely noted. But at the same time, the suggestion that the colonialists might be stealing the land – a suggestion evoked by the painful ironies surrounding the coffin in Shawnadithit's map – is never brought up, all the while the atlas twice mentions that the Beothuk themselves stole (*Historical Atlas*, The Atlantic Realm, Plate 20, and p. 48). For the reviewer Paul Robinson (1987: 5), this stress on Beothuk stealing came two times too many:

> *Somehow I found this explanation of the Beothuk's eradication too simple, too convenient, as if it were expedient to skirt nimbly a topic that most of us choose to ignore. The memory of Shawnadithit, her mother and sisters – the last of a race of people – is too strong in my mind. They are deserving of better treatment, particularly in a book as majestic in its approach to our complete heritage as is the Historical Atlas of Canada.*

Robinson's criticisms – although they ignore the simultaneous attention to harassment in the proffered explanations – are well taken. However, I would argue that it is precisely the question of the *Historical Atlas's* 'majestic' approach to cartographic completeness that is at issue. The nationalizing desire to produce a complete cartography that accords full place to native peoples is brought into crisis by the very disciplined lengths the research and narrative go to secure completeness.[35] This epistemological tension becomes clearer in the second major moment in which a Beothuk presence is acknowledged, namely the final map of the volume, entitled 'Native Canada, *ca.* 1820' (*Historical Atlas of Canada: Volume One,* Plate 69).

In many ways, the map 'Native Canada, *ca.* 1820' is a disruptive way to conclude an atlas subtitled 'From the Beginning to 1800'. Instead of a singularised colonial genesis story that might conclude by depicting the originary capitalist movement of the fur trade and fisheries in 1800, the map evokes a vast multiplicity of native movements. Instead of homogenizing native experiences, it charts diversity. And instead of mapping points of colonial control – which might be readily (mis-)used in court as testimony to the vigour of colonialism and the extinction of native rights – it highlights the complex dynamics of survival in the face of advancing colonial settlement. Here, then, perhaps more than in any other place in the *Historical Atlas,* we find a real radicalisation of the plural '*Des Origines*' in the doubled-up sensitivity of Volume One's French subtitle. Yet the epistemological tension noted above also works here. In the end, the diversity and disruption are all recalled within a coherent map of the whole of Canada – a map that no native cartographers would ever have drawn, let alone imagined as their national community. The various historical native movements – including those of the Beothuk – are charted so comprehensively and pictured so coherently in one map that the final result is their recollection on the single unbroken horizon of the modern Canadian nation. They become not first nations as such, but 'Native Canada, about 1820', a synchronic moment in Canada's progressivist march from archaic past into modern nationhood. Colliding with this coherency, the alienation of Shawnadithit's new-found-land introduces 'a newness that is not part of the progressivist division between past and present, or the archaic and the modern' (Bhabha, 1994: 227). Instead, this newness is, to quote Bhabha (1994: 227) again, 'the "foreign" element that reveals the interstitial'.

Most immediately, we can note that the mapping of encounters in 1820 made by Shawnadithit casts some doubt on the accuracy of the Newfoundland portion of 'Native Canada, *ca.* 1820'. In this portion, the likely seasonal routes taken by the Beothuk from the interior to the sea are shown by small arrows, while larger arrows depict the advance of colonial settlement (Figure 20.4). In the context of the events portrayed by Shawnadithit, the Beothuk's seasonal routes – and particularly the route towards the southwest – seem quite unlikely. Moreover, as Cormack's annotations to Shawnadithit's map suggest, the number of Beothuk around this time was probably about twenty-seven. Such reduced numbers – based in large part on Shawnadithit's testimony – would suggest that, by the end of the 1810s, the whole traditional pattern

---

[35] Alan Green put the problem this way: 'Native peoples are not slighted. . ., [y]et the national scope of the atlas and, *a fortiori*, Cole Harris's nationalistic and Innisian preface, fly in the face of the non-national realities of native history. The pre-history maps all tend to point to the artificiality of any conception of 'Canada' before the advent of the French. . .. [T]here is some contradiction between Cole Harris's desire to do right by the Indians and the patriotic themes that he stresses in summing up Volume 1.' (Review, *Labour/ Le Travail* (1988), 22, 274.).

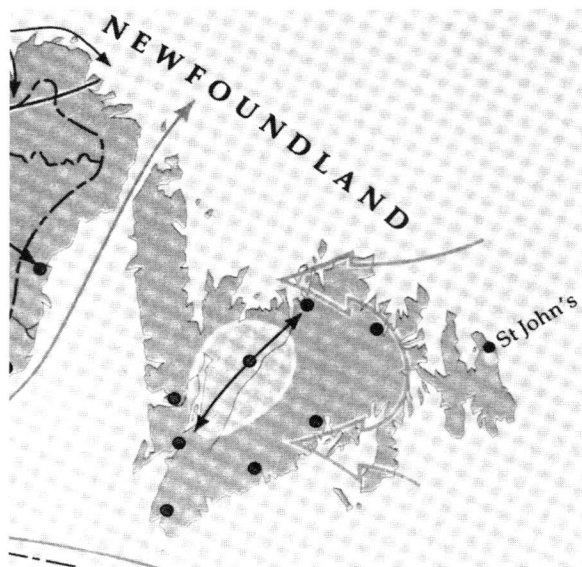

Figure 20.4   A black and white copy of the Newfoundland portion of Plate 69, 'Native Canada, *ca.* 1820' in the *Historical Atlas of Canada, Volume One.*

of life and movement had been equally reduced. By the 1820s, then, survival in the interior – and most immediately in the winter of 1820, surviving the foray of Captain Buchan's party up river – were the more likely concerns. The fact that Shawnadithit mapped these events, and not (as far as we know) some seasonal hunting and gathering patterns, also suggests their priority from the perspective of the colonized. The *Historical Atlas* does not entirely efface such a perspective, and it is only fair to note that it reflects the bleak priority of survival in the face of colonialism with words appended to the map.

'Harassed and barred from the coastal resources, the Beothuk were nearing extinction'. Nonetheless, even with this written qualification, the actual depiction of Beothuk seasonal movements remains.

More significant than the questions of seasonal movements is the issue of how the whole colonial scene is represented. Shawnadithit's map traces out a new-found spatiality traced by the travels of the colonizers and the colonized. At the level of content, it portrays a middle ground – an interstitial space, shared, but travelled through and experienced, differently by the two groups. At the level of the way in which that is represented, Shawnadithit's depiction of the scene is also something of an interstitial work. It employs the paper-and-pencil 'outline-ordered' mapping format of the Europeans with whom she was trying to communicate.

Yet, as I have argued in the preceding three sections, it also reflects a series of Beothuk perspectives that challenge European notions of space, scale, time and abstraction. The *Historical Atlas* perspective on 'Native Canada, *ca.* 1820' – as with the earlier map of burial and habitation sites (Figure 20.3) – does not throw those same notions into question. It directly counters the myth that, by 1800, Canadian history was primarily a matter of European struggle on the land, but it does not abandon the pan-Canadian perspective of the European explorers. Its representation, unlike Shawnadithit's non-national picturing of people, is abstract and detached. It is mapped out on the template of the modern nation. It still usefully represents the grand dynamics of contact – the small arrows of native movement outsized by the large incoming arrows of colonial settlement- – but it does so completely within the framework of modern cartography. There is no space here for the complex in-between moments of contact and conflict represented by Shawnadithit. Indeed, because the whole national scene of 'Native Canada' must be brought together as such on a single sheet, there is not even space – let alone the (possibilities of) non-synchronous time – to include the complexities of exchange, timing and native interpretation presented in Shawnadithit's story map.

Instead, with a flurry of colours and images befitting the European war-room, 'Native Canada, *ca.* 1820' allows the modern citizen reader to take in, with a single glance, a last synchronous slice of Canada's non-modern past before the modernizing, nationalizing impulse of the next two volumes obscures the genealogy of that presence from the present. This contrast with Shawnadithit's contestatory cartography is not meant to belittle the widely acclaimed achievements of Volume One in putting native people on the map. These are achievements that remain important for ongoing struggles to decolonize. However, it is meant to provoke further questions concerning what the editor himself described as the limits of such a good-willed, anti-ethnocentric, but national project as the *Historical Atlas*. Such limits are not easily overcome, and anti-ethnocentrism more generally seems always already limited. It is, therefore, with this more general theme that I will conclude.

20.5 A Conclusion

The silva is savage, the *via rupta* is written, discerned, and inscribed violently as difference, as form imposed on the *hylē*, in the forest, in wood as matter; it is difficult to imagine that access to the possibility of the road map is not at the same time access to writing.

Jacques Derrida (1976: 108)

Derrida's arguments here about Lévi-Strauss's account of the Nambikwara and a 'road-map' are complex. This said, it calls for no special scholar of the Heidegger, whose '*hylē*' he is citing, to understand one of the philosopher's basic points about the unintended ethnocentrism within Lévi-Strauss's display of anti-ethnocentric critique. In *Tristes Tropiques,* Lévi-Strauss had condemned writing as the genesis of everything bad in western civilisation. Romantically, in a supposedly anti-ethnocentric attempt to praise the Nambikwara as better than the West, he proceeded to portray them as living in innocent bliss until the advent of writing.

Derrida's deconstruction points to the ethnocentric assumption in this display of reverse ethnocentrism, reminding his readers that writing comes in more than the modern western alphabetic form, and that it exists even in the ability to make a road through the forest. Access to such a road, he suggests, amounts to a possibility of a road map – a seemingly western invention, but yet, a possibility also alive in the jungle of the Nambikwara (where, in a further twist, a road also practically marked the advent of the west).[36]

Derrida's argument is especially relevant here for a reason that can be understood in three parts. Firstly, as Mignolo (1993:219) has argued:

> *The history of cartography related to the New World runs parallel to the celebration of alphabetic writing and its complicity with historical narratives.*

As narratives like those of Bishop John made clear, the very same gesture that diminished Shawnadithit's map making pencil skills as signs of skill-lessness operated together with an implicit elevation of alphabetic writing as the distinguishing mark of civilization. Secondly, as the writing of Lewis exemplifies, it is the connection between European mapping and the protocols of such alphabetic writing that are generally used as the basis for putting cartography like Cook's higher in Eurocentric hierarchies of progress. Thirdly, and finally, we can see in the critiques others have made of Lewis the same display of reverse ethnocentrism that Derrida critiques in Lévi-Strauss. Denis Wood (1993), for instance, argues that:

> *It is time to acknowledge that people like. . . Malcolm Lewis are simply wrong when they speak of human groups with cognitive abilities less than ours.*

He then proceeds to repeat the same double gesture of Lévi-Strauss. On the one hand, he praises non-western, supposedly 'non-mapping cultures' for their innocence, their difference from the preoccupation with property in the 'ever-growing map making societies' (Wood, 1993: 6). On the other hand, his argument implies that this innocence also amounts to a form of cultural simplicity, a simplicity compared with which the complexity of map making societies (bad as they are) seems more interesting and more productive of 'so unfathomably many more *kinds* of maps' (Wood, 1993: 6, italics in the original).

In response to these various attempts to dichotomize, academicize and, ultimately, I think, neutralize the question of violence in cartographic power relations, Shawnadithit's surveys, like Derrida's deconstructions, ask a resounding 'Why?' Unlike

---

[36] In this context, the context too of the Nambikwara's track or *picada,* one should 'meditate', says Derrida (1976: 107–108), 'upon all of the following together: writing as the possibility of the road and of difference, the history of writing and the history of the road, of the rupture, of the *via rupta,* of the path that is broken, beaten, *fracta,* of the space of reversibility and of repetition traced by the opening, the divergence from, and the violent spacing of nature, of the natural, savage, salvage, forest.' In the knowledge that it is also a violence, but in the interests of making a more singular point, my above account is rather a meditation on some of these parts *apart.*

Derrida, though, they do so from the past, from a space of terror and turmoil, in an alienating but all too earthly voice of resistance.

To many readers, Shawnadithit's observations of the observers may not comprise a particularly promising picture of resistance. As I have pointed out, they were transformed into maps under duress in the context of considerable traumatic change and pressure. Moreover, the way her work has since been treated as a collection of artefacts for supplementing national narratives would seem to mark the nadir of co-opted resistance. They would, in this sense, seem only to vindicate Harley's (1998a: 301) assessment that: 'Maps are pre-eminently a language of power, not of protest.' However, were we to follow Foucault (1981) whom Harley himself claimed to be following here, we would want to insist that the same power relations underpinning the authority of empire are not monopolized or held by its agents, but rather exercised in ways that will always remain open to critical re-appropriation:

> *that is, there is no binary and all-encompassing opposition between rulers and ruled..., no such duality extending from the top down and reacting on more and more limited groups to the very depths of the social body. One must suppose rather that the manifold relationships of force ... are the basis for wide-ranging effects of cleavage that run through the social body as a whole.*

For the contemporary cartographic critic, I do not believe it is valuable, therefore, to neglect the small possibilities for resistant voices that remain. Rather than denying the complexity of maps like those of Shawnadithit and taking the route of self-flagellation, *à la* Lévi-Strauss, we can instead begin the work of re-examining and reworking the cleavages spoken of by Foucault. Spivak (1992), for example, finds a disruptive and alienating-turned-potentially-liberating voice in the final death agony of a woman lying down to die on a national map of India. Douloti, the ultimate subaltern woman, can, even in her death, seem to poke a hole in the post-colonial national myth of India, indicating how people like herself have been abandoned by the national dream and, thereby, in Spivak's (1992: 99) words, making 'visible the fantasmatic nature of a merely hegemonic nationalism'. Shawnadithit's alienating observations of early Canada are far more agentic than this. She actually surveyed the initial advances of what was to become the culture of hegemonic nationalism. We do not have to read her commentary solely in analogical terms. Instead, with black and red pencil, she offers a cartographic critique of a process that mapped her native land as Newfoundland. To find resistance in this, we only have to unpack her work's entombment as national artefact and read it again as a mapping of new-found people and practices. Simply reading the map, we find new-found land, a palimpsest in process.

Acknowledgements

The author would like to thank Barbara Belyea, Dan Clayton, Nick Chrisman, Ed Dahl, Derek Gregory, Cole Harris, Francis Harvey, Bruce Willems-Braun and three anonymous referees for comments on previous drafts.

References

Abele, F. and Stasiulis, D. (1989) Canada as a 'white settler colony': What about natives and immigrants? in *The New Canadian Political Economy* (eds. W. Clement and G. Williams), McGill-Queen's University Press, Montreal, pp. 240–277.

Akerman, J.E. (1993) Blazing a well-worn path: Cartographic commercialism, highway promotion, and automobile tourism in the United States, 1880-1930. *Cartographica*, **30**(1), 10–20.

Anderson, B. (1991) *Imagined Communities: Reflections on the Origin and Spread of Nationalism*, revised edn, Verso, London.

Barthes, R. (1972) *Mythologies* (trans. A. Lavers), Hill and Wang, New York.

Belyea, B.(undated) Inland Journeys, Native Maps: Hudson's Bay Company Exploration 1754–1802, unpublished manuscript.

Belyea, B. (1992a) Images of power: Derrida/Foucault/Harley. *Cartographica*, **29**(2), 1–9.

Belyea, B. (1992b) Amerindian maps: the explorer as translator. *Journal of Historical Geography*, **18**(3), 267–277.

Berton, P. (1976) *My Country*, Penguin, Toronto.

Bhabha, H.K. (1984) Of mimicry and man: The ambivalence of colonial discourse. *October*, **28**, 125–133.

Bhabha, H. (1994) How newness enters the world, in *The Location of Culture* (ed. H. Bhabha), Routledge, New York.

Brody, H. (1988) *Maps and Dreams: Indians and the British Columbia Frontier*, 2nd edn, Douglas and McIntyre, Vancouver.

Carter, P. (1988) *The Road to Botany Bay: An Exploration of Landscape and History*, Knopf, New York.

Chrisman, N.R. (1987) Design of geographic information systems based on social and cultural goals. *Photogrammetric Engineering and Remote Sensing*, **53**(10), 1367–1370.

Clifton, J.A. (1990) The Indian story: A cultural fiction, in *The Invented Indian: Cultural Fictions and Government Policies*, Transaction Publishers, New Brunswick.

Crate, J. (1990) Shawnandithit (Last of the Beothuks). *Canadian Literature*, **124–125**, 17.

Derrida, J. (1976) *Of Grammatology* (trans. G.C. Spivak), Johns Hopkins University Press, Baltimore, MD.

Derrida, J. (1982) The Supplement of Copula: Philosophy before linguistics, in *Margins of Philosophy* (trans. A. Bass), University of Chicago Press, Chicago, pp. 175–205.

Edney, M.H. (1993) The patronage of science and the creation of imperial space: The British mapping of India, 1799-1843. *Cartographica*, **30**(1), 61–67.

Emberly, J.V. (1993) *Thresholds of Difference: Feminist Critique, Native Women's Writings, Postcolonial Theory*, University of Toronto Press, Toronto.

Foucault, M. (1981) *The History of Sexuality: An Introduction* (trans. R. Hurley), Penguin, London.

Gilbert, W. (1992) Beothuk-European relations in Trinity Bay. *Newfoundland Quarterly*, **87** (3), 2–10.

Godard, B. (1990) The politics of representation: Some Native Canadian women writers. *Canadian Literature*, **124/5**, 183–225.

Gregory, D. (1994) *Geographical Imaginations*, Blackwell, Oxford.

Gregory, D. (1995) Between the book and the lamp: Imaginative geographies of Egypt, 1849-50. *Transactions of the Institute of British Geographers NS*, **20**(1), 29–57.

Harley, B. (1988a) Maps, knowledge, and power, in *The Iconography of Landscape* (eds. D. Cosgrove and S. Daniels), Cambridge University Press, Cambridge.

Harley, J.B. (1988b) Silences and secrecy: The hidden agenda of cartography in early modern Europe. *Imago Mundi*, **40**, 57–76.

Harley, J.B. (1990) Cartography, ethics and social theory. *Cartographica*, **27**(2), 1–23.

Harley, B. (1992a) Re-reading the maps of the Columbian encounter. *Annals of the American Association of Geographers*, **82**(3), 522–536.

Harley, J.B. (1992b) Deconstructing the Map, in *Writing Worlds: Discourse, Text and Metaphor in the Representation of Landscape* (eds. T.J. Barnes and J.S. Duncan), Routledge, New York, pp. 231–247.

Harris, C. (1987) *Historical Atlas of Canada: Volume One: From the Beginning to 1800*, University of Toronto Press, Toronto.

Heidenreich, C.E. (1981) Mapping the location of native groups, 1600–1760. *Mapping History/ L'Histoire Par Les Cartes*, 2.

Helgerson, R. (1992) *Forms of Nationhood: The Elizabethan Writing of England*, University of Chicago Press, Chicago.

Helgerson, R. (1993) Nation or estate?: Ideological conflict in the early modern mapping of England. *Cartographica*, **30**(1), 68–74.

Hewson, J. (1978) Beothuk Vocabularies. Technical Papers of the Newfoundland Museum.

Horwood, H. (1959) The people who were murdered for fun. *Maclean's Magazine*, 10 October, 26–44.

Howley, J.P. (1915) *The Beothuks or Red Indians: The Aboriginal Inhabitants of Newfoundland*, Cambridge University Press, Cambridge.

Huggan, G. (1989) Decolonising the map: Post-colonialism, post-structuralism and the cartographic connection. *Ariel*, **20**(4), 116–129.

Huggan, G. (1991) Maps, dreams and the presentation of ethnographic narrative: Hugh Brody's 'Maps and Dreams' and Bruce Chatwin's 'The Songlines'. *Ariel*, **22**(1), 57–69.

Jenness, D. (1934) The vanished red Indians of Newfoundland. *Canadian Geographical Journal*, **8**(1), 26–32.

Katz, C. and Smith, N. (1993) Grounding metaphor: Towards a spatialized politics, in *Place and the Politics of Identity* (eds M. Keith and S. Pile), Routledge, New York.

Kerr, D. and Holdsworth, D.W. (1990) Historical Atlas of Canada, vol. **3**, *Addressing the Twentieth Century 1891–1961*, University of Toronto Press, Toronto.

Kirwin, W. (1992) A note on Beothuk names in Newfoundland. *Onomastica Canadiana*, **74**(1), 39–45.

Kulchyski, P. (1992) Primitive subversions: Totalization and resistance in native Canadian politics. *Cultural Critique*, **21**, 171–195.

Leslie, F. and Upton, L.F. (1977) The extermination of the Beothucks of Newfoundland. *Canadian Historical Review*, **53**(2), 133–153.

Lestringant, F. (1994) *Mapping the Renaissance World: The Geographical Imagination in the Age of Discovery* (trans. D. Fausett), University of California Press, Berkeley, CA.

Lewis, G.M. (1979) The indigenous maps and mapping of North American Indians. *The Map Collector*, **9**, 25–32.

Lewis, G.M. (1993) Metrics, geometries, signs, and language: Sources of cartographic miscommunication between native and Euro-American cultures in North America. *Cartographica*, **30**(1), 98–106.

Marcus, G. and Fischer, M. (1986) *Anthropology as Cultural Critique*, University of Chicago Press, Chicago.

Mignolo, W.D. (1993) Misunderstanding and colonization: The reconfiguration of memory and space. *The South Atlantic Quarterly*, **92**(2), 208–260.

Miller, A.G. (1991) Transformations of time and space: Oaxaca, Mexico, circa 1500–1700, in *Images of Memory: On Remembering and Representation* (eds S. Küchler and W. Melion), Smithsonian Institution Press, Washington, DC, pp. 141–175.

Mitchell, T. (1988) *Colonising Egypt*, Cambridge University Press, Cambridge.

Nealon, J.T. (1992) Exteriority and appropriation: Foucault, Derrida and the discipline of literary criticism. *Cultural Critique*, **21**, 97–119.

O'Dea, F. (1971) *The 17th-Century Cartography of Newfoundland*, University of Toronto Press, Toronto.

Orlove, B. (1991) Mapping reeds and reading maps: The politics of representation in Lake Titicaca. *American Ethnologist*, **18**, 3–38.

Orlove, B. (1993) The ethnography of maps: The cultural and social contexts of cartographic representation in Peru. *Cartographica*, **30**(1), 29–46.

Pastore, R. (1989) The collapse of the Beothuk world. *Acadiensis*, **19**(1), 52–71.

Pastore, R. (1990) Native history in the Atlantic region during the colonial period. *Acadiensis*, **20**(1), 200–25.

Pickles, J. (1995) *Ground Truth: The Social Implications of Geographic Information Systems*, Guilford Press, New York.

Piternick, A.B. (1993) The Historical Atlas of Canada: The project behind the product. *Cartographica*, **30**(4), 21–31.

Rabasa, J. (1985) Allegories of the atlas, in *Europe and its Others*, vol. **2** (eds. F. Barker *et al.*), University of Essex, Colchester.

Raynauld, R. (1984) Les pêcheurs et les colons Anglais n'ont pas exterminé les Béothuks de Terre-Neuve. *Recherches Amérindiennes au Québec*, **19**(1), 45–59.

Robinson, P. (1987) Mapping Canada's early years. *Atlantic Provinces Book Review*, **14**(4).

Rowe, F.W. (1977) *Extinction: The Beothuks of Newfoundland*, McGraw Hill Ryerson, Toronto.

Rundstrom, R. (1990) A cultural interpretation of Inuit map accuracy. *The Geographical Review*, **80**(20), 155–168.

Rundstrom, R. (1991) Mapping, postmodernism, indigenous people and the changing direction of North American cartography. *Cartographica*, **28**(2), 1–12.

Sparke, M. (1994) Writing on patriarchal missiles: The chauvinism of the Gulf War and the limits of critique. *Environment and Planning A*, **26**, 1061–1089.

Spivak, G.C. (1988a) Subaltern studies: Deconstructing historiography, in *Other Worlds: Essays in Cultural Politics*, Routledge, New York, pp. 197–221.

Spivak, G.C. (1988b) Can the subaltern speak? in *Marxism and the Interpretation of Culture* (eds. C. Nelson and L. Grossberg), University of Illinois Press, Urbana, IL.

Spivak, G.C. (1989) A response to The Difference Within: Feminism and Critical Theory, in *The Difference Within: Feminism and Critical Theory* (eds. E. Meese and A. Parker), John Benjamin's Publishing Co., Philadelphia.

Spivak, G.C. (1991) Time and timing: Law and history, in *Chronotypes: The Construction of Time* (eds J. Bender and D.E. Wellbery), Stanford University Press, Stanford.

Spivak, G.C. (1992) Woman in difference: Mahasweta Devi's 'Douloti the Bountiful', in *Nationalisms and Sexualities* (eds. A. Parker, M. Russo, D. Sommer and P. Yaeger), Routledge, New York, pp. 96–117.

Staveley, M. (1987) Newfoundland: Economy and society at the margin, in *Heartland and Hinterland: A Geography of Canada*, 2nd edn (ed. L.D. McCann), Prentice Hall, Scarborough, ON.

Such, P. (1978) *Vanished Peoples: The Archaic, Dorset and Beothuk People of Newfoundland*, NC Press Limited, Toronto.

Tompkins, E.(undated) The Geological Survey of Newfoundland. Museum Notes: Information Sheets from the Newfoundland Museum.

Watson, S. (1994) In race, wilderness, territory and the origins of modern Canadian landscape painting. *Semiotext(e)*, **17**, 93–104.

Winichakul, T. (1994) *Siam Mapped: A History of the Geo-Body of a Nation*, University of Hawaii Press, Honolulu.

Winter, K. (1975) *Shananditti: The Last of the Beothuks*, J.J. Douglas Ltd, Vancouver.

Wood, D. (1993) Maps and mapmaking. *Cartographica*, **30**(1), 1–19.

Wood, D. and Fels, J. (1992) *The Power of Maps*, Guilford Press, New York.

Young, R. (1990) *White Mythologies: Writing History and the West*, Routledge, New York.

Zubaida, S. (1990) Exhibitions of power. *Economy and Society*, **19**(3), 359–375.

21

The Look of Surveillance Returns

Reflection Essay: *Between Demythologizing and Deconstructing the Map*

Matthew Sparke

Department of Geography, University of Washington, USA

21.1 Influences

Since the time when my article was first published, critical cartographic studies have advanced considerably 'beyond the binaries' that it originally sought to challenge (Del Casino and Hanna, 2006). Here, in this reflection paper, my goal is to build on these advances in enquiry into cartographic representation by exploring how they relate to more recent idealistic and voluntaristic suggestions about moving beyond geographic representations and traditional maps altogether. The rather different possibilities of both 'non-representational theory' and 'voluntary geographic information' are thereby re-framed with some re-presentations of my own original argument. To set the scene, though, these reflections begin by revisiting an unsettled binary between roots and routes that was an important inspiration of my article, and which now affords a biographical introduction into its critical geography. In short, I begin by reflecting on how three bio-graphical roots of the article can now be retraced as geo-graphical routes too (hyphens intended, however unsettlingly).

The first route was my own movement into and through Canada, studying at the time as a British graduate student at the University of British Columbia (UBC) in Vancouver and conducting my PhD research at, amongst other places, the national archives in Ottawa (where I first came across Shawnadithit's maps in Howley's book). The second

---

route, was my interest in the adaptation of French post-structuralist theory by post-colonial critics, especially Gayatri Chakravorty Spivak, whose reflections on the representation of the geo-graphic agency of subaltern women in India offered anti-colonial insight when re-routed again into the Canadian context (Spivak, 1976, 1985, 1988). And the third was the particular inspiration of arguments by Brain Harley, Denis Wood and John Fels on the need to consider the power relations implicated in and moving through maps. It is an honour to have my article reprinted here alongside their important interventions, not least of all because my title and the notion of shuttling between demythologization and deconstruction was derived from their terminology. I want now, therefore, to explain why they became such significant signposts for my own post-structuralist-turned-post-colonial journey into Canadian history, before proceeding to reflect on how my article relates to more recent debates over non-representation, representation and political participation with maps, GIS and Web-enabled VGI (Volunteered Geographic Information, or 'Neo-Geo' as it's sometimes called) cartography.

Harley's work was enormously important at the time because of the way it demonstrated how one could examine maps in terms of Michel Foucault's arguments about the concatenation of 'power-knowledge'. His 'Deconstructing the Map' article (Chapter 16) sought thus to make sense of maps as representations of knowledge *about* the world that were deeply tied to relations of power *in* the world. To demonstrate these worldly ties, Harley drew on his encyclopaedic understanding of diverse cartographic archives: showing, amongst other things, their huge importance in the geographical development of European colonialism. I am still impressed by the ways in which he and other historians of cartography such as Matthew Edney (Chapter 18) and James Akerman (2009) deliver such detailed and lively accounts of the ways maps are thereby implicated in imperial power, and, to some extent, my article reflected a desire to detail some similar dynamics *vis-à-vis* the Canadian cartographic archive.

At the same time, Denis Wood and John Fels showed that it was possible to make these sorts of arguments about the worldliness of cartographic power-knowledge in ways that appealed to a wider audience outside of academia. Their article (Chapter 14), and their book *The Power of Maps* clearly spoke to a huge public interest in the role maps play in everyday life. Moreover, their pithy, often poetic sound bites seemed to me to speak truth to power without ever sacrificing the possibility of contributing to intellectual innovation too. One of their most obvious contributions in this respect was to wider debates over the myth-making movements of concealment and revelation that Roland Barthes and others had connected to travel writing and landscape representations more generally. And, because of my own intellectual travels, I found this especially inspiring.

As an undergraduate I had learned a great deal from David Harvey's Marxian analyses of urban representations (including Haussmann's re-mapping of Paris) and their class-interested political-economy (Harvey, 1985). I had also read cultural critiques by other geographers who developed demythologization arguments in ways that showed that the discipline might have something to contribute to what was then the fast-expanding field of cultural studies (Anderson, 1988; Duncan and Duncan, 1988; Gregory and Ley, 1988; Jackson, 1989). Coming to UBC (partly as a result of an interest in this line of work)

I was also introduced by Gerry Pratt and Aruna Srivastava to a wide range of feminist scholarship that took the critique of vision, 'see-it-all-from-above' god tricks and omniscient representation forward in ways that clearly showed how maps were bound up with masculinist, as well as, imperial power-knowledge (Haraway, 1989; Kolodny, 1975; Rose, 1993). And, meanwhile, I was also influenced by historical geographers in Vancouver (a city itself named after a colonial explorer) who sought to study colonial archives as spatial history – that is as 'routes' versus 'roots' (*à la* Carter, 1987) – and who thereby traced active native geographies obscured by wilderness myths of so-called British Columbia (Clayton, 2000; Galois, 1994; Harris, 2002). Thus, while Harley did not draw on exactly the same theoretical repertoire, and while neither did Woods and Fels address all of these Marxist, feminist and post-colonial concerns, the critical cartographers' 'deconstruction' and 'demythologization' of maps still seemed to me to be intellectually resonant with wider contemporary efforts to expand critical theorizing in geography.

The resonance noted, there was also intellectual dissonance too. In this respect it is ironic to describe the interventions of Harley, Wood and Fels as 'signposts', because their key terms actually pointed in opposite directions to the ways I used them for orientation in my article. While Harley spoke of 'deconstructing' the map, my own point was that his work really represented a form of demythologization: an attempt, in short, to debunk cartography's claim to uninterested objectivity by showing the powerful interests in which it operated. Conversely, it seemed to me, that while Woods and Fels drew on Barthesian arguments about semiotic demythologization, the pragmatic spirit of their project had more in common with Derrida's affirmative approach to deconstruction (which Harley by contrast rejected as overly textualist and relativizing). They were thus more interested in how maps produce particular con-solidations of truth rather than in suggesting that we can identify once and for all what the real truth is behind the biases and cover-ups of a particular map. In other words, while Harley performed a version of cartographic ideology critique by pulling away the mask of objectivity and revealing the social and political structures behind maps, Woods and Fels pointed to the need to address the epistemological structures of maps themselves, the way they go on framing, consolidating and thereby re-making real world truths that they are assumed merely to record and represent.

I do not want to overstate these differences here. Clearly, the two sorts of critical moves share much in common. But for me, following Barbara Belyea's critique of Harley (Belyea, 1992), and learning from Spivak's explanation of Derrida's *'writing'* for literary and social theorists more generally (Spivak, 1976), it was important not to throw out the nuance of deconstruction with the wishy-washy bathwater of relativistic postmodernism. Deconstruction, it seemed to me, had a more robust role to play in keeping critique honest about its own limits, of always asking – as Spivak (1985) had herself done in her deconstruction of subaltern studies – critical questions about how the reinsertion of a neglected agent or account into the historical–geographical record risked essentializing the identity of the agent or reifying the representation of the account. Moreover, reading Spivak's own arguments alongside José Rabasa's post-structuralist-turned-post-colonial reading of colonial mapping, it became very clear that a rigorous deconstructive approach offered something that you could not find in ideology critique alone: namely, a heightened sensitivity to the epistemological possibilities of counter-hegemonic cartographic production. If the map could be, in Rabasa's (1985: 17) terms, 'a signalling tool for scrambling previous territorializations', it might not only be a coactive co-product of colonial power-knowledge, but also an anti-colonial assemblage of power-knowledge too. Rabasa's suggestion that maps were, in this sense, geographical palimpsests was particularly provocative and, for the same reason, it is worth re-quoting again here. 'The transposition of the image of the palimpsest', he argued, 'becomes an illuminative metaphor for understanding geography as a series of erasures and overwritings that have transformed the world. The imperfect erasures are, in turn, a source of hope for the reconstitution or reinvention of the world from native points of view' (Rabasa, 1985: 37).

It was just such a native point of view that I sought to highlight in the counter-hegemonic cartography of Shawnadithit. Demythologizers could debunk the limits of Cook's cartography and other colonial maps of Newfoundland all they liked, I thought, but without the deconstructive idea of the map as a palimpsest we might not register the re-territorializing re-mapping of the same space as New-Found-Land from a native point of view. Shawnadithit's maps – which are all now available online (Shawna-dithit, 2009) – had generally been represented in the Canadian canon as anthropological relics of a pre-historic people. My argument was that they could be read instead as native mapping that returned the look of colonial surveillance and, as it were, anthropologized colonial history while redrawing the island's geography from an ab-original perspective (i.e. both native and moving away from 'original' European assertions of new-ness).

Although I cited Homi Bhabha in arguing that the look of surveillance returns resistantly, my points about Shawnadithit's story maps were also re-routed through a more embodied account of ab-original agency. I remain wary of Bhabha's tendency to read off post-colonial possibilities from just the poetics of performative word play (Bhabha, 2004). Anti-colonial agency is a historical reality, and one does not necessarily need supplementary suppositions about mimicry and hybridity to notice it (Mitchell, 1997). Indeed, much of the First Nations struggle going on in Canada (both now and when I wrote the article) has demonstrated this repeatedly, often in ways that also offer a sobering geographic displacement of Bhabha's locutions about location too (Sparke, 2005; and for an important ongoing struggle that foregrounds a native map –as part of the Kétuskéno Declaration against Albertan oil sands destruction of native homelands – see Beaver Lake Nation, 2009). Returning to the debates in cartography via Spivak's and Rabasa's post-colonial re-routings of post-structuralism, I wanted to focus on the politics of context rather than just the poetics of text. The demythologization moment was therefore vital. The basic facts of native geographic agency obscured by colonial myths of Newfoundland had to be recognized. Yet, at the same time, my aim was to show that deconstruction could usefully supplement demythologization by providing a way of remaining alert to how anti-colonial knowledge production re-territorialized the cartographic overlays of the colonialists while remaining subject to further re-territorialization and re-colonization itself.

I had already written about the 'white mythology' of colonial overwriting in a review essay examining the relationship between deconstruction and post-colonial critique more generally (Sparke, 1994). So it was building on this earlier work that I attempted to show empirically how the white man's colonial map could become re-readable as a palimpsest of power-knowledge in which counter-hegemonic cartographies were – in Derrida's terms – 'active and stirring' too. This was an argument which, I further suggested *contra* Harley, was also compatible with Foucault's own insistence on studying power in terms of its capillary circulation through social relations, including those of resistance as much as of dominance and governance. Today, moves like this away from sovereign or juridical conceptualizations of power continue to enable non-romanticizing accounts of resistance in (and through) critical geography (Sparke, 2008). As such, they also intersect with new concerns about non-represen-tation, performativity and participation that have further shaped debates in critical cartography in the period since the mid 1990s. It is to these new intersections on the routes between demythologization and deconstruction that these reflections now turn.

21.2 Intersections

In retrospect, my article's theorization of a post-structuralist-turned-post-colonial argument 'between deconstruction and demythologization' seems to have disappeared up its own subtleness. Notwithstanding my attempts to cite the relevant literatures and demonstrate the requisite theoretical *pouvoir-savoir*, the intervention has had little impact or recognition in post-structuralist/post-colonial theory, nor for that matter in Canadian historiography. It may have been too Canadian for the former audience, and not Canadian enough for the latter, but, whatever the case, the article has not travelled far outside of critical cartographic studies within geography (Wikipedia, 2009). Even within this narrower subfield it has hardly become a chart topper (only 23 cites in total as listed on *Google Scholar* in 2009). It was noticed enough to be republished in an edited book entitled *Places Through the Body* (Sparke, 1998). But, despite this reprinting, it has still been left out of an authoritative online bibliography of work on the history of cartography (Edney, 2007), and, to add mis-representation to non-representation, has been treated in one its few moments in the citational sun as an example of 'non-representational theory' in critical cartographic studies (Perkins, 2004).

Given the relative non-representation of the article in citational circles, it might seem unfortunately ironic as well as ungrateful to complain about the association with non-representational theory. Chris Perkins in fact provides a great summary of the article's main argument, and usefully links it to ongoing work by others on the cultural-political contexts of cartographic texts. I can also understand why he connects the article's attention to Shawnadithit's action-orientated and affect-laden mapping to the argu-ments by Nigel Thrift and his followers about the need to track the active and affective aspects of social and cultural life – movements, tastes, feelings, and so on – that are excessive to or otherwise elude disciplinary writes-turned-rites of representation. However, as Del Casino and Hanna (2006: 37) argue in their creative critique of binary oppositions dividing representational maps from spatial practices: 'it is not necessary to conflate all theories of practice and performance with non-representational theory'. Moreover, the additional attention given in my article to colonial violence, dispossession and resistance seems to me to be exactly the sort of attention to worldly oppression and the struggles it engenders that non-representational theorists prefer to leave non-represented. This may well be done in the idealistic hope that non-representation will make the violence and political problems simply go away. But I would submit that it undermines resistance. It is hard to map-back and produce 'rival' cartographies if one runs away from the challenges of representation altogether. No doubt this is why some of the most innovative work in critical cartographic and cultural studies would therefore seem to replace non-representational idealism with a much more critical attention to the representational relays in which and through which cartographies of resistance develop (Basu, 2009; Brown and Knopp, 2008; Camp, 2002; Crampton and Krygier, 2006; De Leeuw, 2009; Harris and Hazen, 2006, 2009; Kitchin and Dodge, 2007; McKittrick, 2006; Nah, 2006; Radcliffe, 2009; St. Martin, 2009). However, given the increasing influence of non-representational theory, it also seems idealistic to just ignore it in the hopes that it will disappear in a puff of neo-Hegelian self-contradiction. What then are its implications for critical cartographic studies such as those represented by my article?

Non-representational geographers might protest that they are not apolitical, that, unlike the German idealists critiqued by Marx and Engels (1970), they are interested instead in noticing aspects of politics that others ignore, and that they do this by tracing the ways in which action and affect are manipulated in the lived geographies of everyday life. However, as Clive Barnett (2008) has argued, there is a philosophical sleight of hand performed in such non-representational representations about manipulated affect, a sort of idealization of politics that makes normative claims about what is wrong with the world (and what could be right) under the cover of ontological arguments about affect. While I find Barnett's own argument too ontologically axiomatic itself, his basic point translates well into other less stipulative styles of criticism. To put the argument in a more deconstructive way, for example, the non-representationalists want to have their non-representational cake while still eating and regurgitating all sorts of political commitments that remain unexamined and unexplained as politics precisely because they become (non-)represented as ontology as affect.[1] The cake's ingredients are never fully disclosed, but seem to consist of a remix of anti-writing metaphysics baked with Deleuzian and Freudian ingredients rather than any apparent awareness of Derrida's deconstruction of the scene of writing in Freud, nor yet of Irigaray's unfolding of affect in and against Lacanian lack (on the latter, see Thien, 2005). And, philosophical affect aside, from the explicitly political points of view of Marxist, feminist and post-colonial geography, the apolitical mixing is also a problem. Indeed, it might further be interpreted as 'problematic' – in Althusser's adaptation of the term – because it is symptomatic of an 'advanced liberal' subject formation – in the self-descriptive sense of Nikolas Rose (2007). That is to say, it redefines critical politics around the individ-ualistic and largely libertarian anxieties of privileged elites (or at least aspirant elites) whose main worries in life are not body counts and the annihilation of friends and family, but rather the sorts of calculations about bodies that turn risk surveillance into

---

[1] To put this still more playfully, paraphrasing the American comic singer Tom Lehrer: if people feel they cannot represent, the very least they can do is stop representing the feeling in endless articles! http://en.wikipedia.org/wiki/That_Was_the_Year_That_Was.

something risky even for straight white men of property (Sparke, 2006; see also Braun, 2007). Looking at the look of surveillance as the manipulation of affect amidst neoliberalism is in this sense quite different to examining how the look of surveillance returns as an affective mode of ab-original agency amidst imperialism.

These political differences between my article's argument and the crypto-normative contradictions of non-representational theory are important to emphasize because my own work has also been criticized by Barnett for what he sees as a similar philosophical sleight– a 'slippage', he calls it, between deconstruction and normative critique (Barnett, 2007: 503). Moreover, with his definition of 'foundationalism' in the latest *Dictionary of Human Geography*, Barnett goes so far as to condemn Spivak too for attempting to 'finesse' thinking like a sceptic and acting like a foundationalist (Barnett, 2009: 262). Disturbingly, he even accuses her, along with Judith Butler and a host of others, of being 'deceitful' because of a 'tendency to misconstrue what is at stake in issues of foundationalism' (2009: 262). I do not want to dwell here on the strange mix of metaphysics and moralism in this accusation, nor on the way Barnett himself finesses extraordinarily complex debates amongst globally-respected theorists to make a sceptical judgement with what reads like a foundationalist faith in a very local, almost private, philosophical correctness. Instead, I want to suggest that in thereby misconstruing Spivak, Barnett ignores arguments she makes about representation that actually offer a compelling alternative to the problems he identifies in non-representational theory. These arguments are worth highlighting here, because I think they also offer support for the sorts of collective or at least collaborative work being advocated by Casino, Hanna, Kitchin, Dodge and others in their invitations to critical cartographers to examine maps as representational practices.

Some of Spivak's most well known arguments about representation are made in her 'Can The Subaltern Speak?' essay. This essay was an important inspiration for developing my critique of Shawnadithit's silencing as a quaint artefact in Canadian colonial history (Spivak, 1988), and attuned as it is to the limits of ethnocentric theory – notably, Foucault's focus on European spaces of discipline that actually took form amidst spaces of empire – the essay also offers an antidote to the parochial pattern in 'non-rep' repatriations of affect as that which affects 'us' (us, in this case, being the same advanced liberals noted above). Answering her essay title's question in the negative, Spivak famously critiques Deleuze and Foucault for suggesting that critical scholars should let the oppressed and marginalized speak for themselves. She well understands the dangers of misrepresentation within dominant structures of power-knowledge, but she argues that to advocate non-representation is an abdication of critical representational responsibility, and, moreover, a denial of the ways in which we go on representing whether we say so or not. We cannot not represent, Spivak argues, and yet, as we do so, we must also always subject our representations to persistent critique for what they ignore or misrepresent. As a way of explaining this approach, she points to the two German words for representation– *Darstellung* (to represent as in a portrait) and *Vetretung* (to represent as in a political debate). She argues that we need to recognize how these two senses of representation –the pictorial sense and the political proxy sense –each brings the other into a productive form of crisis. Every political representation has to be aware of the limits of representation in the picturing sense (and

what it might therefore ignore or misrepresent), and every picturing form of representation has to be made accountable to its political repercussions (and how it might
have a political outcome whether intended or not). There is no denying the limits and
problems of pictorial representation here, but there is equally no attempt to say political
representation is completely reducible to these limits and problems.

It seems to me that these arguments about the double relay of representation are
especially useful for critical cartographers as we continue to confront the challenges of
studying maps as politically-consequential representational practices. On the one hand,
they help us avoid ontologizing anti-essentialism. This is the sort of move to which
Barnett objects and that tends to ascribe worldly outcomes or political judgements to
purely epistemic structures and positions (such as the non-representationality of affect,
for example). On the other hand, Spivak's arguments about representation do not close-
off (as Barnett's decree about 'deceit' would seem to do) the ongoing production of
explanations, arguments and, yes, maps that simultaneously make themselves vulnerable
to critique for moments of misrepresentation, over-writing and exclusionary enframing.
This is precisely what so much of the work in critical cartography now appears to be
doing (Kitchin and Dodge, 2007). It does not write-off maps as always already doomed to
misrepresentation, nor yet suggest that the ways in which maps operate as practices and
performances can be deduced purely on the basis of an anti-essentialist axiomatics or
apolitical deconstructionism. Instead, it leads to a way of understanding the production
and use of maps as political that simultaneously suggests we can leverage the lessons of
deconstruction to open moments of cartographic representation to an iterative and
collaborative process of critique, re-presentation and, thus, ultimately, re-mapping too
(and in this spirit of critical collaboration I would suggest that Kitchin and Dodge so
emphasize the performativity of maps that they risk downplaying the dominative power/
knowledge some maps such as military maps still encode vis-a-vis others such as native
maps). For the faithful few who have found metaphysical meaning in some select seminal
works of western Philosophy, talk of critical collaboration may sound like a populist
finesse or even a deceit. But readers should judge for themselves, and instead of seeking
sanctuary amidst the fortifications of a new foundationalism, the protection against
deception and misrepresentation would here come in the form of a commitment to
working collaboratively, creatively *and* critically on an ongoing basis, always making the
moments of cartographic truth production and performance open to the World Wide
Web, so to speak, of critical engagement.

Contemporary examples of critical cartographic production and engagement are
now many, and they include amongst their most exciting innovations increasing work
with the actual Web of 'Web 2.0' mapping: the so-called Neo-Geo map making
revolution that has been enabled in part by the Internet, the explosion in geo-tagged
online data, and the parallel development of map 'mash-up' technologies such as
Google's My Maps. In certain respects, these contemporary cartographic developments
can be seen as a new chapter in PPGIS or 'participatory GIS' (see Elwood and Ghose, this
volume; and Elwood, 2006), especially insofar as they index another historical shift
away from the state holding a monopoly on the production of cartographic truth
(Elwood, 2008). Yet while all sorts of subaltern social subjects can and do use Web-
enabled technologies to produce new maps that can speak a new cartographic truth to

power (Elwood, 2010), and while this can surely be seen as an enabling expansion of cartographic citizenship beyond the ranks of ESRI-enabled Apollonian elites (Kingsbury and Jones, 2009), there are also obviously dangers of technological utopianism and biopolitical abuse attending the new possibilities for demystifying map making (making Elwood and Ghose's cautions about PPGIS all the more relevant). Here again deconstruction can critically supplement demythologization. Likewise, comparisons with the scene of Shawnadithit volunteering her story maps to Cormack can prompt critical reflection on the recapture and cooptation of geo-graphic agency through the neoliberal and neocolonial reterritorialization of the Neo-Geo revolution.

For one thing, the grim colonial circumstances of Shawnadithit's not-really-voluntary map-making point to the ways in which broader patterns of violence and dispossession might overshadow the production of Volunteered Geographic Information (VGI being a term used to name the sorts of spatial data generated by Web-users). Reflecting on this parallel we can note that while it may function as an enabling technology for map making by 'citizen scientists' (Goodchild, 2007), much VGI is generated and circulated on online commercial networks that are simultaneously colonizing and reducing the meaning of citizenship to the neoliberal common denominators of market choice, entrepreneurial calculation and consumer training (and this is to say nothing of the ways in which military data collection also surveys these new e-spaces much as the British marines once surveyed Newfoundland; see also Basu, 2009).

Given that Shawnadithit finally died due to an infectious disease, one telling contemporary example of this neoliberalization of VGI technology today is the iPhone application *Outbreaks Near Me* (HealthMap, 2009; Harmon, 2009). Launched in the midst of the H1N1 pandemic in September 2009, this geoweb-enabled application is an illustration at one level of the democratisation of public health disease surveillance (Figure 21.1). But, at another level, it just as clearly exemplifies the way in which such cartographic 'democratisation' is only beneficial for a privileged few who can buy into the new market-mediated surveillance system as a way of managing their personal risk exposure and protecting their own individual bodies. Here we see not an imperial liberal like Cormack making an anthropological survey of a dying culture, nor mobile marines marching with bodies across native land, but rather 'advanced liberal' kinetic elites who move across borders and through new found lands of flatness with embodied privilege (Sparke, 2006, 2009).

The inequalities of today's 'accumulation by dispossession' (Harvey, 2005) become still more disconcerting when we turn to consider the contemporary sorts of alienation that might mirror Shawnadithit's brutal alienation and the genocide of the Beothuk in Newfoundland. What happens, for instance, when online user-generated spatial data reported or otherwise relayed by citizen scientists on the Web subsequently becomes used in campaigns against those deemed to be 'non-citizen' threats to national security? Is there not a danger of crowd-sourcing turning thus into mob-sourcing (e.g. the rise of Spiritual Mapping and so-called Prayer Maps such as *Repent Amarillo*, 2010) or, at the very least, of elites using the tools to further entrench their privilege (Burrows, 2008)? To be sure, there are many cases of the new map-making tools working to contest such contemporary forms of alienation (for a useful survey see Crampton, 2009). The online *Atlas of Torture* (www.univie.ac.at/bimtor/countrymap) is one such example, as too is

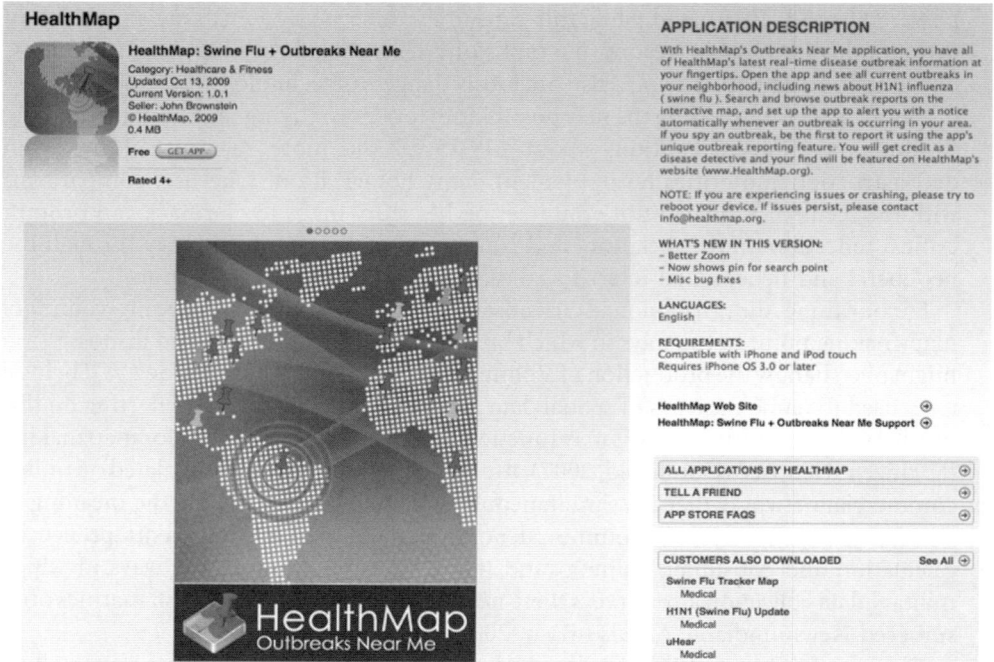

Figure 21.1   Outbreaks Near Me. (Source: Author screenshot of HealthMap, 2009).

the work of Trevor Paglen that uses online data to track the sorts of conjoint neocolonial-neoliberal violence represented by the CIA's outsourcing of torture (Paglen, 2007, 2009; Paglen and Thompson, 2006). In this sense Neo-Geo technologies can enable critical work for human rights just like, and sometimes alongside, formal GIScience (Madden and Ross, 2009). However, at the same time as they might enable the representation of political violence and genocide, it would be naïve to ignore the ways in which pre-existing asymmetries in access to the Internet and computer technologies will go on skewing the distribution of benefits and costs of Web 2.0 map making. This is not to suggest at all that Neo-Geo technologies are always and everywhere neoliberal and neocolonial. Nor does it mean that we can simply read-off the ways that they do serve such interests and imperatives simply on the basis of their epistemological organisation. Instead, it suggests an ongoing need to keep moving between demythologization and deconstruction as map-making becomes more accessible and democratised and as the look of surveillance – as in the *Atlas of Torture* – returns once more to survey the surveyors of a new world disorder.

Further Reading

Del Casino, V.J. and Hanna, S.P. (2006) Beyond the 'binaries': A methodological intervention for interrogating maps as representational practices. *ACME: An International E-Journal for Critical Geographies*, **4**(1), 34–56, http://www.acme-journal.org/vol4/VDCSPH.pdf. (This is a great read for quickly catching up on where the debates in critical cartography have gone from the mid 1990s onwards.).

Crampton, J. (2009) Cartography: performative, participatory, political. *Progress in Human Geography*, **33**, 840–848. (Another useful overview of recent work on the power-knowledge nexus in mapping.).

Elwood, S. (2010) Geographic information science: emerging research on the societal implications of the geospatial web. Progress in Human Geography. in press (This progress report provides a terrific entry-point into the fast developing debates on the Neo-Geo revolution and geoweb).

Paglen, T. (2009) *Blank Spots on the Map: The Dark Geography of the Pentagon's Secret World*, Dutton, New York. (As reported on the *Colbert Report* (www.colbertnation.com), the look of surveillance returns here to survey the US military and its geographies of disappearance.).

Pickles, J. (2004) *A History of Spaces: Cartographic Reason, Mapping, and the Geo-coded World*, Routledge, New York. (An erudite introduction into the modern emergence of cartographic reason.).

References

Akerman, J.R. (2009) *The Imperial Map*, University of Chicago Press, Chicago, IL.

Anderson, K. (1988) Cultural hegemony and the race-definition process in Chinatown, Vancouver: 1880–1980. *Environment and Planning D: Society and Space*, **6**, 127–149.

Barnett, C. (2009) Foundationalism, in *The Dictionary of Human Geography* (eds D. Gregory, R. Johnston, G. Pratt *et al.*.), John Wiley & Sons, Inc., New York, pp. 261–262.

Barnett, C. (2007) Review of in the space of theory: postfoundational geographies of the nation-state by M Sparke. *Environment and Planning A*, **39**, 502–504.

Barnett, C. (2008) Political affects in public space: normative blind-spots in non-representational ontologies. *Transactions of the Institute of British Geographers NS*, **33**, 186–200.

Basu, R. (2009) Phronesis through GIS: Exploring political spaces of education. *Professional Geographer*, **61**(4), 481–492.

Beaver Lake Nation (2009) *Beaver Lake Cree Nation versus Canada and Alberta*, www.raventrust.com/media/BLCNPressKit.pdf.

Belyea, B. (1992) Images of power: derrida/foucault/harley. *Cartographica*, **29**(2), 1–9.

Bhabha, H. (2004) *The Location of Culture*, Routledge, New York.

Braun, B. (2007) Biopolitics and the molecularization of life. *Cultural Geographies*, **14**(1), 6–28.

Brown, M. and Knopp, L. (2008) Queering the map: the productive tensions of colliding epistemologies. *Annals of the Association of American Geographers*, **98**, 40–58.

Burrows, R. (2008) Geodemographics and the construction of differentiated neighbourhoods, in *Cohesion in Crisis? New Dimensions of Diversity and Difference* (eds J. Flint and D. Robinson), Policy Press, Bristol, pp. 219–237.

Camp, S. (2002) The pleasures of resistance: enslaved women and body politics in the plantation South, 1830–1861. *Journal of Southern History*, **68**, 533–540.

Carter, P. (1987) *The Road to Botany Bay: An Essay in Spatial History*, Faber, London.

Clayton, D. (2000) *Islands of Truth: The Imperial Fashioning of Vancouver Island*, UBC Press, Vancouver.

Crampton, J. (2009) Cartography: performative, participatory, political. *Progress in Human Geography*, **33**, 840–848.

Crampton, J. and Krygier, J. (2006) An Introduction to critical cartography. *ACME: An International E-Journal for Critical Geographies*, **4**(1), 11–33, www.acme-journal.org/vol4/JWCJK.pdf.

De Leeuw, S. (2009) 'I am outraged': Media, racism, and colonial narratives of 'equality' in response to the Nisga'a Treaty of Northwestern British Columbia, Canada. *WicazoSa Review: A Journal of Native American Studies*, **23**(2).

Del Casino, V.J. and Hanna, S.P. (2000) Representations and identities in tourism map spaces. *Progress in Human Geography*, **24**(1), 23–46.

Del Casino, V.J. and Hanna, S.P. (2006) Beyond the 'binaries': A methodological intervention for interrogating maps as representational practices. *ACME: An International E-Journal for Critical Geographies*, **4**(1), 34–56, www.acme-journal.org/vol4/VDCSPH.pdf.

Duncan, J. and Duncan, N. (1988) (Re)reading the landscape. *Environment and Planning D: Society and Space*, **6**, 117–126.

Edney, M. (2007) Recent trends in the history of cartography: A selective, annotated bibliography to the English-language literature. *Coordinates*, Series B, No. 6, http://purl.oclc.org/coordinates/b6.htm.

Elwood, S. (2006) Beyond cooptation or resistance: Urban spatial politics, community organizations, and GIS-based spatial narratives. *Annals of the Association of American Geographers*, **96**(2), 323–341.

Elwood, S. (2008) Volunteered geographic information: Future research directions motivated by critical, participatory, and feminist GIS. *GeoJournal*, **72**, 173–183.

Elwood, S. (2010) Geographic information science: emerging research on the societal implications of the geospatial web. Progress in Human Geography, in press.

Elwood, S. and Ghose, R. (2001) PPGIS in Community Development Planning: Framing the Organizational Context. *Cartographica*, **38**(3/4), 19–33.

Galois, R. (1994) *Kwakwaka'wakw Settlements, 1775–1920: A Geographical Analysis and Gazetteer*, with contributions by Jay Powell and Gloria Cranmer Webster on behalf of the U'mista Cultural Centre, Alert Bay, BC, UBC Press, Vancouver.

Goodchild, M.F. (2007) Citizens as sensors: The world of volunteered geography. *GeoJournal*, **69**(4), 211–221.

Gregory, D. and Ley, D. (1988) Culture's geographies. *Environment and Planning D: Society and Space*, **1**, 115–116.

Haraway, D. (1989) *Primate Visions: Gender, Race, and Nature in the World of Modern Science*, Routledge, New York.

Harmon, K. (2009) Slick mobile app tracks H1N1, other outbreaks near you. *Scientific American*, 1 September 2009, www.scientificamerican.com/blog/60-second-science/post.cfm?id=sick-mobile-app-tracks-h1n1-other-o-2009-09-01.

Harris, R.C. (2002) *Making Native Space: Colonialism, Resistance, and Reserves In British Columbia*, UBC Press, Vancouver.

Harris, L.M. and Hazen, H.D. (2006) Power of maps: (Counter)mapping for conservation. *ACME: An International E-Journal for Critical Geographies*, **4**(1), 99–130, www.acme-journal.org/vol4/LMHHDH.pdf.

Harris, L. and Hazen, H. (2009) Rethinking maps from a more-than-human perspective: nature-society, mapping, and conservation territories, in *Rethinking Maps* (eds M. Dodge, R. Kitchin and C. Perkins), Routledge, London, pp. 50–67.

Harvey, D. (1985) *Consciousness and the Urban Experience: Studies in the History and Theory of Capitalist Urbanization*, John Hopkins University Press, Baltimore, MD.

Harvey, D. (2005) *A Brief History of Neoliberalism*, Oxford University Press, New York.

HealthMap (2009) *Outbreaks Near Me*, www.healthmap.org/iphone/.

Jackson, P. (1989) *Maps of Meaning: An Introduction to Cultural Geography*, Unwin Hyman, London.

Kitchin, R. and Dodge, M. (2007) Rethinking maps. *Progress in Human Geography*, **3**(3), 331–344.

Kingsbury, P. and Jones, J.P. (2009) Walter Benjamin's dionysian adventures on google earth. *Geoforum*, **40**, 502–513.

Kolodny, A. (1975) *The Lay of the Land: Metaphor as Experience and History in American Life and Letters*, University of North Carolina Press, Chapel Hill, NC.

Madden, M. and Ross, A. (2009) Genocide and GIScience: Integrating personal narratives and geographic information science to study human rights. *The Professional Geographer*, **61**(4), 508–526.

Marx, K. and Engels, F. (1970) *The German Ideology*, International Publishers, New York.

McKittrick, K. (2006) *Demonic Grounds: Black Women and the Cartographies of Struggle*, University of Minnesota Press, Minneapolis, MN.

Mitchell, K. (1997) Different diasporas and the hype of hybridity. *Environment and Planning D: Society and Space*, **15**(5), 533–553.

Nah, A. (2006) (Re)mapping indigenous 'race'/place in postcolonial peninsular Malaysia. *Geografiska Annaler Series B-Human Geography*, **88**, 285–297.

Paglen, T. (2009) *Blank Spots on the Map: The Dark Geography of the Pentagon's Secret World*, Dutton, New York.

Paglen, T. (2007) Groom lake and the imperial production of nowhere, in *Violent Geographies. Fear, Terror, and Political Violence* (eds D. Gregory and A. Pred), Routledge, New York, pp. 237–254.

Paglen, T. and Thompson, A.C. (2006) *Torture Taxi: On the Trail of the CIA's Rendition Flights*, Melville House Publishing, Hoboken, NJ.

Perkins, C. (2004) Cartography – cultures of mapping: power in practice. *Progress in Human Geography*, **28**, 381–391.

Rabasa, J. (1985) Allegories of the Atlas, in *Europe and its Other*, vol 2 (ed. F. Barker), University of Essex, Colchester, pp. 1–16.

Radcliffe, S. (2009) National maps, digitalisation and neoliberal cartographies: transforming nation-state practices and symbols in postcolonial Ecuador. *Transactions of the Institute of British Geographers NS*, **34**(4), 426–444.

Repent Amarillo (2010) Prayer Map accessed at http://maps.google.com/maps/ms?ie=UTF8& hl=en&msa=0&msid=108090264584267789065.00048c5908228b028d687&ll=35.214771,-101.810303&spn=0.268152,0.676346&z=11

Rose, G. (1993) *Feminism and Geography: The Limits of Geographical Knowledge*, Polity Press, Cambridge.

Rose, G. (2007) *The Politics of Life Itself: Biomedicine, Power, and Subjectivity in the Twenty-First Century*, Princeton University Press, Princeton, NJ.

Shawnadithit, (2009) *The Drawings of Shanawdithit*, as recorded Howley, J.P. (1915) *The Beothucks or Red Indians: The Aboriginal Inhabitants of Newfoundland*, www.mun.ca/rels/native/beothuk/beo2gifs/texts/shana2.html.

Sparke, M. (1994) White mythologies and anemic geographies. *Environment and Planning D: Society and Space*, **12**(1), 105–123.

Sparke, M. (1998) Mapped bodies and disembodied maps: (Dis)placing cartographic struggle in colonial Canada, in *Places Through the Body* (eds H. Nast and S. Piles), Routledge, New York, pp. 305–336.

Sparke, M. (2005) *In the Space of Theory: Postfoundational Geographies of the Nation-State*, University of Minnesota Press, Minneapolis, MN.

Sparke, M. (2006) A neoliberal nexus: Citizenship, security and the future of the border. *Political Geography*, **25**(2), 151–180.

Sparke, M. (2008) Political geographies of globalization (3): resistance. *Progress in Human Geography*, **32**(1), 1–18.

Sparke, M. (2009) On denationalization as neoliberalization: Biopolitics, class interest and the incompleteness of citizenship. *Political Power and Social Theory*, **20**, 287–300.

Spivak, G.C. (1976) Translator's preface, in *Of Grammatology* (ed. J. Derrida), Johns Hopkins University Press, Baltimore, MD.

Spivak, G.C. (1985) Subaltern Studies: Deconstructing Historiography, in *Subaltern Studies: Writings on South Asian History and Society, Volume Four* (ed. R. Guha), Oxford University Press, Delhi.

Spivak, G.C. (1988) Can the subaltern speak? in *Marxism and the Interpretation of Culture* (eds C. Nelson and L. Grossberg), University of Illinois Press, Urbana, IL, pp. 271–313.

St. Martin, K. (2009) Toward a cartography of the commons: Constituting the political and economic possibilities of place. *Professional Geographer*, **61**(4), 493–507.

Thien, D. (2005) After or beyond feeling?: A consideration of affect and emotion in geography. *Area*, **37**(4), 450–456.

Wikipedia, (2009) *Shanawdithit*, http://en.wikipedia.org/wiki/Shanawdithit.

Index

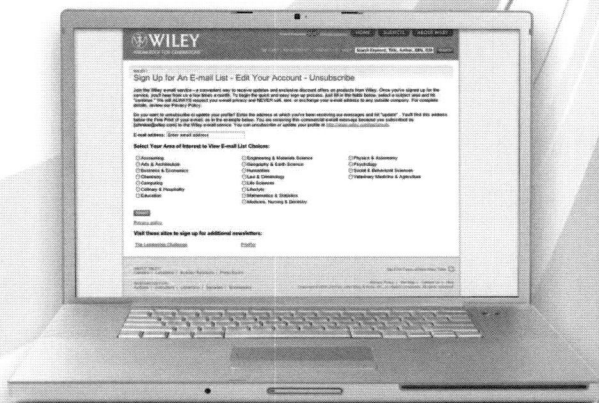